Plant in Challenging Environments

Volume 7

This book series provides recent advancements in wide areas related to higher plants and how they adapt / evolve under environmental changes in a scenario of climate change. It investigates plants under the complementary point of views, including agronomy aspects (vegetables and fruits), nutrition and health (food security), "omics," epigenetics, contamination by heavy metals, environmental stresses (salinity, drought, high and low temperatures), interaction with beneficial or pathogenic microorganisms, and application of exogenous molecules (nitric oxide, melatonin, chitosan, silicon, etc.) to palliate negative effects. It also includes changes due to climatic condition (high/low rainfall) taking into account that the climate change is often the reason why plants evolve in a challenging environment.

This book series also covers molecular-/cellular-level responses of plants under different climatic reasons. Families of molecules derived from hydrogen peroxide (H2O2), nitric oxide (NO) and hydrogen sulfide (H2S) designated as reactive oxygen, nitrogen and sulfur species (ROS, RNS and RSS, respectively) are included since, depending on the production level, they function both as signal molecules and as a mechanism of response against adverse/changing environmental conditions that can produce multiple cellular damages, alter the redox state or even trigger cell death. During these ensued metabolic processes, some anti-oxidative/oxidative enzymes are also disturbed or triggered abruptly, but there are adequate mechanisms of regulation/homeostasis in the different subcellular compartments to keep these enzymes under control.

In the last decades, the progression in this field has been enormous, but still there is so much in this field to understand the plethora of phenomena behind.

Sanjib Kumar Panda • Mather A. Khan
Editors

Plant Functional Genomics for Abiotic Stress Resilience

Springer

Editors
Sanjib Kumar Panda
Department of Biochemistry
Central University of Rajasthan
Bandarsindri, Kishangarh, Ajmer
Rajasthan, India

Mather A. Khan
Bond Life Sciences Center, Division of Plant Science and Technology
University of Missouri
Columbia, MO, USA

ISSN 2730-6194 ISSN 2730-6208 (electronic)
Plant in Challenging Environments
ISBN 978-3-032-01703-1 ISBN 978-3-032-01704-8 (eBook)
https://doi.org/10.1007/978-3-032-01704-8

This Springer imprint is published by the registered company Springer Nature Switzerland AG
The registered company address is: Gewerbestrasse 11, 6330 Cham, Switzerland

Preface

Land plants have evolved trailblazing systems to cope with severe stressful environmental situations during evolution. Plants have multifaceted molecular systems to retort and acclimate to abiotic stress, including drought, cold, heat, salt, and heavy metal stress, via propagating induced signals to sensory mechanisms to improve tolerance and survival. Under stressful environmental conditions, land plants have developed unique systems to propagate environmental stress signals from sensing tissues to appropriate targets to achieve optimal growth maintenance and tolerance to stress conditions.

In this book, *Plant Functional Genomics for Abiotic Stress Resilience*, studies have been comprehensively integrated to understand complex plant responses to abiotic stress. Understanding the networks of stress perception, signal transduction, and defensive responses under adverse environmental conditions can be explicitly achieved through the amalgamation of diverse omics data and functional genomics tools to fortify crop stress resilience. Omics approaches such as transcriptomics, metabolomics, genomics, proteomics, and epigenomics have revolutionized our understanding by integrating genetic engineering techniques to engineer plant traits, paving the way for the development of next-generation crops. This book not only highlights these technological breakthroughs but also accentuates the status of integrating multi-omics approaches to gain a holistic understanding of plant stress responses.

The chapter in the book follows the evolutionary and dynamic approaches of plant adaptation strategies integrated with new revolutionized techniques as an approach to enhance our understanding to harness this knowledge for crop improvement. As a student or researcher, this book will serve you as a valuable resource that encompasses our understanding from the foundational concepts to the most advanced stages, transitioning from natural phenomena to today's cutting-edge approaches in abiotic stress management.

We would like to extend our gratitude to all the contributors for their invaluable expertise and dedication, and the readers for their interest in advancing plant sciences, hoping for a more advanced and sustainable agricultural future.

Bandarsindri, Kishangarh, Ajmer, Rajasthan, India — Sanjib Kumar Panda
Columbia, MO, USA — Mather A. Khan

Contents

Chapter 1
Unveiling Abiotic Stress Sensing and Regulatory Pathways for Enhanced Crop Resilience

Divya Gupta, Mather A. Khan, Renu Kumari, Asha Kumari, Mayur Patel, and Sanjib Kumar Panda

Abstract Abiotic stresses, their impact on plants, and their regulation and mitigation adopted by plants are the focus of studies from several decades. Understanding these relatable questions can provide opportunities to improve crop yield and productivity which are the key requirement in global scenarios to improve the hunger index with the growing population. Here we focus on the of plants' sensing and regulatory mechanisms under various abiotic stresses. In this, several integrated pathways have been studied which involve Ca^{+2} ion signaling, phytohormones, TFs (Transcription factors), and sensory mechanisms. The regulatory mechanism altogether regulates the regulation of stress-perceptive genes to improve stress tolerance. The study can be helpful to marker-assisted breeding to grow better and more stress-resilient varieties of different crops.

Keywords Abiotic stress · Plants · Regulatory mechanism · Productivity

1.1 Navigating Plant Stress: From Molecular Responses to Crop Enhancement

Plants are immobile and unlike animals, they cannot choose to avoid stressful environments by moving from there. So, they have to adapt complicated but fine strategies to deal with stressful environments. The common steps to be followed in stressful conditions are (i) sensing, through receptors or proteins on the plasma membrane (ii) regulating stress by increasing calcium ion concentration, kinase/

D. Gupta · R. Kumari · A. Kumari · M. Patel · S. K. Panda (✉)
Plant Functional Genomics & Molecular Biology Laboratory, Department of Biochemistry, Central University of Rajasthan, Ajmer, India
e-mail: sanjib.panda@curaj.ac.in

M. A. Khan
Bond Life Sciences Center, Division of Plant Science and Technology, University of Missouri, Columbia, MO, USA

S. K. Panda, M. A. Khan (eds.), *Plant Functional Genomics for Abiotic Stress Resilience*, Plant in Challenging Environments 7,
https://doi.org/10.1007/978-3-032-01704-8_1

phosphatase activity, different TFs, etc., and (iii) expression of stress-regulatory genes to augment tolerance (Dixit and Jayabaskaran 2014; Zhu 2016). The regulation at the molecular level also led to changes at the morphological level like stomatal closure, curly and yellowish leaves, reduced size and number of leaves, poor grain yield, etc. (Hasanuzzaman and Fujita 2022; Bhardwaj et al. 2023). We have focused on how plants sense and regulate stress, key secondary molecules like lipid molecules, ROS, calcium ion, H_2O_2, etc. to process downstream signaling, the role of phytohormones, and regulation of stress-responsive genes (Demidchik et al. 2002; Scrase field and knight 2003; Xu et al. 2024). In osmotic stress, stress regulation is chiefly regulated in an ABA-dependent and independent manner (Seok et al. 2023). Under heat stress, HsfA1 (heat stress transcription factor A1) actively regulates plant response to enhance tolerance (Ohama et al. 2017). In cold stress, ICE1 (Inducer of CBF Expression 1) controls cold-responsive genes (Zhu et al. 2016). In metal stress, PCs and MTs (phytochelatins and metallothioneins) conjugate with metal and are stored in a vacuole to harness the negative impact of metal toxicity at the cellular level (Hodson 2012). This comparative study can suggest an overview of the involvement of several genes that can be targeted by gene editing technologies to improve crop production and nutritional values and can also improve important traits by the marker-assisted system of conventional breeding (Ashraf et al. 2012; El-mounadi et al. 2020; Ahmad 2023).

1.2 Unraveling the Drought Dilemma: Understanding Plant Responses to Water Scarcity

Drought occurs when natural water resources are depleted and become more frequent due to, global warming, deforestation, unpredicted rainfall, and alteration in the monsoon pattern (Brown et al. 2018; Konapala et al. 2020; Aryal et al. 2020; Waseem et al. 2023). These ecological changes lead to lower humidity in the surrounding air and decrease soil moisture under high temperatures, resulting in water deficit conditions that negatively impact normal plant growth and lead to physiological, morphological, and biochemical changes in the plant (Fig. 1.1) (Álvarez et al. 2023). The plant has a complex stress perception and regulatory mechanism to avoid or tolerate drought stress conditions. After drought stress perception, various signaling Cascades regulate their inter- and intra-dependent functioning in the plant and various modifications like epigenetic changes and post-translational modification can take place. The plant response can comprise four adaptations: avoidance, tolerance, escape, and recovery (Shelake et al. 2022).

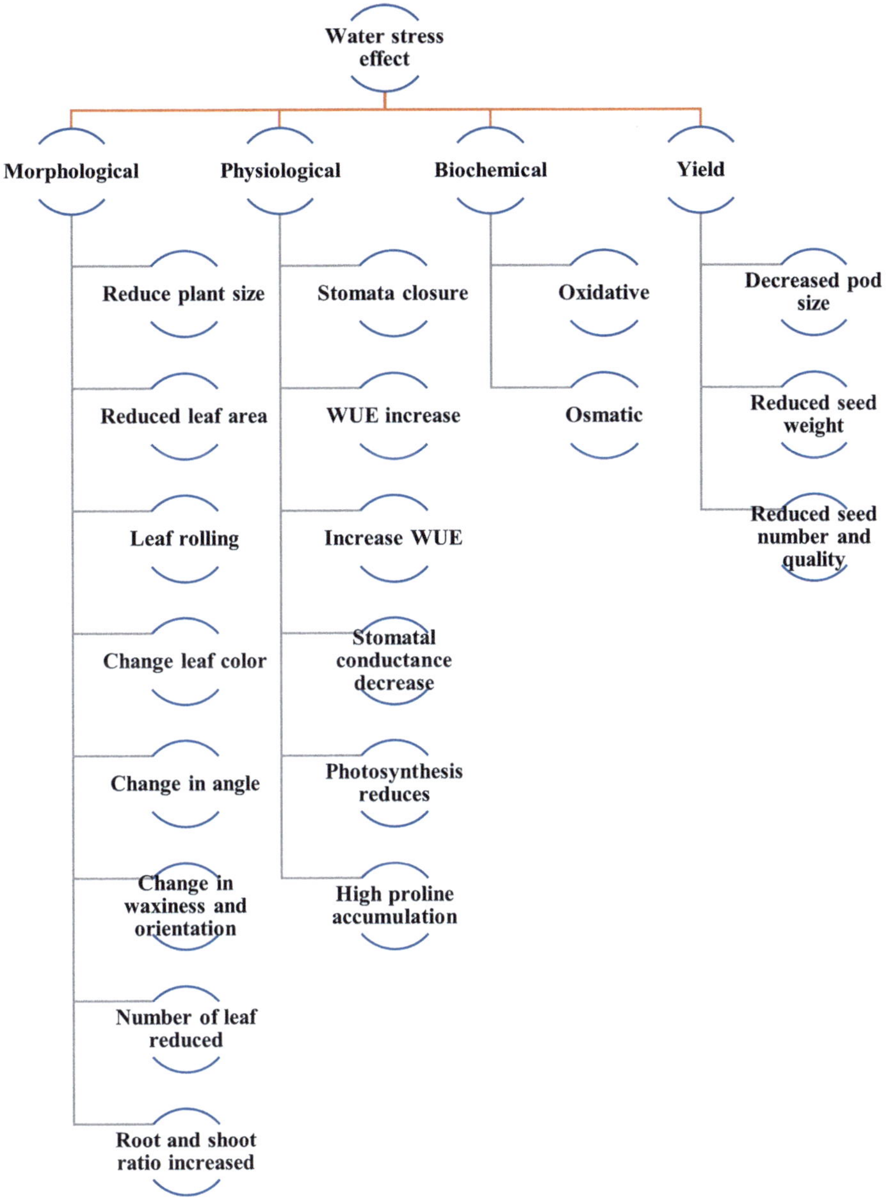

Fig. 1.1 Flowchart of the impact of drought stress. Here illustrated the impact of drought stress at morphological, biochemical, physiological and yield level

1.2.1 Deciphering Molecular Mechanisms of Water Stress Sensing: ABA Signaling

Abscisic acid (ABA) accumulation under drought stress is the key hormone that regulates the stress response in plants (Hsiao 1973), and induces downstream signaling and stomata closure (Fig. 1.2) (Schroeder et al. 2001; Yamaguchi-Shinozaki and Shinozaki 2006; Postiglione and Muday 2023). ABA signaling path consists of three factors PYL/PYR/RCAR (pyrabactin resistance (like)/regulatory component of ABA receptor) (Park et al. 2009; Ma et al. 2009a, b), PP2Cs (Meyer et al. 1994; Park et al. 2009), and SnRK2 (Fujii and Zhu 2009). In the presence of ABA, PYL/PYR/RCAR inhibit PP2C, which normally represses the activity of SnRK or other leading kinases, and triggers regulation in regards to stress (Chen et al. 2020; Zhu 2016; Qi et al. 2018; Gong et al. 2020; Chen et al. 2020). SnRKs have been separated into three subfamilies SnRK1, SnRK2, and SnRK3 (Hrabak et al. 2003) and SnRK2 is further sectioned into groups (group I, II, and III). Here, the group I listed kinases that are ABA independent, group II kinases are activated

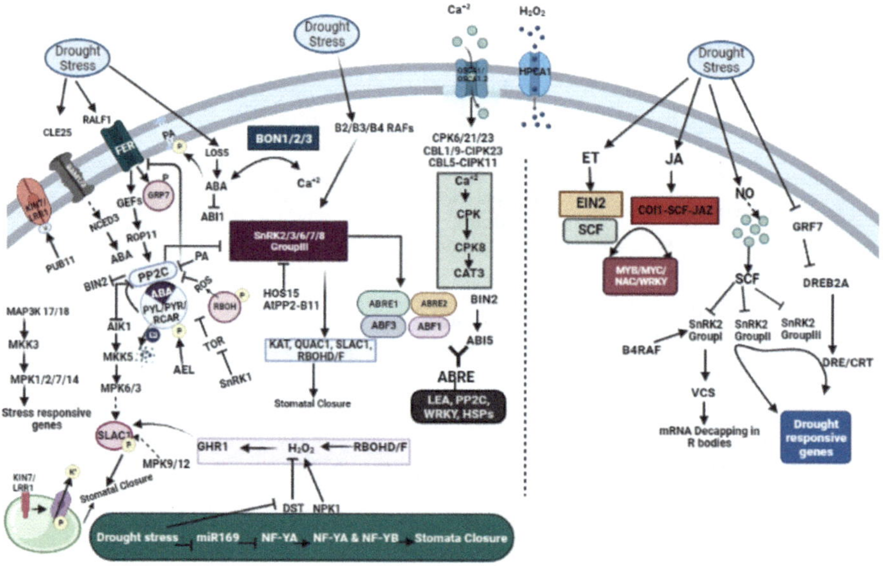

Fig. 1.2 Cross-talk signalling under drought stress regulation. *FER* FERONIA, *BON* BONZAI, *PP2-B11* PHLOEM PROTEIN 2-B11, *HOS15* HIGH EXPRESSION OF OSMOTICALLY RESPONSIVE GENES 15, *AEL* ARABIDOPSIS EARLY FLOWERING 1-LIKE, *AIK1* ABA-INSENSITIVE PROTEIN KINASE 1, *TOR* TARGET OF RAPAMYCIN, *SnRK2* SNF1-related protein kinase 2, *ROP11* RHO-RELATED PROTEIN FROM PLANTS 11, *GRP7* FER–GLYCINE RICH PROTEIN 7, *KIN* KINASE7, *NCED3* NINE-CIS- EPOXYCAROTENOID DIOXYGENASE 3, *KAT1* POTASSIUM CHANNEL IN ARABIDOPSIS THALIANA 1, *BIN2* BRASSINOSTEROID INSENSITIVE2, *RBOHD* RESPIRATORY BURST OXIDASE HOMOLOG D, *CLE25* CLAVATA3/ESR-RELATED25, *DST* DROUGHT AND SALT TOLERANCE, *QUAC1* QUICK-ACTIVATING ANION CHANNEL 1

either independent of ABA or weakly by ABA, and group III kinases are stimulated in the existence of ABA (Kulik et al. 2011; Rachowka et al. 2023). Under standard growth circumstances, the TOR complex (target of rapamycin) phosphorylates PYL/PYR/RCAR to disrupt interaction with PP2C, inhibited by SnRK2 under drought stress by phosphorylating its RAPTOR component (Wang et al. 2018). SnRK2s also form a complex with SnRK1 to prevent its interaction with TOR to promote normal plant growth under ideal conditions. However, SnRK1 inhibits the TOR complex under stress to delay growth and trigger a stress response (Belda-Palazon et al. 2020). SnRK2.6 (OST1) promotes downstream signaling under drought stress such as phosphorylation and activation of ABF1, AREB1/ABF2, ARBE2/ABF4, and ABF3 to manage the control of drought-responsive genes expression (Fig. 1.2). OST1 also regulates the SLAC1 or QUAC1 anion channel to reduce phosphorylation at Ser-120 to reduce transpiration via stomata closure (Geiger et al. 2009; Lee et al. 2009; Imes et al. 2013).

Recent studies highlighted some factors accountable for the attenuation of ABA signaling by deprivation of SnRK2.3 by SCF E3 ligase, PP2- B11, and dilapidation of OST1 by HOS15 E3 ligase (Cheng et al. 2017, Ali et al. 2019). Similarly, SnRK2-the interacting calcium sensor also negatively regulates both ABA-dependent/independent SnRK2 (Bucholc et al. 2011). Similarly, AEL and EL1, the Arabidopsis homologs of rice CASEIN KINASE1 (CK1), promote the degradation of PYL by phosphorylation and affect ABA signaling (Fig. 1.2).

The involvement of several mitogen-activated protein kinases has also been reported in ABA signaling in regard to drought stress. AIK1- MKK5- MPK6, a Cascade intricates in ABA-arbitrated regulation which involves primary root growth and stomata closure (Li et al. 2017a, b). Similarly, MPK3K17/18- MKK3- MPK1/2/7/14 positively regulates stress-responsive genes and is activated by the SnRK2 gene family responsive to ABA signaling (Danquah et al. 2015).

1.2.2 Exploring Upstream Signaling Components

Besides downstream signaling, numerous components involved in upstream signaling have been studied so far (Fig. 1.2) (Yuan et al. 2014; Chen et al. 2020; Grison et al. 2019). Like activation of OSCA1 and Ca^{+2} ion channels under osmotic stress, to regulate calcium-dependent Cascade regulation (Yuan et al. 2014; Fu and Yang 2023). Similarly, the immune repressor protein, BONZAI (BON1/2/3) is also responsible for calcium ion signaling and accumulation of ABA (Chen et al. 2020). In abiotic stress situations, waves of calcium ion signaling act as a secondary messenger by activating calcium-dependent protein kinases (CPKs) (Shi et al. 2018). Activated CPKs phosphorylate ABF and ABI TFs to order gene expression in regards to drought-like LEA, RD29B, etc. (Zhang et al. 2020). In an ABA-independent way, H_2O_2 under stress conditions, interacts with HPCA1-GHR1 in the presence of calcium ion which activates kinase and leads to phosphorylation of SLAC1 for stomata closure (Zhang et al. 2022).

Another group of kinase in abiotic stress is RLK which senses signals from their surroundings like neighboring cells, environment, and pathogens, followed by signal transmission to regulate downstream signaling (Shiu and Bleecker 2003; Becraft 2002; Shiu et al. 2004; De Smet et al. 2009). Here under drought stress, CLE25 peptide and other hydraulic signals are transported by the xylem towards shoots (Korver et al. 2018) and sensed by BAM1 and BAM3 and induce expression of NCED enzyme to positively regulate ABA biosynthesis. Under drought stress, the interaction between FERONIA (FER) receptor kinase and RALF1, activation of ROP11/ARAC10, and GTPase to promote ABI2 activation (Yu et al. 2012) which can also inhibit FER as feedback inhibition for growth of plants (Fig. 1.2) (Liao et al. 2017; Chen et al. 2016).

1.2.3 Beyond ABA: Drought Stress Regulation

Apart from ABA, ethylene and Jasmonic acid also participate in drought stress regulation. For example, gene silencing of the ethylene biosynthesis gene (ACS6) in maize was designated to upsurge grain yield concerning drought stress (Fig. 1.2) (Habben et al. 2014). Similarly, AUXIN REGULATED GENE INVOLVED IN ORGAN SIZE (ARGOS) family gene, which negatively regulates ethylene, was reported to enhance grain yield and forbearance to drought stress due to water shortage conditions (Shi et al. 2017). In the JA signaling pathway, JAZ1, a negative regulator in rice, exhibited reduced drought forbearance (Fu et al. 2017). Similarly, bHLH148 in rice performs as a positive transcriptional up regulator for OsDREB1 and OsJAZ about drought conditions (Seo et al. 2011). These lessons indicate the status of phytohormones in drought stress regulation. In the ABA independent pathway, groups I and II of the SnRK2 family play notable roles. Here Group I comprises kinases not stimulated by ABA and Group II kinases do not activate or are weakly triggered by ABA (Kulik et al. 2011). Here SnRK2 Group I gets activated in the presence of B4 Raf kinase under drought conditions (Fabregas et al. 2020; Takahashi et al. 2018; Soma et al. 2020) in processing bodies (Soma et al. 2020) and then Group I SnRK2 cooperate with VARICOSE, an mRNA decapping activator and regulate mRNA for post-transcriptional modification (Soma et al. 2020).

In gene regulation under ABA-independent signaling, DREB2A plays an important role under drought stress. It is negatively controlled by GRF7 under normal circumstances. Suppression of GRF7 via knockdown and knockout reported enhanced expression of DREB2A under normal conditions (Kim et al. 2012). Further DREBs interact with DRE/CRT for the processing of downstream signaling (Nakashima et al. 2009).

1.3 Salt Stress in Plants: Understanding the Impact and Mechanisms of Sodium Ion Sensing

Salt toxicity and its accumulation led to reduce plant growth and development (Singh et al. 2021). Similar to drought conditions, several physiological, biochemical, and metabolic changes take place in plants exposed to salt. The key identifiers of excessive salt conditions are osmotic stress, an initial-stage response, and ionic stress, an afterward-stage response (Munns and Tester 2008). Osmotic stress reduces the photosynthetic rate and ionic toxicity steps into inhibition of leaf opening and causes early leaf senescence, resulting in reduced growth and development (Farooq et al. 2017). Salt stress sensing and its regulation take place in a cross-talk manner where the mechano-sensor senses salt stress and signals are received by receptors and transporters and further downstream signaling takes place in the presence of RLKs (receptor-like kinases), phosphatase, etc. followed by regulation of salt-responsive genes to improve salt tolerance.

1.3.1 Salt Stress Signaling in Plants: Understanding Responses to Sodium Imbalance

In response to hyperosmotic conditions, mechanosensitive ion channels get activated. A hyperosmolarity-gated calcium-permeable ion channel (OSCA1), located in the plasma membrane, senses hyperosmolarity, and osca1 mutants displayed abrupt calcium ion signaling under osmotic stress (Fig. 1.3) (Yuan et al. 2014). Similarly, under sodium ion stress MOCA1 (monocation-induced calcium ion increase1) or inositol phosphoryl ceramide glucuronosyl transferase (IPUT1) forms glycosyl inositol phosphoryl ceramide (GIPC) by transferring glucuronic acid from UDP- glucuronic acid to inositol phosphoryl ceramide (Rennie et al. 2014; Jiang et al. 2019;). Here, GIPC in regards to salt stress binds to Na^+ and triggers Ca^{+2} transport to the cytoplasm (Jiang et al. 2019). Similarly, receptor-like kinase, FERONIA (FER), confined on the plasma membrane, senses cell wall damage under salt stress. MSL and MCA (mechanosensitive ion channel) family proteins, sense reduced turgor pressure under salt stress. MSL10 leads to a spike in Ca^{+2} transport and accumulation of ROS (Basu and Haswell 2020), by sensing cell swelling. HPCA1 as an H_2O_2 sensor, induce Ca^{+2} influx in regards to salt stress (Wu et al. 2020). SOS1, Na^+/K^+ antiporter, in Arabidopsis, act as salt sensor (Shi et al. 2002). Enhanced Ca^{+2} ion, ROS, and other secondary messengers are decoded by Ca^{+2}-dependent kinases, phosphatases, Ca^{+2}-dependent proteins, etc. (Dodd et al. 2010). Enhanced Ca^{+2} in cytoplasm received by SOS3 and SCaBP8/CBL10 and promoter activation of SOS2, a Ser/Thr protein kinase, by releasing from self-inhibition (Liu and Zhu 1998; Ishitani et al. 2000; Guo et al. 2001; Quan et al. 2007) and localize stimulated SOS2 on the PM (Halfter et al. 2000; Guo et al. 2001; Quan et al. 2007). Now triggered SOS2, phosphorylates and triggers SOS1 (Quan

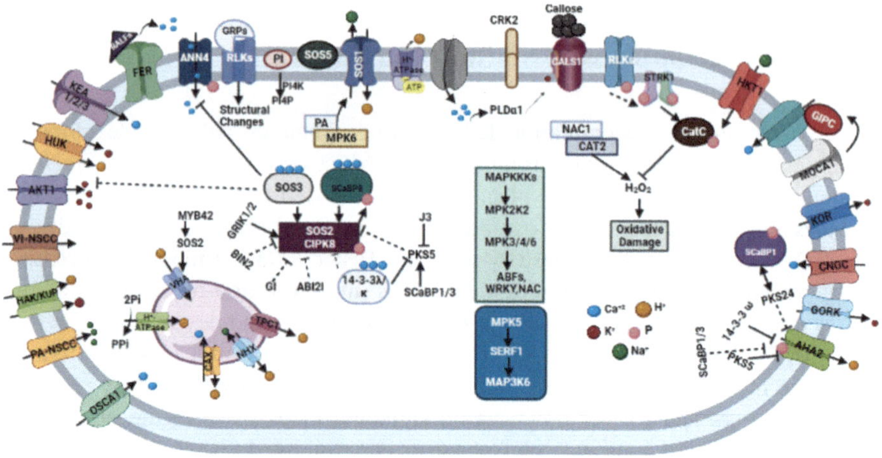

Fig. 1.3 Salt stress mitigation strategies adopted by plants. *KEA* K^+ exchange antiporter, *HUK-HUK, HKT* high potassium affinity transporter, *CDPK* calcium-dependent protein kinases, *CBL* calcineurin B-like proteins, *CIPK* CBL-interacting protein kinases, *MAPK* mitogen-activated protein kinase, *AKT* K^+ inward rectifying channel, *NSCC* nonselective cation channels, *KOR* outward-rectifying K^+channels, *HAK/KUP* potassium transporters, *OSCA1* hyperosmolality-induced $[Ca2+]_i$ increase 1, *FER* Feronia, *ANN4* ANNEXIN4, *SOS* salt overly sensitive, *CRK2* CYS-RICH RECEPTOR-LIKE KINASE 2, *CALS* CALLOSE SYNTHASE 1, *STRK1* salt tolerance receptor-like cytoplasmic kinase 1, *GIPC* glycosyl inositol phosphoryl ceramide, *MOCA1* monocation-induced $[Ca^{2+]}_i$ increases 1, *CNGC* cyclic nucleotide-gated ion channel, *GORK* guard cell outward potassium channel, *AHA2* H+-ATPase, *GRIK* GEMINIVIRUS REP INTERACTION KINASE1, *SCaBP8* SOS3-LIKE CALCIUM BINDING PROTEIN8, *BIN2* BRASSINOSTEROID-INSENSITIVE 2, *ABI2* PP2C ABA INSENSITIVE2, *CAX* H+ /Ca2+ antiporter, *NHX* K+/H+ exchangers, *TPC1* TWO PORE CHANNEL1

et al. 2007). Under hyperosmotic conditions, the SOS2-SOS3 composite also regulates NHX and H^+-ATPase on the vacuole (Qiu et al. 2004; Batelli et al. 2007). Under usual circumstances, SOS2 kinase action is repressed by BIN2, 14-3-3 (phospho ser/thr binding protein), GIGANTEA (GI), and PKS5 proteins by direct/indirect interaction (Kim et al. 2013; Zhou et al. 2014; Yang et al. 2019). GI and 14-3-3 are further tarnished by the 26S proteasomal degradation pathway under salt stress (Kim et al. 2013; Tan et al. 2016). Similar to SOS, HKT1, and HAK4, sodium transporter, regulate salt stress by phosphorylation (Uozumi et al. 2000; Ren et al. 2005; Munns et al. 2012; Zhang et al. 2019).

1.3.2 *Role of Phytohormones: Regulatory Mechanism Under Salt Stress*

Ethylene biosynthesis primes to salt forbearance reduction by enhancing the uptake of sodium ions and ROS accretion. Further, ethylene also regulates other phytohormones concerning salt stress. OsEIL1/2 acts as a key regulator and regulates HKT, a sodium ion transporter, OsYUC8, to promote auxin biosynthesis, and inhibit GY1 to regulate JA biosynthesis. Similarly, MHZ4/5 acts as a mediator between ethylene and ABA regulation in response to salt stress. In combination, these interactions inhibit root growth, ROS accumulation, Na^+/K^+ uptake, etc., and negatively regulate salt forbearance (Qin and Huang 2020). In regards to salt stress, ABA relates with PYL/PYR/RCAR and inhibits PP2C, and activates SnRK2 phosphorylation. Activated SnRK2 regulates ABRE/ABF, WRKY, NAC, MYC, etc., and adjusts the expression of salt-stress regulatory genes to advance salt acceptance such as LEA, RBOH/D, etc. AREB activates the expression of BAM1/AMY3 which damages starch into sugar and produces sugar-procured osmolytes (Fig. 1.4) (Thalmann et al. 2016).

1.3.3 Mitogen-Activated Kinases in Salt Stress Regulation

Mitogen-activated kinases also regulate salt stress like other kinases. In regard to salt stress, MKK2 phosphorylates and activates MPK4/6 and regulates salt-responsive genes (Teige et al. 2004). Similarly, MPK5 induces MAPKK6 and MAPK5 expression by phosphorylation of SALT RESPONSIVE ETHYLENE RESPONSE FACTOR 1(SERF1) in the existence of salt stress (Schmidt et al. 2013). In Arabidopsis, MPK3/4/6 phosphorylates HSFA4A and regulates the expression of HSP17.6A to progress salt forbearance (Perez-Salamo et al. 2014; Andrasi et al. 2019). On the contrary, MPK3/6 activated by SIT1, promotes ethylene production which negatively regulates salt forbearance (Li et al. 2014).

Thus, in salt-excessive conditions, a complex regulatory mechanism coordinates salt sufferance, and still exploitation and understanding of the mechanisms remain to make a clear understanding to promote the breeding of salt-resilient crops.

1.4 Heat Stress Effects on Plant Physiology: Disruption of Growth and Development

A temperature rise is specifically the major factor to disturb the plant life cycle, growth, and progress (Fitter and Fitter 2002). The consequence of high temperature (HT) varies in diverse plant species at different developmental stages. High temperatures disrupt photosynthesis, lead to the accretion of ROS, misfolding of proteins, and organelle malfunctioning (Kotak et al. 2007; Wahid et al. 2007; Hasanuzzaman et al. 2013). An increase of 5–6 °C (around 28 °C) in temperature, promotes root growth and early flowering, reduces seed production, and plants become vulnerable to pathogenic infection (Alcazar and Parker 2011; Gray and Brady 2016). An increase in temperature also leads to thermomorphogenesis resulting in hypocotyl

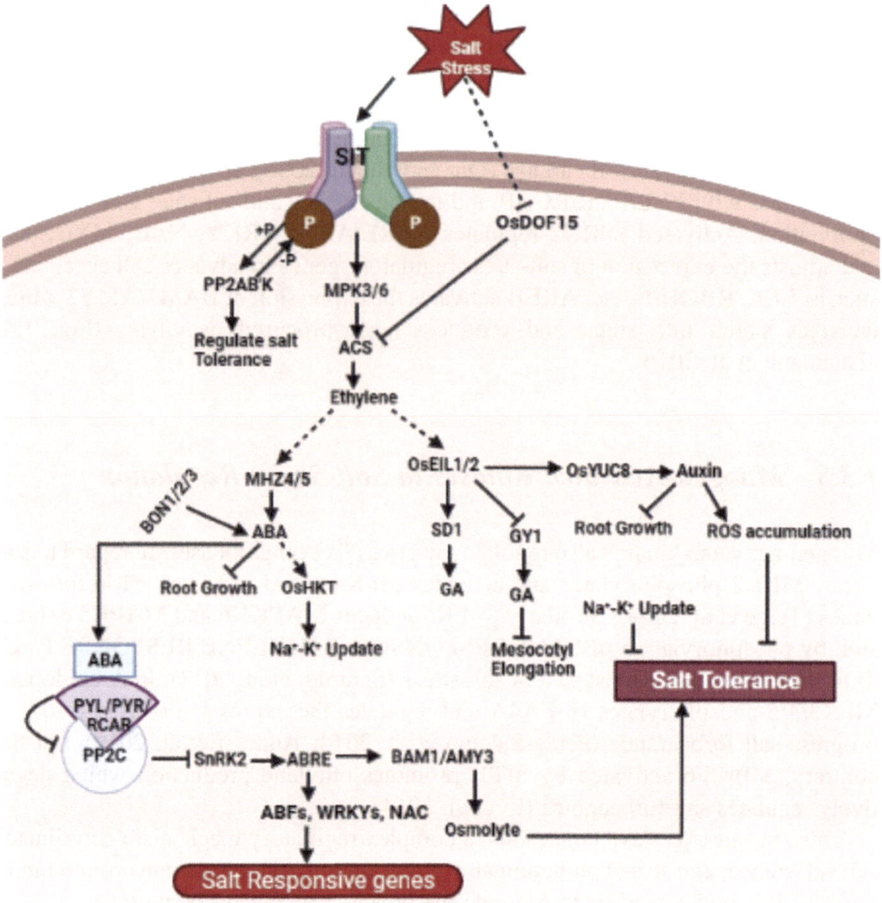

Fig. 1.4 Role of phytohormones in the regulation of salt-responsive genes. *PP2B'k* protein phosphatase 2A (PP2A) B′κ subunit, *DOF15* DNA-binding with one finger 15, *MHZ* maohzi, *YUC8 YUCCA* flavin monooxygenases 8, *BAM1* β-Amylase1, *ACS* 1-aminocyclopropane-1-carboxylate synthase, *ABA* Abscisic acid, *PYL/PYR/RCAR* Pyrabactin resistance, pyrabactin resistance-like, and the regulatory component of ABA receptors, *PP2C* type 2C protein phosphatases

elongation, a decrease in leaf thickness, and a lessening in stomatal density (Quint et al. 2016; Ibanez et al. 2017). An upsurge in temperature from 28 to 37 °C leads to heat shock in plants and affects the reproduction system, leading to poor fertilization and early miscarriage of embryos (Hedhly 2011; Sage et al. 2015). A rise in temperature up to 40 °C or higher causes cell death due to severe structural injuries (Li et al. 2018). To understand the thermotolerance and thermosensing mechanism, their regulation needs to be focused.

1.4.1 Heat Stress Perception and Regulatory Mechanisms in Plants

High-temperature regulatory mechanisms are intriguing, and the response and regulation chiefly evolve in a PIF4-dependent (phytochrome interacting factors 4) (Fig. 1.5) and HSFA1-dependent manner (Fig. 1.6). PIF4, a bHLH-type transcription factor (TF), is intricated with photomorphogenesis and is also a key manager of thermomorphogenesis (Ding et al. 2020). Under normal conditions, red light activated PhyB (phytochrome B) and performed 26S-mediated proteasomal degeneration of PIF4 by phosphorylation (Lorrain et al. 2008). At high-temperature, PhyB reverts from an active state (Pfr) to an inactive state (Pr) in the shade and allows the accretion of PIF4, which promotes the elongation of cells (Jung et al. 2016; Legris et al. 2016). Histone variants H2A.Z also mediate thermosensing in Arabidopsis in the daytime (Kumar and Wigge 2010). Under PIF4 dependent response H2A.Z eviction is mediated by SPT4-1 and SPT4-2, RNA Poly II elongation factor, INO80 complex, and HSFA1S (HSF family TFs) as they function upstream of H2A.Z (Cortijio et al. 2017; Xue et al. 2021; Perrella et al. 2022).

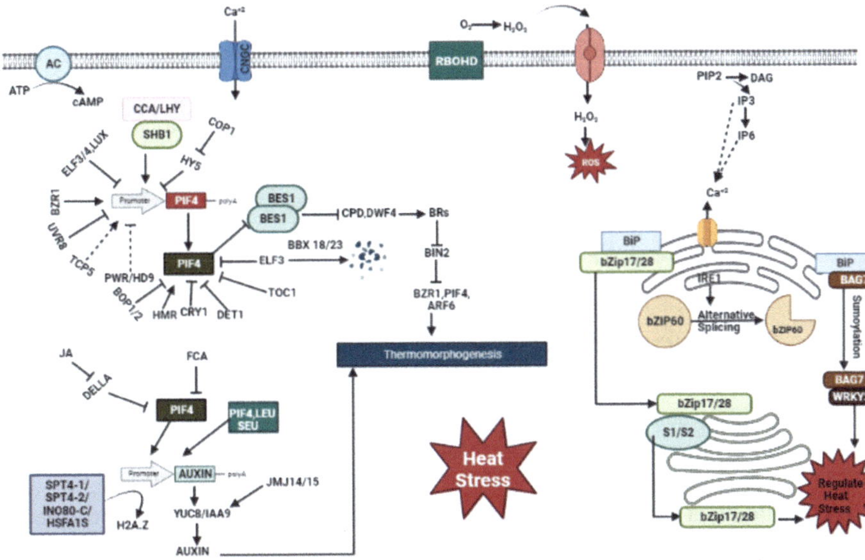

Fig. 1.5 Role of PIF4 in heat stress regulation. *RBOHD* H2O2-producing enzymes, *CCA* CIRCADIAN CLOCK ASSOCIATED 1, *LHY* LATE ELONGATED HYPOCOTYL, *SHB1* short hypocotyl under blue 1, *TCP* TEOSINTE BRANCHED 1/CYCLOIDEA/PCF, *SEU-* SEUSS, *LEU* LEUING, *COP1* E3 ubiquitin ligase, *SPT-* RNA POLYMERASE II (POLII) elongation factors, *PWR* POWERDRESS, LUX LUX ARRYTHMO, *ELF3* EARLY FLOWERING 3, *BOP1/2* BLADE-ON-PETIOLE 1, *HMR* HEMERA, *TOC1* TIMING OF CAB EXPRESSION 1, *IRE* INOSITOL-REQUIRING ENZYME 1, *BES1* BRI1-EMS-SUPPRESSOR 1, *CPD* CONSTITUTIVE PHOTOMORPHOGENESIS AND DWARFISM, *DWF* DWARF4, *BAG7* B-cell lymphoma2 (Bcl-2) -associated athanogene 7, *BiP* BINDING PROTEIN)

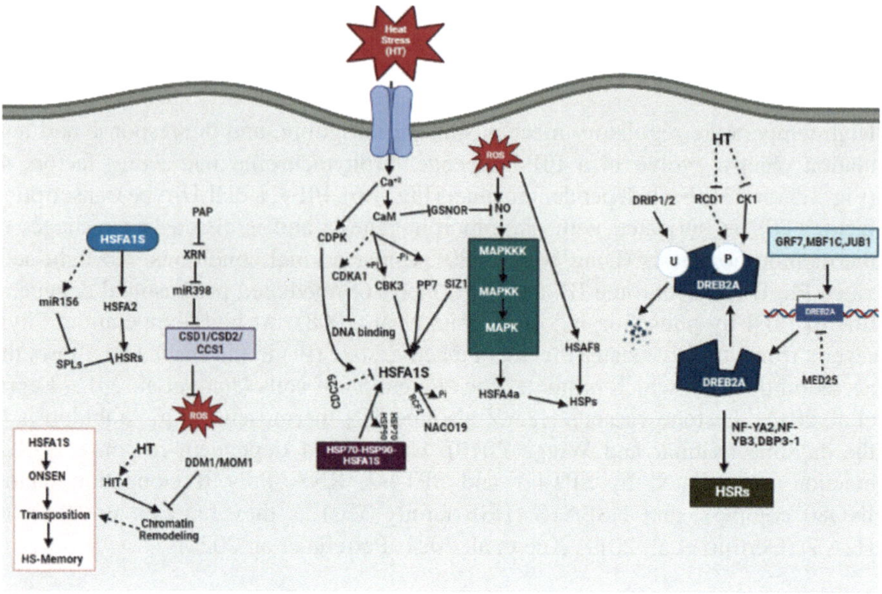

Fig. 1.6 Role of HSFA1 in heat stress regulation. *CDPK* calcium-dependent protein kinase, *SPL* SQUAMOSA-PROMOTER BINDING-LIKE, *GSNOR* nitrosylation, S-nitroso glutathione reductase, *GRF7* GROWTH-REGULATING FACTOR 7, *MBF1c* MULTIPROTEIN BRIDGING FACTOR 1C, *JUB1* JUNGBRUNNEN1, *SIZ1* SUMO E3 ligase SAP AND MIZ1 DOMAIN-CONTAINING LIGASE1, *PP7*CaM-binding protein phosphatase 7, *CSD1* COPER/ZINC SUPEROXIDE DISUTASE 1, *CCS* d COPPER CHAPERONE FOR SOD1, *DREB* DEHYDRATION RESPONSIVE ELEMENT BINDING 2A, *DDM1* DECREASE IN DNA METHYLATION 1, *MOM1* MORPHEUS' MOLECULE 1, *RCF2* C-REPEAT BINDING FACTOR gene expression 2

1.4.2 Phytohormes as Heat Stress Regulators

Under HT, PIF4 activates themomorphogenesis-responsive genes like YUC8 and IAA19, Auxin biosynthesis, and BR biosynthesis genes (Fig. 1.5). The PIF4-SEU-LEU complex endorses the expression of YUC8 and IAA19 via H3K4me3 methylation (Agulnick et al. 1996). Correspondingly, PIF4 was reported to be intricated in interaction with the promoter of BR biosynthesis-responsive genes like DWARF4 and BRASSINOSTEROID-6-OXIDASE 2 (BR6OX2) and also complex BZR1-ARF6-PIF4 control cell elongation (Wei et al. 2017; Oh et al. 2012, 2014). On the contrary, the DELLA protein, the GA signaling component, was reported as a negative regulator of thermomorphogenesis by interacting with complex BZR1-ARF6-PIF4 (Koini et al. 2009). BIN2, the BR signaling component, is also reported as a negative controller of thermomorphogenesis (Zhu et al. 2017; Li and Nam 2002; Peng et al. 2008). Under normal condition BES1, homodimer, inhibit the BR biosynthesis-responsive gene. Under HT, PIF4 prevents the formation of a homodimer of BES1 and regulates BR biosynthesis (Martinez et al. 2018).

1.4.3 Transcriptional and Post-transcriptional Regulation of PIF4

Studies have revealed the transcriptional and post-transcriptional regulation of PIF4 via different key regulators (Fig. 1.5). Transcription of PIF4 has been negatively controlled by Evening complex (EC) comprised of ELF4-EF3-LUX (Ezer et al. 2017; Nusinow et al. 2011). Similarly, TOC1 (TIMING OF CAB EXPRESSION) was studied to be involved in PIF4 inhibition in the evening to suppress thermoresponsive growth (Zhu et al. 2016). The study has shown the fact an increase in temperature leads to plant growth but increases susceptibility to pathogenic infection, although the mechanism is vague (Alcazar and Parker 2011). Gangappa et al. (2017) reported downregulation of defense-responsive genes in PIF4 overexpressing plant and reported enhanced resistance to *Pseudomonas syringae* DC3000 infection in pif4 mutant concerning wild type. As a concluding remark, the PhyB-PIF4 module is important to balance both plant growth and defense under HTs (Gangappa et al. 2017).

1.4.4 Molecular Regulation of HSF and DREB Family Members Under Heat Stress

HSAF1s, HSF family members, act as key regulators to activate HSR genes and other thermotolerance-responsive genes (Ohama et al. 2017). In normal conditions, HSP70 and HSP90 inhibit the activity of HSFA1 but HT leads to the separation of HSP70 and HSP90 from HSFA1 to enhance thermotolerance (Yamada et al. 2007; Hahn et al. 2011). Studies have also focused on the critical role of DREB2A under HT (Liu et al. 1998). Several regulatory components are involved in DREB2A regulation. Under non-stress conditions, GRF7 inhibits DREB2A, on contrary HSFA1s, JUB1 and MBF1c induce DREB2A under HT (Sakuma et al. 2006; Wu et al. 2012; Suzuki et al. 2011). Similarly, DRIP1/2 regulates DREB2A by 26S proteasomal dilapidation under usual conditions (Qin et al. 2008). RCD1 negatively regulates DREB2A, degraded under HT (Vainonen et al. 2012). NRD (NEGATIVE REGULATORY DOMAIN), a domain of DREB2A, is phosphorylated by CK1 and promotes its deprivation concerning normal condition (Mizoi et al. 2019). DREB2A action is also promoted by NF-YB3-DPB3-1-NF-YA2 trimer under HT (Sato et al. 2014; Sato et al. 2016).

1.4.5 Integrated Regulatory Network of Heat Stress Response in Plants

In heat stress regulation, miRNA also plays an important regulatory component (Fig. 1.6). Like, miR398 targets CSD1/CSD2/CCS1, ROS scavengers, to induce ROS production to promote the regulation of HSFA1 positively (Guan et al. 2013). Similarly, the miR156-SP2 complex regulates the memory of HT (Stief et al. 2014). Similarly, under HT, induction of tocopherol (Vit E) leads to the accumulation of PAP which inhibits XRN and promotes miR398 biogenesis to enhance HT (Fang et al. 2019). Heat stress memory is also mediated by transposons. DDM and MOM1 repress transposon via chromatin remodeling. On the other hand, ONSEN, a retrotransposon, target of HSFA1 and HSFA2 gets activated under HT (Ito et al. 2011; Cavark et al. 2014). Heat stress triggers misfolded protein which triggers the unfolded protein response (UPR) in ER (Howell 2013). Under HT, bZIP60 mRNA splicing is mediated by IRE1, an RNA splicing factor, localized on ER membrane. IRE1 excises bZIP transmembrane domain and mediates its translocation into the nucleus (Deng et al. 2011).

Being an important factor of the sensing mechanism of HT and induction of heat stress signal for downstream signaling, calcium ion, and ROS have been highlighted in recent studies to promote heat stress response (Liu et al. 2005; Sangwan et al. 2002; Suri and Dhindsa 2008; Saidi et al. 2009; Volkov et al. 2006). Calcium ion incorporation is mediated by calcium ion channels like CNGC, which open in response to enhanced cyclic nucleotide under HT (Gao et al. 2012; Finka et al. 2012; Tunc-Ozdemir et al. 2013). Further, IP3 and IP6 promote the emancipation of calcium ions from ER to the cytosol through ER-localized Ca^{+2} channels (Zhang et al. 2009). Similarly, NADPH oxidase, RBOHB/D actively generate ROS under HT (Konigshofer et al. 2008; Wang et al. 2014), and further ROS causes accretion of NO (Wang et al. 2014) and regulates HSPs to improve thermotolerance.

1.5 Cold Stress Responses in Plants: Mechanisms and Adaptation Strategies

Cold stress can be alienated into two categories based on the temperature ranges of chilling stress (0–15 °C) and freezing stress (>0 °C) (Guo et al. 2018; Liu et al. 2018a, b). Chilling stress affects plants by affecting membrane solubility (Orvar et al. 2000), reducing ROS scavenging enzyme activity (Siddiqui and Covicchioli 2006; Ruelland et al. 2009), and by affecting protein synthesis and gene expression regulation due to secondary structures of RNA (Ruelland et al. 2009; Rajkowitsch et al. 2007). However, freezing promotes extracellular ice nucleation causes severe cell dehydration, and leads to irreversible cell damage (Dowgert and Steponkus 1984; Pearce 2001). For acclimatization under cold stress, tolerance mechanisms are adopted by plants such as an increase in the production of proline and soluble sugars, acting as osmolytes (Ruelland et al. 2009), the accretion of defensive proteins like LEA (late embryogenesis abundant), antifreeze proteins (AFPs), and cold shock proteins (CSPs) to enhance cold tolerance (Ruelland et al. 2009).

1.5.1 Cold Stress Perception and Regulation in Plants

Vicissitudes in membrane flexibility are a remarkably considerable factor under heat and cold stress which leads to Ca^{+2} influx and start regulation of stress-responsive genes (Gong et al. 1998; Zhu 2016; Knight et al. 1996). As, the membrane fluidity changes in rice, COLD1, a G- protein regulator, coupled with RGA1, spikes Ca^{+2} influx (Ma et al. 2015). Similarly, OsCIPK7 regulates Ca^{+2} influx under stress. ANN1 and CNGC trigger Ca^{+2} influx concerning cold and are activated by OST1 and OsAPK8 respectively (Wang et al. 2020; Liu et al. 2020).

1.5.2 Reactive Oxygen and Nitrogen Species (RONS) as Cold Stress Regulator

Adding to calcium ion signaling, ROS, and nitric oxide are also important regulatory factors involved in cold tolerance. Conversion of GSH to GSSG activates the H_2O_2 scavenging system (Kocsy et al. 2001) in the existence of ABA, calcium ions, and salicylic acid to regulate cold-responsive genes (Fig. 1.7). Researchers have also emphasized the role of nitric acid as a nitric reductase dependent NO concentration reportedly involved in cold stress acclimatization (Costa-Broseta et al. 2018; Zhao et al. 2009; Cantrel et al. 2011). Although some of the studies reported NO as a negative controller in concern to cold (Abat and Deswal 2009; Sehrawat et al. 2013). Under cold stress, S-nitrosylated proteins have also marked up their involvement in primary regulatory mechanisms like photosynthesis, other defense mechanisms, etc. (Puyaubert et al. 2014). Cantrel et al. (2011) premeditated the reduced expression of CBF1/3 and its regulons under poor nitric oxide concentration (Fig. 1.7).

1.5.3 CBF-Dependent Regulatory Strategies Under Cold Stress

Regulation of cold tolerance can be maintained in both CBF-dependent and independent styles. The expression of CBFs is controlled by several TFs. Ca^{+2} mediated calmodulin binding transcriptional activator (CAMATA 3 and 5) was studied as a regulator of CBF1/2 under steep decline in temperature (Fig. 1.8) (Kidokoro et al. 2017). ICE1 and ICE2 are positively intricated in the instruction of the expression of

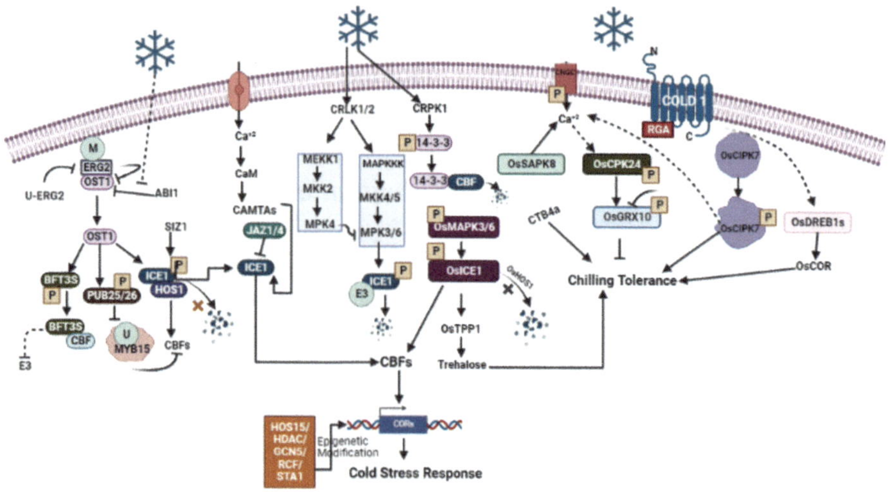

Fig. 1.7 Cross talk of phytohormones for cold stress tolerance. *BR* Brassinosteroid, *ETH* Ethylene, *PhyB* Phytochrome B, *PIF* PHYTOCHROME INTERACTING FACTORS, *CBF* C-repeat Binding Factor, *GSH* reduced glutathione, *GSSG* oxidized glutathione, *BZR* BRASSINAZOLE RESISTANT1, *BES* BRI1-EMS-SUPPRESSOR1, *BIN2* BRASSINOSTEROID INSENSITIVE2, *HOS* HIGH EXPRESSION OF OSMOTICALLY RESPONSIVE GENE15, *PRR* PSEUDO RESPONSE REGULATORS, *COR* cold responsive

CBF concerning cold stress. ICE1 undergoes post-translational modifications like ubiquitination, phosphorylation, and sumoylation to induce cold-responsive genes under downstream signaling (Ding et al. 2019a, b). ICE1 is ubiquitinated by HOS1, an E3 ligase, OST1 regulates ICE1 phosphorylation and SIZ1 is responsible for its sumoylation (Miura et al. 2007; Ding et al. 2015, 2019a, b). Under cold stress, OST1 disrupts the interaction between ICE1 and HO1 and promotes its stability (Fig. 1.8) (Ding et al. 2015). Similarly, OsMPK3 was also reported to enhance ICE1 by releasing HOS1 by phosphorylation (Zhang et al. 2017). By inhibiting the MKK4/5-MPK3/6 complex, MEKK1-MKK2-MPK4 positively regulates the cold stress by phosphorylating ICE1 via releasing HOS1 (Zhao et al. 2017) and improves the interaction between ICE1 and OsTPP1 to enhance trehalose biosynthesis for freezing tolerance (Zhao et al. 2017). In some studies, MYB-type TF was reported as a negative regulator of cold stress tolerance as inhibits expression of the CBE gene (Agarwal et al. 2006) In studies, OsMYB3R-2 was reported to be intricated in the induction of OsCyB1-1 and OsDREB1s genes to positively control cold tolerance (Ma et al. 2009a, b) but on the contrary, OsMYBS3 inhibits expression of OsDREB1B, as a negative regulator (Su et al. 2010).

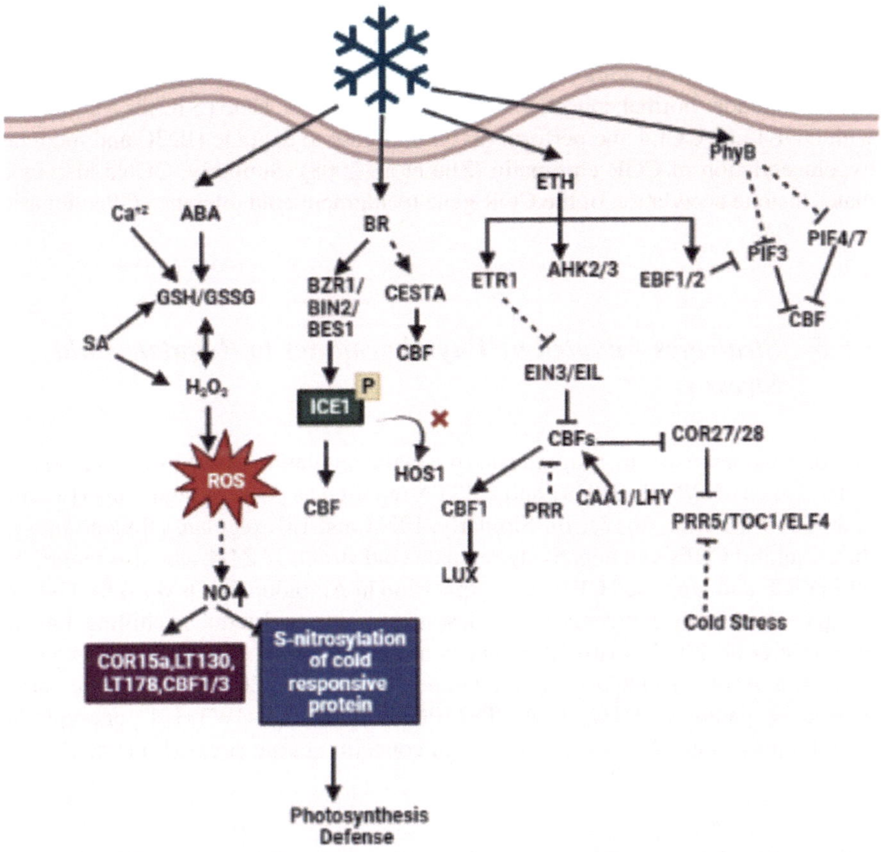

Fig. 1.8 Regulatory mechanism of cold stress regulation at the cellular level. *CRLK1* CALCIUM/CALMODULIN-REGULATED RECEPTOR-LIKE KINASE1, *CRPK1* COLD-RESPONSIVE PROTEIN KINASE1, *CBFs* C-REPEAT/DRE BINDING FACTORs, *RGA* RICE G PROTEIN-ALPHA SUBUNIT 1, *EGR2* CLADE-E GROWTH-REGULATING 2, *PUB25* PLANT U-BOX 25, *CTB4a* COLD TOLERANCE AT BOOTING STAGE 4a, *OsGrx10* OsCPK24–GLUTATHIONE- DEPENDENT THIOLTRANSFERASE 10, *STA1* STABILIZED 1, *RCF1* REGULATOR OF CBF GENE EXPRESSION 1, *CAMTA* calmodium-binding transcriptional activator 3

1.5.4 Noncoding RNA and Epigenetic Modification and Cold Stress

In the CBE-COR pathway, epigenetic regulation also regulates gene expression concerning cold stress. The role of miRNAs has been verified by several studies. In the study upregulation of miR393 and downregulation of miR319c and miR398a were reported about cold stress (Sunkar and Zhu 2004). Overexpression of miR397 in Arabidopsis reported enhanced acceptance of cold stress (Dong and Pei 2014). A lncRNA, SVALKA, resulted in enhanced expression after 4 h of cold tolerance while reduced CBF1 expression was observed (Kindgren et al. 2018). For histone modification under cold stress, AtHDA6 positively regulated cold tolerance (To et al. 2011). In Arabidopsis, HOS15-HD2C regulates hypoacetylation of COR chromatin under normal conditions. But with cold stress, HOS15 forms a compound with DDB1 and CUL4 and performs as an E3 ligase to degrade HD2C and modulate hyperacetylation of COR chromatin (Zhu et al. 2008). Similarly, GCN5 also modulates histone acetylation of the COR gene to augment cold tolerance (Vlachonasios et al. 2003).

1.5.5 Strategies Adapted by Phytohormones to Regulate Cold Stress

Phytohormones also play a significant role in the regulation of cold stress-responsive genes. Like BIN2/BZR1/BES1 and CESTA reported as positive regulators (Eremina et al. 2016; Li et al. 2017a, b). Similarly, EIN3 and EIL regulate ethylene biosynthesis, inhibit CBFs and negatively regulate cold stress. JAZ1/4 was also reported to inhibit ICE and repressed CBF gene expression in Arabidopsis. OsMYB30- OsJAZ9 complex reportedly represses β- amylase expression and reduces chilling forbearance (Lv et al. 2017). OsbZIP73 makes a complex with OsbZIP71 and regulates ROS and ABA to enhance cold tolerance (Liu et al. 2018a, b). In the auxin-dependent pathway, LTG, LOW-TEMPERATURE GROWTH 1, encrypts for Casein kinase and adjusts growth of plant concerning cold stress (Lu et al. 2014).

1.5.6 Cold Stress Response: Interplay with Circadian Clock and Photoperiod

The circadian clock and cold stress retaliation are also closely related. CCA1, LHY, and PRRs are essential components to maintain the circadian clock. CCA1 and LHY act as positive regulators in response to cold by regulating CBFs and COR genes (Dong et al. 2011, Nakamichi et al. 2009). In contrast, PRR5/7/9 acts as a negative controller about freezing stress (Nakamichi et al. 2009). Similar to the circadian clock, photoperiod and light has also been shown as regulatory factors with cold stress. Studies suggested that short-day plants are more cold tolerant than long-day plants (Lee and Thomashow 2012). The study reported PIF4 and PIF7 as negative regulators of cold stress under long-day situations (Kidokoro et al. 2009; Lee and Thomashow 2012). Similarly, PIF3 also represses CBF gene expression but is inhibited by EBF1 and EBF2 to improve cold tolerance. COR gene regulation has also been studied independently of CBFs. Similarly, The JA receptor, NPR1, under cold stress, monomerizes and triggers HSF1A after its translocation to the nucleus and thus triggers COR gene expression (Olate et al. 2018).

1.6 Heavy Metal Induced Stress in Plants: Mechanisms and Adaptation Strategies

Among different abiotic stresses, metal toxicity is also an important growth and development limiting factor in plants. Metal toxicity leads to numerous physical, biochemical, and physiological vicissitudes that hinder root growth, cause changes in leaf color, and result in poor nutrient uptake. At the physiological level, it induces ROS production, blocks necessary functional groups, and acts as a substitute for important regulatory ions (Schutzendubel and Polle 2002). From the cell wall to the molecular level several approaches are approved by plants to enhance tolerance in regards to metal stress such as (i) metal ions binding to the root cell wall; (ii) PM acting as a barrier on the origin of size and charge of metal ions to restrict their movement across the membrane; (iii) use of metal transporter to efflux ions from cytoplasm; (iv) chelate metal ion by forming a compound with PCs or MTs to act as organic acids; and (v) store metal ions in the vacuole to avoid toxicity by reducing their concentration in the cytosol.

1.6.1 Metal Stress Regulation in Plants: Signaling and Responses

Initial symptoms of metal ion toxicity are osmatic stress, dehydration, oxidative stress, etc. with poor nutrient uptake, which is similar to other abiotic stresses (Chen et al. 2001; Yadav 2010; Rucinska-Sobkowiak 2016). Under metal stress, calcium ion-dependent regulation, MAPK kinase signaling, and hormonal signaling are the key regulatory components to regulate downstream signaling to enhance plant tolerance (Fig. 1.9). Calcium ion signaling regulates metal ion toxicity via modulating physiological and biochemical changes and induces antioxidant enzymes and other stress-receptive genes (Ahmad et al. 2015). The correlation between CDPK and MAPK has been validated concerning stress (Takahashi et al. 2011; Wurzinger et al. 2011; Opedenakker et al. 2012). Xie et al. (2014) reported phosphorylation of MPK5 in rice mediated by CPK18. Similarly, MKK3 and Ca/CaM together activate MPK8 and negatively regulate RBOHD (Takahashi et al. 2011). So, the interconnected regulation is reportedly crucial to maintain redox homeostasis and ion equilibrium in cells under metal stress. Enhanced ROS production leads to the initiation of the MAPK Cascade. MEKK1/MKK4/6/MPK3/6, a MAPK Cascade regulates both biotic and abiotic stress-responsive regulation in Arabidopsis (Jalmi and Sinha 2015; Pitzschke et al. 2009). Asai et al. (2008) reported that OXI1-MPK6 positively regulates ROS as a feedback response. Similarly, MEKK1/MKK4/MPK3/6 also stimulates ROS production (Kovtun et al. 2000), which describes MAPK Cascade activation by ROS production.

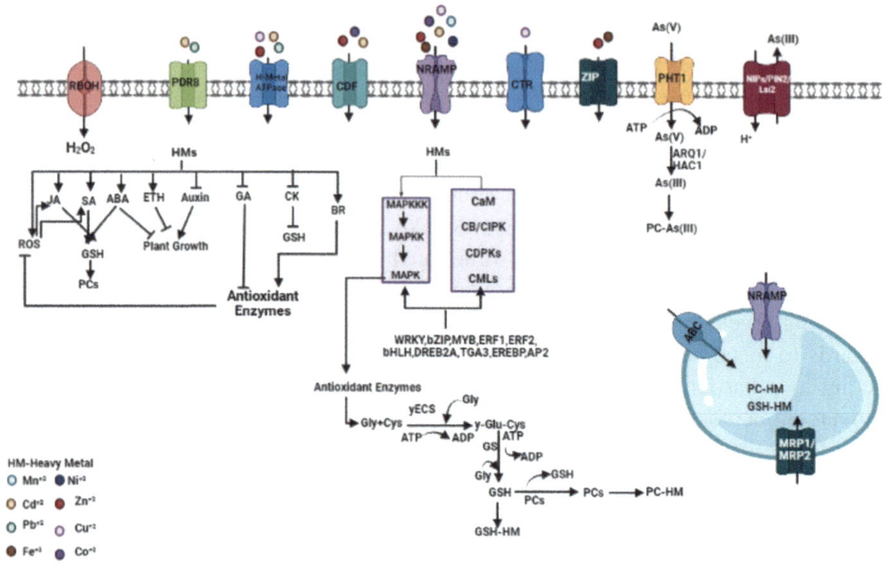

Fig. 1.9 Metal ion toxicity regulation at the cellular level. *PDR8* ABC transporter, *NIP* Al-malate transporter, *ZIP* zinc-iron permease, *NRAMP* naturally resistant associated macrophage protein, *CDF* cation diffusion facilitator, *TGA3* TGA1a-related gene, *PHT1* Pi/As(V) transporters, *PIN2/NIP/Lsi2* auxin transporters, *EREBP* ethylene-responsive element binding proteins, *ARQ1/HAC1* arsenate reductases

1.6.2 Phytohormones as Key Regulators Under Metal Toxicity

Phytohormones play an important role to enhance metal toxicity tolerance by (1) Inhibiting plant growth at an early stage of toxicity (2) By regulating antioxidant production (3) Metal ion chelation and accumulation (Nguyen et al. 2021). In phytohormones regulation also MAPK Cascade plays an important role. An interplay between auxin, cytokinin, and MAPKs was reported where MKK4/5 and MPK3/6 acted as key players to regulate the expression of PIN 1b/9 in rice (Singh et al. 2015). But the engrossment of MAPK Cascade in auxin-response is still unclear as under Cd stress MAPK has reported its involvement as a negative regulator of auxin-responsive genes like YUCCA, PIN, IAA, etc. (Zhao et al. 2014). Like auxin, MAPK Cascade also regulates ethylene response but in heavy metal stress, it is still unclear (Opdenakker et al. 2012). MPK3/6 phosphorylates ACS2/6 and enhances ethylene fabrication (Liu and Zhang 2004; Li et al. 2012). Similarly, MKK9/MPK3/6 was studied to regulate EIN3, a key TF in ethylene production, downstream of CTR1 (ethylene receptor) (Xu et al. 2008; Yoo and Sheen 2008). In a current study also, a link was suggested between MPK3/6 responsive ROS creation and ethylene response in Arabidopsis under Cd stress (Schallingen et al. 2015).

1.6.3 Transporter-Mediated Metal Toxicity Regulation in Plants

To reduce metal ion toxicity, the efflux of excessive ions into apoplast and chelation by phytochelatin (PCs), metallothioneins (MTs), organic acids, and amino acids in the cytoplasm are the key responses where conjugate of metal and ligand complexes are carried to the vacuole (Verbruggen et al. 2009; Sharma and Dietz 2006; Mendoza-Cozatl et al. 2011). To maintain ion homeostasis, several transporters have been identified and roughly grouped into ZIP, NRAMP, ABC, and CDF transporter (Singh et al. 2015; Park et al. 2012). ZIP transports zinc, iron, cadmium, and manganese (Eide et al. 1996). In the study, MPK3/6 addressed their involvement in the expression of IRT1 (a ZIP family member), FRO2, and FIT as mpk3/6 mutated lines exhibited their downregulation (Ye et al. 2015). NRAMP, another metal transporter, usually upregulates under iron, cadmium, and manganese deficiency and is located on plastid and vacuole membranes (Thomine and Schroeder 2013). ABC transporter is classified into 8 subfamilies, involved in diverse subcellular processes (Park et al. 2012). AtPDR8, an ABC transporter, provided resistance against Cd and Pb (Kim et al. 2007). Similarly, AtABCC1/C2 transport PC-HMs complex (Jalmi et al. 2018). Additionally, chelating compounds show a significant character in metal detoxification (Rauser 1999; Cobbett 2000; Clemens 2001; Hall 2002). Both MT and PC are cysteine-rich polypeptides. Their key strategy for detoxification is to exploit the characteristic of heavy metal to bind to the thiol group of protein (Peuke and Rennenberg 2005). Under metal stress, glutathione (GSH), also plays a vibrant role in defense, synthesized in the presence of two ATP in each step which is catalyzed by γ- glutamyl cysteine synthetase (γ-ECS) and glutathione synthetase (May et al. 1998; Foyer et al. 2001; Noctor et al. 1998). GSH translocate and sequester metal ion by forming ligand GSH-HM and improve metal ion toxicity in plants.

1.7 Conclusion and Future Prospectives: Paving the Way for Resilient Crops and Global Food Security Under Abiotic Stress Condition

Our knowledge in terms of understanding the regulatory mechanism and signaling is very fragmented. The connection between the most known signaling cascades and sensors still needs to be found. Omics approaches including proteomics, genomics, and metabolomics are currently widely adopted to understand the crosstalk between signaling cascades. The review summarizes the regulatory mechanisms and involvement of the stress-perceptive genes from up to down stream signaling to improve stress tolerance under different abiotic stress conditions. The understanding and detailed analysis can help us to improve crops in terms of enhanced senescence, productivity, yield, etc. by using genetic engineering and gene editing techniques as the accuracy is high and precise. So, a detailed analysis of the multifaceted signaling of the plant under different stresses at the molecular level is key to producing stress-tolerant crops to achieve the global target of anticipating hunger of the growing population.

References

Abat JK, Deswal R (2009) Differential modulation of S-nitrosoproteome of Brassica juncea by low temperature: change in S-nitrosylation of Rubisco is responsible for the inactivation of its carboxylase activity. Proteomics 9(18):4368–4380

Agarwal M, Hao Y, Kapoor A, Dong CH, Fujii H, Zheng X, Zhu JK (2006) A R2R3 type MYB transcription factor is involved in the cold regulation of CBF genes and in acquired freezing tolerance. J Biol Chem 281(49):37636–37645

Agulnick AD, Taira M, Breen JJ, Tanaka T, Dawid IB, Westphal H (1996) Interactions of the LIM-domain-binding factor Ldbl with LIM homeodomain proteins. Nature 384(6606):270–272

Ahmad M (2023) Plant breeding advancements with "CRISPR-Cas" genome editing technologies will assist future food security. Front Plant Sci 14:1133036

Ahmad A, Hadi F, Ali N (2015) Effective phytoextraction of cadmium (Cd) with increasing concentration of total phenolics and free proline in Cannabis sativa (L) plant under various treatments of fertilizers, plant growth regulators and sodium salt. Int J Phytoremed 17(1):56–65

Alcázar R, Parker JE (2011) The impact of temperature on balancing immune responsiveness and growth in Arabidopsis. Trends Plant Sci 16(12):666–675

Ali A, Kim JK, Jan M, Khan HA, Khan IU, Shen M, Park J, Lim CJ, Hussain S, Baek D, Wang K (2019) Rheostatic control of ABA signaling through HOS15-mediated OST1 degradation. Mol Plant 12(11):1447–1462

Álvarez S, Acosta-Motos JR, Sánchez-Blanco MJ (2023) Morphological performance and seasonal pattern of water relations and gas exchange in Pistacia lentiscus plants subjected to salinity and water deficit. Front Plant Sci 14:1237332

Andrási N, Rigó G, Zsigmond L, Pérez-Salamó I, Papdi C, Klement E, Pettkó-Szandtner A, Baba AI, Ayaydin F, Dasari R, Cséplő Á (2019) The mitogen-activated protein kinase 4-phosphorylated heat shock factor A4A regulates responses to combined salt and heat stresses. J Exp Bot 70(18):4903–4918

Aryal JP, Sapkota TB, Khurana R, Khatri-Chhetri A, Rahut DB, Jat ML (2020) Climate change and agriculture in South Asia: adaptation options in smallholder production systems. Environ Dev Sustain 22(6):5045–5075

Asai S, Ohta K, Yoshioka H (2008) MAPK signaling regulates nitric oxide and NADPH oxidase-dependent oxidative bursts in Nicotiana benthamiana. Plant Cell 20(5):1390–1406

Ashraf M, Akram NA, Foolad MR (2012) Marker-assisted selection in plant breeding for salinity tolerance. In: Plant salt tolerance: methods and protocols, pp 305–333

Basu D, Haswell ES (2020) The mechanosensitive ion channel MSL10 potentiates responses to cell swelling in Arabidopsis seedlings. Curr Biol 30(14):2716–2728

Batelli G, Verslues PE, Agius F, Qiu Q, Fujii H, Pan S, Schumaker KS, Grillo S, Zhu JK (2007) SOS2 promotes salt tolerance in part by interacting with the vacuolar H+-ATPase and upregulating its transport activity. Mol Cell Biol 27(22):7781–7790

Becraft PW (2002) Receptor kinase signaling in plant development. Annu Rev Cell Dev Biol 18(1): 163–192

Belda-Palazon B, Adamo M, Valerio C, Ferreira LJ, Confraria A, Reis-Barata D, Rodrigues A, Meyer C, Rodriguez PL, Baena-González E (2020) A dual function of SnRK2 kinases in the regulation of SnRK1 and plant growth. Nature Plants 6(11):1345–1353

Bhardwaj R, Lone JK, Pandey R, Mondal N, Dhandapani R, Meena SK, Khan S (2023) Insights into morphological and physio-biochemical adaptive responses in mungbean (Vigna radiata L.) under heat stress. Front Genet 14:1206451

Brown S, Nicholls RJ, Lázár AN, Hornby DD, Hill C, Hazra S, Appeaning Addo K, Haque A, Caesar J, Tompkins EL (2018) What are the implications of sea-level rise for a 1.5, 2 and 3° C rise in global mean temperatures in the Ganges-Brahmaputra-Meghna and other vulnerable deltas? Reg Environ Chang 18:1829–1842

Bucholc M, Ciesielski A, Goch G, Anielska-Mazur A, Kulik A, Krzywińska E, Dobrowolska G (2011) SNF1-related protein kinases 2 are negatively regulated by a plant-specific calcium sensor. J Biol Chem 286(5):3429–3441

Cantrel C, Vazquez T, Puyaubert J, Rezé N, Lesch M, Kaiser WM, Dutilleul C, Guillas I, Zachowski A, Baudouin E (2011) Nitric oxide participates in cold-responsive phosphosphingolipid formation and gene expression in Arabidopsis thaliana. New Phytol 189(2):415–427

Cavrak VV, Lettner N, Jamge S, Kosarewicz A, Bayer LM, Mittelsten Scheid O (2014) How a retrotransposon exploits the plant's heat stress response for its activation. PLoS Genet 10(1): e1004115

Chen CT, Chen LM, Lin CC, Kao CH (2001) Regulation of proline accumulation in detached rice leaves exposed to excess copper. Plant Sci 160(2):283–290

Chen J, Yu F, Liu Y, Du C, Li X, Zhu S, Wang X, Lan W, Rodriguez PL, Liu X, Li D (2016) FERONIA interacts with ABI2-type phosphatases to facilitate signaling cross-talk between abscisic acid and RALF peptide in Arabidopsis. Proc Natl Acad Sci 113(37):E5519–E5527

Chen K, Gao J, Sun S, Zhang Z, Yu B, Li J, Xie C, Li G, Wang P, Song CP, Bressan RA (2020) BONZAI proteins control global osmotic stress responses in plants. Curr Biol 30(24): 4815–4825

Cheng C, Wang Z, Ren Z, Zhi L, Yao B, Su C, Liu L, Li X (2017) SCFAtPP2-B11 modulates ABA signaling by facilitating SnRK2. 3 degradation in Arabidopsis thaliana. PLoS Genet 13(8): e1006947

Clemens S (2001) Molecular mechanisms of plant metal tolerance and homeostasis. Planta 212: 475–486

Cobbett CS (2000) Phytochelatins and their roles in heavy metal detoxification. Plant Physiol 123(3):825–832

Cortijo S, Charoensawan V, Brestovitsky A, Buning R, Ravarani C, Rhodes D, van Noort J, Jaeger KE, Wigge PA (2017) Transcriptional regulation of the ambient temperature response by H2A. Z nucleosomes and HSF1 transcription factors in Arabidopsis. Mol Plant 10(10):1258–1273

Costa-Broseta Á, Perea-Resa C, Castillo MC, Ruíz MF, Salinas J, León J (2018) Nitric oxide controls constitutive freezing tolerance in Arabidopsis by attenuating the levels of osmoprotectants, stress-related hormones and anthocyanins. Sci Rep 8(1):9268

Danquah A, de Zélicourt A, Boudsocq M, Neubauer J, Frei dit Frey N, Leonhardt N, Pateyron S, Gwinner F, Tamby JP, Ortiz-Masia D, Marcote MJ (2015) Identification and characterization of an ABA-activated MAP kinase cascade in Arabidopsis thaliana. Plant J 82(2):232–244

De Smet I, Voss U, Jürgens G, Beeckman T (2009) Receptor-like kinases shape the plant. Nat Cell Biol 11(10):1166–1173

Demidchik V, Bowen HC, Maathuis FJ, Shabala SN, Tester MA, White PJ, Davies JM (2002) Arabidopsis thaliana root non-selective cation channels mediate calcium uptake and are involved in growth. Plant J 32(5):799–808

Deng Y, Humbert S, Liu JX, Srivastava R, Rothstein SJ, Howell SH (2011) Heat induces the splicing by IRE1 of a mRNA encoding a transcription factor involved in the unfolded protein response in Arabidopsis. Proc Natl Acad Sci 108(17):7247–7252

Ding Y, Li H, Zhang X, Xie Q, Gong Z, Yang S (2015) OST1 kinase modulates freezing tolerance by enhancing ICE1 stability in Arabidopsis. Dev Cell 32(3):278–289

Ding Y, Lv J, Shi Y, Gao J, Hua J, Song C, Gong Z, Yang S (2019a) EGR 2 phosphatase regulates OST 1 kinase activity and freezing tolerance in Arabidopsis. EMBO J 38(1):e99819

Ding Y, Shi Y, Yang S (2019b) Advances and challenges in uncovering cold tolerance regulatory mechanisms in plants. New Phytol 222(4):1690–1704

Ding Y, Shi Y, Yang S (2020) Molecular regulation of plant responses to environmental temperatures. Mol Plant 13(4):544–564

Dixit AK, Jayabaskaran C (2014) Calcium signaling during abiotic stress in plants: roles of calcium dependent protein kinases. In: Physiology of nutrition and environmental stresses on crop productivity, p 348

Dodd AN, Kudla J, Sanders D (2010) The language of calcium signaling. Annu Rev Plant Biol 61: 593–620

Dong CH, Pei H (2014) Over-expression of miR397 improves plant tolerance to cold stress in Arabidopsis thaliana. J Plant Biol 57:209–217

Dong MA, Farré EM, Thomashow MF (2011) Circadian clock-associated 1 and late elongated hypocotyl regulate expression of the C-repeat binding factor (CBF) pathway in Arabidopsis. Proc Natl Acad Sci 108(17):7241–7246

Dowgert MF, Steponkus PL (1984) Behavior of the plasma membrane of isolated protoplasts during a freeze-thaw cycle. Plant Physiol 75(4):1139–1151

Eide D, Broderius M, Fett J, Guerinot ML (1996) A novel iron-regulated metal transporter from plants identified by functional expression in yeast. Proc Natl Acad Sci 93(11):5624–5628

El-Mounadi K, Morales-Floriano ML, Garcia-Ruiz H (2020) Principles, applications, and biosafety of plant genome editing using CRISPR-Cas9. Front Plant Sci 11:56

Eremina M, Unterholzner SJ, Rathnayake AI, Castellanos M, Khan M, Kugler KG, May ST, Mayer KF, Rozhon W, Poppenberger B (2016) Brassinosteroids participate in the control of basal and acquired freezing tolerance of plants. Proc Natl Acad Sci 113(40):E5982–E5991

Ezer D, Jung JH, Lan H, Biswas S, Gregoire L, Box MS, Charoensawan V, Cortijo S, Lai X, Stöckle D, Zubieta C (2017) The evening complex coordinates environmental and endogenous signals in Arabidopsis. Nature Plants 3(7):1–12

Fàbregas N, Yoshida T, Fernie AR (2020) Role of Raf-like kinases in SnRK2 activation and osmotic stress response in plants. Nat Commun 11(1):6184

Fang X, Zhao G, Zhang S, Li Y, Gu H, Li Y, Zhao Q, Qi Y (2019) Chloroplast-to-nucleus signaling regulates microRNA biogenesis in Arabidopsis. Dev Cell 48(3):371–382

Farooq M, Gogoi N, Hussain M, Barthakur S, Paul S, Bharadwaj N, Migdadi HM, Alghamdi SS, Siddique KH (2017) Effects, tolerance mechanisms and management of salt stress in grain legumes. Plant Physiol Biochem 118:199–217

Finka A, Cuendet AFH, Maathuis FJ, Saidi Y, Goloubinoff P (2012) Plasma membrane cyclic nucleotide gated calcium channels control land plant thermal sensing and acquired thermotolerance. Plant Cell 24(8):3333–3348

Fitter AH, Fitter RSR (2002) Rapid changes in flowering time in British plants. Science 296(5573): 1689–1691

Foyer CH, Theodoulou FL, Delrot S (2001) The functions of inter-and intracellular glutathione transport systems in plants. Trends Plant Sci 6(10):486–492

Fu H, Yang Y (2023) How plants tolerate salt stress. Curr Issues Mol Biol 45(7):5914–5934

Fu J, Wu H, Ma S, Xiang D, Liu R, Xiong L (2017) OsJAZ1 attenuates drought resistance by regulating JA and ABA signaling in rice. Front Plant Sci 8:2108

Fujii H, Zhu JK (2009) Arabidopsis mutant deficient in 3 abscisic acid-activated protein kinases reveals critical roles in growth, reproduction, and stress. Proc Natl Acad Sci 106(20):8380–8385

Gangappa SN, Berriri S, Kumar SV (2017) PIF4 coordinates thermosensory growth and immunity in Arabidopsis. Curr Biol 27(2):243–249

Gao F, Han X, Wu J, Zheng S, Shang Z, Sun D, Zhou R, Li B (2012) A heat-activated calcium-permeable channel–Arabidopsis cyclic nucleotide-gated ion channel 6–is involved in heat shock responses. Plant J 70(6):1056–1069

Geiger D, Scherzer S, Mumm P, Stange A, Marten I, Bauer H, Ache P, Matschi S, Liese A, Al-Rasheid KA, Romeis T (2009) Activity of guard cell anion channel SLAC1 is controlled by drought-stress signaling kinase-phosphatase pair. Proc Natl Acad Sci 106(50):21425–21430

Gong M, van der Luit AH, Knight MR, Trewavas AJ (1998) Heat-shock-induced changes in intracellular Ca2+ level in tobacco seedlings in relation to thermotolerance. Plant Physiol 116(1):429–437

Gong Z, Xiong L, Shi H, Yang S, Herrera-Estrella LR, Xu G, Chao DY, Li J, Wang PY, Qin F, Li J (2020) Plant abiotic stress response and nutrient use efficiency. Sci China Life Sci 63:635–674

Gray SB, Brady SM (2016) Plant developmental responses to climate change. Dev Biol 419(1): 64–77

Grison MS, Kirk P, Brault ML, Wu XN, Schulze WX, Benitez-Alfonso Y, Immel F, Bayer EM (2019) Plasma membrane-associated receptor-like kinases relocalize to plasmodesmata in response to osmotic stress. Plant Physiol 181(1):142–160

Guan Q, Lu X, Zeng H, Zhang Y, Zhu J (2013) Heat stress induction of mi R398 triggers a regulatory loop that is critical for thermotolerance in Arabidopsis. Plant J 74(5):840–851

Guo Y, Halfter U, Ishitani M, Zhu JK (2001) Molecular characterization of functional domains in the protein kinase SOS2 that is required for plant salt tolerance. Plant Cell 13(6):1383–1400

Guo X, Liu D, Chong K (2018) Cold signaling in plants: insights into mechanisms and regulation. J Integr Plant Biol 60(9):745–756

Habben JE, Bao X, Bate NJ, DeBruin JL, Dolan D, Hasegawa D, Helentjaris TG, Lafitte RH, Lovan N, Mo H, Reimann K (2014) Transgenic alteration of ethylene biosynthesis increases grain yield in maize under field drought-stress conditions. Plant Biotechnol J 12(6):685–693

Hahn A, Bublak D, Schleiff E, Scharf KD (2011) Crosstalk between Hsp90 and Hsp70 chaperones and heat stress transcription factors in tomato. Plant Cell 23(2):741–755

Halfter U, Ishitani M, Zhu JK (2000) The Arabidopsis SOS2 protein kinase physically interacts with and is activated by the calcium-binding protein SOS3. Proc Natl Acad Sci 97(7):3735–3740

Hall JÁ (2002) Cellular mechanisms for heavy metal detoxification and tolerance. J Exp Bot 53(366):1–11

Hasanuzzaman M, Fujita M (2022) Plant responses and tolerance to salt stress: Physiological and molecular interventions. Int J Mol Sci 23(9):4810

Hasanuzzaman M, Nahar K, Alam MM, Roychowdhury R, Fujita M (2013) Physiological, biochemical, and molecular mechanisms of heat stress tolerance in plants. Int J Mol Sci 14(5): 9643–9684

Hedhly A (2011) Sensitivity of flowering plant gametophytes to temperature fluctuations. Environ Exp Bot 74:9–16

Hodson MJ (2012) Metal toxicity and tolerance in plants. Biochemist 34(5):28–32

Howell SH (2013) Endoplasmic reticulum stress responses in plants. Annu Rev Plant Biol 64:477–499

Hrabak EM, Chan CW, Gribskov M, Harper JF, Choi JH, Halford N, Kudla J, Luan S, Nimmo HG, Sussman MR, Thomas M (2003) The Arabidopsis CDPK-SnRK superfamily of protein kinases. Plant Physiol 132(2):666–680

Hsiao TC (1973) Plant responses to water stress. Annu Rev Plant Physiol 24(1):519–570

Ibañez C, Poeschl Y, Peterson T, Bellstädt J, Denk K, Gogol-Döring A, Quint M, Delker C (2017) Ambient temperature and genotype differentially affect developmental and phenotypic plasticity in Arabidopsis thaliana. BMC Plant Biol 17(1):1–14

Imes D, Mumm P, Böhm J, Al-Rasheid KA, Marten I, Geiger D, Hedrich R (2013) Open stomata 1 (OST 1) kinase controls R–type anion channel QUAC 1 in Arabidopsis guard cells. Plant J 74(3):372–382

Ishitani M, Liu J, Halfter U, Kim CS, Shi W, Zhu JK (2000) SOS3 function in plant salt tolerance requires N-myristoylation and calcium binding. Plant Cell 12(9):1667–1677

Ito H, Gaubert H, Bucher E, Mirouze M, Vaillant I, Paszkowski J (2011) An siRNA pathway prevents transgenerational retrotransposition in plants subjected to stress. Nature 472(7341): 115–119

Jalmi SK, Sinha AK (2015) ROS mediated MAPK signaling in abiotic and biotic stress-striking similarities and differences. Front Plant Sci 6:769

Jalmi SK, Bhagat PK, Verma D, Noryang S, Tayyeba S, Singh K, Sharma D, Sinha AK (2018) Traversing the links between heavy metal stress and plant signaling. Front Plant Sci 9:12

Jiang Z, Zhou X, Tao M, Yuan F, Liu L, Wu F, Wu X, Xiang Y, Niu Y, Liu F, Li C (2019) Plant cell-surface GIPC sphingolipids sense salt to trigger Ca2+ influx. Nature 572(7769):341–346

Jung JH, Domijan M, Klose C, Biswas S, Ezer D, Gao M, Khattak AK, Box MS, Charoensawan V, Cortijo S, Kumar M (2016) Phytochromes function as thermosensors in Arabidopsis. Science 354(6314):886–889

Kidokoro S, Maruyama K, Nakashima K, Imura Y, Narusaka Y, Shinwari ZK, Osakabe Y, Fujita Y, Mizoi J, Shinozaki K, Yamaguchi-Shinozaki K (2009) The phytochrome-interacting factor PIF7 negatively regulates DREB1 expression under circadian control in Arabidopsis. Plant Physiol 151(4):2046–2057

Kidokoro S, Yoneda K, Takasaki H, Takahashi F, Shinozaki K, Yamaguchi-Shinozaki K (2017) Different cold-signaling pathways function in the responses to rapid and gradual decreases in temperature. Plant Cell 29(4):760–774

Kim DY, Bovet L, Maeshima M, Martinoia E, Lee Y (2007) The ABC transporter AtPDR8 is a cadmium extrusion pump conferring heavy metal resistance. Plant J 50(2):207–218

Kim JS, Mizoi J, Kidokoro S, Maruyama K, Nakajima J, Nakashima K, Mitsuda N, Takiguchi Y, Ohme-Takagi M, Kondou Y, Yoshizumi T (2012) Arabidopsis GROWTH-REGULATING FACTOR7 functions as a transcriptional repressor of abscisic acid–and osmotic stress–responsive genes, including DREB2A. Plant Cell 24(8):3393–3405

Kim WY, Ali Z, Park HJ, Park SJ, Cha JY, Perez-Hormaeche J, Quintero FJ, Shin G, Kim MR, Qiang Z, Ning L (2013) Release of SOS2 kinase from sequestration with GIGANTEA determines salt tolerance in Arabidopsis. Nat Commun 4(1):1352

Kindgren P, Ard R, Ivanov M, Marquardt S (2018) Transcriptional read-through of the long non-coding RNA SVALKA governs plant cold acclimation. Nat Commun 9(1):4561

Knight H, Trewavas AJ, Knight MR (1996) Cold calcium signaling in Arabidopsis involves two cellular pools and a change in calcium signature after acclimation. Plant Cell 8(3):489–503

Kocsy G, Tóth B, Berzy T, Szalai G, Jednákovits A, Galiba G (2001) Glutathione reductase activity and chilling tolerance are induced by a hydroxylamine derivative BRX-156 in maize and soybean. Plant Sci 160(5):943–950

Koini MA, Alvey L, Allen T, Tilley CA, Harberd NP, Whitelam GC, Franklin KA (2009) High temperature-mediated adaptations in plant architecture require the bHLH transcription factor PIF4. Curr Biol 19(5):408–413

Konapala G, Mishra AK, Wada Y, Mann ME (2020) Climate change will affect global water availability through compounding changes in seasonal precipitation and evaporation. Nat Commun 11(1):3044

Königshofer H, Tromballa HW, Löppert HG (2008) Early events in signalling high-temperature stress in tobacco BY2 cells involve alterations in membrane fluidity and enhanced hydrogen peroxide production. Plant Cell Environ 31(12):1771–1780

Korver RA, Koevoets IT, Testerink C (2018) Out of shape during stress: a key role for auxin. Trends Plant Sci 23(9):783–793

Kotak S, Larkindale J, Lee U, von Koskull-Döring P, Vierling E, Scharf KD (2007) Complexity of the heat stress response in plants. Curr Opin Plant Biol 10(3):310–316

Kovtun Y, Chiu WL, Tena G, Sheen J (2000) Functional analysis of oxidative stress-activated mitogen-activated protein kinase cascade in plants. Proc Natl Acad Sci 97(6):2940–2945

Kulik A, Wawer I, Krzywińska E, Bucholc M, Dobrowolska G (2011) SnRK2 protein kinases—key regulators of plant response to abiotic stresses. Omics 15(12):859–872
Kumar SV, Wigge PA (2010) H2A. Z-containing nucleosomes mediate the thermosensory response in Arabidopsis. Cell 140(1):136–147
Lee CM, Thomashow MF (2012) Photoperiodic regulation of the C-repeat binding factor (CBF) cold acclimation pathway and freezing tolerance in Arabidopsis thaliana. Proc Natl Acad Sci 109(37):15054–15059
Lee SC, Lan W, Buchanan BB, Luan S (2009) A protein kinase-phosphatase pair interacts with an ion channel to regulate ABA signaling in plant guard cells. Proc Natl Acad Sci 106(50): 21419–21424
Legris M, Klose C, Burgie ES, Rojas CCR, Neme M, Hiltbrunner A, Wigge PA, Schäfer E, Vierstra RD, Casal JJ (2016) Phytochrome B integrates light and temperature signals in Arabidopsis. Science 354(6314):897–900
Li J, Nam KH (2002) Regulation of brassinosteroid signaling by a GSK3/SHAGGY-like kinase. Science 295(5558):1299–1301
Li G, Meng X, Wang R, Mao G, Han L, Liu Y, Zhang S (2012) Dual-level regulation of ACC synthase activity by MPK3/MPK6 cascade and its downstream WRKY transcription factor during ethylene induction in Arabidopsis. PLoS Genet 8(6):e1002767
Li CH, Wang G, Zhao JL, Zhang LQ, Ai LF, Han YF, Sun DY, Zhang SW, Sun Y (2014) The receptor-like kinase SIT1 mediates salt sensitivity by activating MAPK3/6 and regulating ethylene homeostasis in rice. Plant Cell 26(6):2538–2553
Li H, Ye K, Shi Y, Cheng J, Zhang X, Yang S (2017a) BZR1 positively regulates freezing tolerance via CBF-dependent and CBF-independent pathways in Arabidopsis. Mol Plant 10(4):545–559
Li K, Yang F, Zhang G, Song S, Li Y, Ren D, Miao Y, Song CP (2017b) AIK1, a mitogen-activated protein kinase, modulates abscisic acid responses through the MKK5-MPK6 kinase cascade. Plant Physiol 173(2):1391–1408
Li B, Gao K, Ren H, Tang W (2018) Molecular mechanisms governing plant responses to high temperatures. J Integr Plant Biol 60(9):757–779
Liao H, Tang R, Zhang X, Luan S, Yu F (2017) FERONIA receptor kinase at the crossroads of hormone signaling and stress responses. Plant Cell Physiol 58(7):1143–1150
Liu Y, Zhang S (2004) Phosphorylation of 1-aminocyclopropane-1-carboxylic acid synthase by MPK6, a stress-responsive mitogen-activated protein kinase, induces ethylene biosynthesis in Arabidopsis. Plant Cell 16(12):3386–3399
Liu J, Zhu JK (1998) A calcium sensor homolog required for plant salt tolerance. Science 280(5371):1943–1945
Liu Q, Kasuga M, Sakuma Y, Abe H, Miura S, Yamaguchi-Shinozaki K, Shinozaki K (1998) Two transcription factors, DREB1 and DREB2, with an EREBP/AP2 DNA binding domain separate two cellular signal transduction pathways in drought-and low-temperature-responsive gene expression, respectively, Arabidopsis. Plant Cell 10(8):1391–1406
Liu HT, Sun DY, Zhou RG (2005) Ca2+ and AtCaM3 are involved in the expression of heat shock protein gene in Arabidopsis. Plant Cell Environ 28(10):1276–1284
Liu C, Ou S, Mao B, Tang J, Wang W, Wang H, Cao S, Schläppi MR, Zhao B, Xiao G, Wang X (2018a) Early selection of bZIP73 facilitated adaptation of japonica rice to cold climates. Nat Commun 9(1):3302
Liu J, Shi Y, Yang S (2018b) Insights into the regulation of C-repeat binding factors in plant cold signaling. J Integr Plant Biol 60(9):780–795
Liu XH, Lyu YS, Yang W, Yang ZT, Lu SJ, Liu JX (2020) A membrane-associated NAC transcription factor OsNTL3 is involved in thermotolerance in rice. Plant Biotechnol J 18(5): 1317–1329
Lorrain S, Allen T, Duek PD, Whitelam GC, Fankhauser C (2008) Phytochrome-mediated inhibition of shade avoidance involves degradation of growth-promoting bHLH transcription factors. Plant J 53(2):312–323

Lu G, Wu FQ, Wu W, Wang HJ, Zheng XM, Zhang Y, Chen X, Zhou K, Jin M, Cheng Z, Li X (2014) Rice LTG 1 is involved in adaptive growth and fitness under low ambient temperature. Plant J 78(3):468–480

Lv Y, Yang M, Hu D, Yang Z, Ma S, Li X, Xiong L (2017) The OsMYB30 transcription factor suppresses cold tolerance by interacting with a JAZ protein and suppressing β-amylase expression. Plant Physiol 173(2):1475–1491

Ma Q, Dai X, Xu Y, Guo J, Liu Y, Chen N, Xiao J, Zhang D, Xu Z, Zhang X, Chong K (2009a) Enhanced tolerance to chilling stress in OsMYB3R-2 transgenic rice is mediated by alteration in cell cycle and ectopic expression of stress genes. Plant Physiol 150(1):244–256

Ma Y, Szostkiewicz I, Korte A, Moes D, Yang Y, Christmann A, Grill E (2009b) Regulators of PP2C phosphatase activity function as abscisic acid sensors. Science 324(5930):1064–1068

Ma Y, Dai X, Xu Y, Luo W, Zheng X, Zeng D, Pan Y, Lin X, Liu H, Zhang D, Xiao J (2015) COLD1 confers chilling tolerance in rice. Cell 160(6):1209–1221

Martínez C, Espinosa-Ruíz A, de Lucas M, Bernardo-García S, Franco-Zorrilla JM, Prat S (2018) PIF 4-induced BR synthesis is critical to diurnal and thermomorphogenic growth. EMBO J 37(23):e99552

May MJ, Vernoux T, Leaver C, Montagu MV, Inzé D (1998) Glutathione homeostasis in plants: implications for environmental sensing and plant development. J Exp Bot 49(321):649–667

Mendoza-Cózatl DG, Jobe TO, Hauser F, Schroeder JI (2011) Long-distance transport, vacuolar sequestration, tolerance, and transcriptional responses induced by cadmium and arsenic. Curr Opin Plant Biol 14(5):554–562

Meyer K, Leube MP, Grill E (1994) A protein phosphatase 2C involved in ABA signal transduction in Arabidopsis thaliana. Science 264(5164):1452–1455

Miura K, Jin JB, Lee J, Yoo CY, Stirm V, Miura T, Ashworth EN, Bressan RA, Yun DJ, Hasegawa PM (2007) SIZ1-mediated sumoylation of ICE1 controls CBF3/DREB1A expression and freezing tolerance in Arabidopsis. Plant Cell 19(4):1403–1414

Mizoi J, Kanazawa N, Kidokoro S, Takahashi F, Qin F, Morimoto K, Shinozaki K, Yamaguchi-Shinozaki K (2019) Heat-induced inhibition of phosphorylation of the stress-protective transcription factor DREB2A promotes thermotolerance of Arabidopsis thaliana. J Biol Chem 294(3):902–917

Munns R, Tester M (2008) Mechanisms of salinity tolerance. Annu Rev Plant Biol 59:651–681

Munns R, James RA, Xu B, Athman A, Conn SJ, Jordans C, Byrt CS, Hare RA, Tyerman SD, Tester M, Plett D (2012) Wheat grain yield on saline soils is improved by an ancestral Na+ transporter gene. Nat Biotechnol 30(4):360–364

Nakamichi N, Kusano M, Fukushima A, Kita M, Ito S, Yamashino T, Saito K, Sakakibara H, Mizuno T (2009) Transcript profiling of an Arabidopsis PSEUDO RESPONSE REGULATOR arrhythmic triple mutant reveals a role for the circadian clock in cold stress response. Plant Cell Physiol 50(3):447–462

Nakashima K, Ito Y, Yamaguchi-Shinozaki K (2009) Transcriptional regulatory networks in response to abiotic stresses in Arabidopsis and grasses. Plant Physiol 149(1):88–95

Nguyen TQ, Sesin V, Kisiala A, Emery RN (2021) Phytohormonal roles in plant responses to heavy metal stress: implications for using macrophytes in phytoremediation of aquatic ecosystems. Environ Toxicol Chem 40(1):7–22

Noctor G, Arisi ACM, Jouanin L, Kunert KJ, Rennenberg H, Foyer CH (1998) Glutathione: biosynthesis, metabolism and relationship to stress tolerance explored in transformed plants. J Exp Bot 49(321):623–647

Nusinow DA, Helfer A, Hamilton EE, King JJ, Imaizumi T, Schultz TF, Farré EM, Kay SA (2011) The ELF4–ELF3–LUX complex links the circadian clock to diurnal control of hypocotyl growth. Nature 475(7356):398–402

Oh E, Zhu JY, Wang ZY (2012) Interaction between BZR1 and PIF4 integrates brassinosteroid and environmental responses. Nat Cell Biol 14(8):802–809

Oh E, Zhu JY, Bai MY, Arenhart RA, Sun Y, Wang ZY (2014) Cell elongation is regulated through a central circuit of interacting transcription factors in the Arabidopsis hypocotyl. elife 3:e03031

Ohama N, Sato H, Shinozaki K, Yamaguchi-Shinozaki K (2017) Transcriptional regulatory network of plant heat stress response. Trends Plant Sci 22(1):53–65
Olate E, Jiménez-Gómez JM, Holuigue L, Salinas J (2018) NPR1 mediates a novel regulatory pathway in cold acclimation by interacting with HSFA1 factors. Nature Plants 4(10):811–823
Opdenakker K, Remans T, Vangronsveld J, Cuypers A (2012) Mitogen-activated protein (MAP) kinases in plant metal stress: regulation and responses in comparison to other biotic and abiotic stresses. Int J Mol Sci 13(6):7828–7853
Örvar BL, Sangwan V, Omann F, Dhindsa RS (2000) Early steps in cold sensing by plant cells: the role of actin cytoskeleton and membrane fluidity. Plant J 23(6):785–794
Park SY, Fung P, Nishimura N, Jensen DR, Fujii H, Zhao Y, Lumba S, Santiago J, Rodrigues A, Chow TFF, Alfred SE (2009) Abscisic acid inhibits type 2C protein phosphatases via the PYR/PYL family of START proteins. Science 324(5930):1068–1071
Park J, Song WY, Ko D, Eom Y, Hansen TH, Schiller M, Lee TG, Martinoia E, Lee Y (2012) The phytochelatin transporters AtABCC1 and AtABCC2 mediate tolerance to cadmium and mercury. Plant J 69(2):278–288
Pearce RS (2001) Plant freezing and damage. Ann Bot 87(4):417–424
Peng P, Yan Z, Zhu Y, Li J (2008) Regulation of the Arabidopsis GSK3-like kinase BRASSINOSTEROID-INSENSITIVE 2 through proteasome-mediated protein degradation. Mol Plant 1(2):338–346
Pérez-Salamó I, Papdi C, Rigó G, Zsigmond L, Vilela B, Lumbreras V, Nagy I, Horváth B, Domoki M, Darula Z, Medzihradszky K (2014) The heat shock factor A4A confers salt tolerance and is regulated by oxidative stress and the mitogen-activated protein kinases MPK3 and MPK6. Plant Physiol 165(1):319–334
Perrella G, Bäurle I, van Zanten M (2022) Epigenetic regulation of thermomorphogenesis and heat stress tolerance. New Phytol 234(4):1144–1160
Peuke AD, Rennenberg H (2005) Phytoremediation with transgenic trees. Z Naturforsch C 60:199–207
Pitzschke A, Djamei A, Bitton F, Hirt H (2009) A major role of the MEKK1–MKK1/2–MPK4 pathway in ROS signalling. Mol Plant 2(1):120–137
Postiglione AE, Muday GK (2023) Abscisic acid increases hydrogen peroxide in mitochondria to facilitate stomatal closure. Plant Physiol 192(1):469–487
Puyaubert J, Fares A, Rézé N, Peltier JB, Baudouin E (2014) Identification of endogenously S-nitrosylated proteins in Arabidopsis plantlets: effect of cold stress on cysteine nitrosylation level. Plant Sci 215:150–156
Qi J, Song CP, Wang B, Zhou J, Kangasjärvi J, Zhu JK, Gong Z (2018) Reactive oxygen species signaling and stomatal movement in plant responses to drought stress and pathogen attack. J Integr Plant Biol 60(9):805–826
Qin H, Huang R (2020) The phytohormonal regulation of Na+/K+ and reactive oxygen species homeostasis in rice salt response. Mol Breed 40(5):47
Qin F, Sakuma Y, Tran LSP, Maruyama K, Kidokoro S, Fujita Y, Fujita M, Umezawa T, Sawano Y, Miyazono KI, Tanokura M (2008) Arabidopsis DREB2A-interacting proteins function as RING E3 ligases and negatively regulate plant drought stress–responsive gene expression. Plant Cell 20(6):1693–1707
Qiu QS, Guo Y, Quintero FJ, Pardo JM, Schumaker KS, Zhu JK (2004) Regulation of vacuolar Na+/H+ exchange in Arabidopsis thaliana by the salt-overly-sensitive (SOS) pathway. J Biol Chem 279(1):207–215
Quan R, Lin H, Mendoza I, Zhang Y, Cao W, Yang Y, Shang M, Chen S, Pardo JM, Guo Y (2007) SCABP8/CBL10, a putative calcium sensor, interacts with the protein kinase SOS2 to protect Arabidopsis shoots from salt stress. Plant Cell 19(4):1415–1431
Quint M, Delker C, Franklin KA, Wigge PA, Halliday KJ, Van Zanten M (2016) Molecular and genetic control of plant thermomorphogenesis. Nature Plants 2(1):1–9

Rachowka J, Anielska-Mazur A, Bucholc M, Stephenson K, Kulik A (2023) SnRK2. 10 kinase differentially modulates expression of hub WRKY transcription factors genes under salinity and oxidative stress in Arabidopsis thaliana. Front Plant Sci 14:1135240

Rajkowitsch L, Chen D, Stampfl S, Semrad K, Waldsich C, Mayer O, Jantsch MF, Konrat R, Bläsi U, Schroeder R (2007) RNA chaperones, RNA annealers and RNA helicases. RNA Biol 4(3):118–130

Rauser WE (1999) Structure and function of metal chelators produced by plants: the case for organic acids, amino acids, phytin, and metallothioneins. Cell Biochem Biophys 31:19–48

Ren ZH, Gao JP, Li LG, Cai XL, Huang W, Chao DY, Zhu MZ, Wang ZY, Luan S, Lin HX (2005) A rice quantitative trait locus for salt tolerance encodes a sodium transporter. Nat Genet 37(10): 1141–1146

Rennie EA, Ebert B, Miles GP, Cahoon RE, Christiansen KM, Stonebloom S, Khatab H, Twell D, Petzold CJ, Adams PD, Dupree P (2014) Identification of a sphingolipid α-glucuronosyltransferase that is essential for pollen function in Arabidopsis. Plant Cell 26(8):3314–3325

Rucińska-Sobkowiak R (2016) Water relations in plants subjected to heavy metal stresses. Acta Physiol Plant 38:1–13

Ruelland E, Vaultier MN, Zachowski A, Hurry V (2009) Cold signalling and cold acclimation in plants. Adv Bot Res 49:35–150

Sage TL, Bagha S, Lundsgaard-Nielsen V, Branch HA, Sultmanis S, Sage RF (2015) The effect of high temperature stress on male and female reproduction in plants. Field Crop Res 182:30–42

Saidi Y, Finka A, Muriset M, Bromberg Z, Weiss YG, Maathuis FJ, Goloubinoff P (2009) The heat shock response in moss plants is regulated by specific calcium-permeable channels in the plasma membrane. Plant Cell 21(9):2829–2843

Sakuma Y, Maruyama K, Qin F, Osakabe Y, Shinozaki K, Yamaguchi-Shinozaki K (2006) Dual function of an Arabidopsis transcription factor DREB2A in water-stress-responsive and heat-stress-responsive gene expression. Proc Natl Acad Sci 103(49):18822–18827

Sangwan V, Örvar BL, Beyerly J, Hirt H, Dhindsa RS (2002) Opposite changes in membrane fluidity mimic cold and heat stress activation of distinct plant MAP kinase pathways. Plant J 31(5):629–638

Sato H, Mizoi J, Tanaka H, Maruyama K, Qin F, Osakabe Y, Morimoto K, Ohori T, Kusakabe K, Nagata M, Shinozaki K (2014) Arabidopsis DPB3-1, a DREB2A interactor, specifically enhances heat stress-induced gene expression by forming a heat stress-specific transcriptional complex with NF-Y subunits. Plant Cell 26(12):4954–4973

Sato H, Todaka D, Kudo M, Mizoi J, Kidokoro S, Zhao Y, Shinozaki K, Yamaguchi-Shinozaki K (2016) The A rabidopsis transcriptional regulator DPB 3-1 enhances heat stress tolerance without growth retardation in rice. Plant Biotechnol J 14(8):1756–1767

Schellingen K, Van Der Straeten D, Remans T, Loix C, Vangronsveld J, Cuypers A (2015) Ethylene biosynthesis is involved in the early oxidative challenge induced by moderate Cd exposure in Arabidopsis thaliana. Environ Exp Bot 117:1–11

Schmidt R, Mieulet D, Hubberten HM, Obata T, Hoefgen R, Fernie AR, Fisahn J, San Segundo B, Guiderdoni E, Schippers JH, Mueller-Roeber B (2013) SALT-RESPONSIVE ERF1 regulates reactive oxygen species–dependent signaling during the initial response to salt stress in rice. Plant Cell 25(6):2115–2131

Schroeder JI, Kwak JM, Allen GJ (2001) Guard cell abscisic acid signalling and engineering drought hardiness in plants. Nature 410(6826):327–330

Schutzendubel A, Polle A (2002) Plant responses to abiotic stresses: heavy metal-induced oxidative stress and protection by mycorrhization. J Exp Bot 53(372):1351–1365

Scrase-Field SA, Knight MR (2003) Calcium: just a chemical switch? Curr Opin Plant Biol 6(5): 500–506

Sehrawat A, Abat JK, Deswal R (2013) RuBisCO depletion improved proteome coverage of cold responsive S-nitrosylated targets in Brassica juncea. Front Plant Sci 4:342

Seo JS, Joo J, Kim MJ, Kim YK, Nahm BH, Song SI, Cheong JJ, Lee JS, Kim JK, Choi YD (2011) OsbHLH148, a basic helix-loop-helix protein, interacts with OsJAZ proteins in a jasmonate signaling pathway leading to drought tolerance in rice. Plant J 65(6):907–921

Seok HY, Lee SY, Sarker S, Bayzid M, Moon YH (2023) Genome-wide analysis of stress-responsive genes and alternative splice variants in Arabidopsis roots under osmotic stresses. Int J Mol Sci 24(19):14580

Sharma SS, Dietz KJ (2006) The significance of amino acids and amino acid-derived molecules in plant responses and adaptation to heavy metal stress. J Exp Bot 57(4):711–726

Shelake RM, Kadam US, Kumar R, Pramanik D, Singh AK, Kim JY (2022) Engineering drought and salinity tolerance traits in crops through CRISPR-mediated genome editing: targets, tools, challenges, and perspectives. Plant Commun 3:100417

Shi H, Quintero FJ, Pardo JM, Zhu JK (2002) The putative plasma membrane Na+/H+ antiporter SOS1 controls long-distance Na+ transport in plants. Plant Cell 14(2):465–477

Shi J, Gao H, Wang H, Lafitte HR, Archibald RL, Yang M, Hakimi SM, Mo H, Habben JE (2017) ARGOS 8 variants generated by CRISPR-Cas9 improve maize grain yield under field drought stress conditions. Plant Biotechnol J 15(2):207–216

Shi S, Li S, Asim M, Mao J, Xu D, Ullah Z, Liu G, Wang Q, Liu H (2018) The Arabidopsis calcium-dependent protein kinases (CDPKs) and their roles in plant growth regulation and abiotic stress responses. Int J Mol Sci 19(7):1900

Shiu SH, Bleecker AB (2003) Expansion of the receptor-like kinase/Pelle gene family and receptor-like proteins in Arabidopsis. Plant Physiol 132(2):530–543

Shiu SH, Karlowski WM, Pan R, Tzeng YH, Mayer KF, Li WH (2004) Comparative analysis of the receptor-like kinase family in Arabidopsis and rice. Plant Cell 16(5):1220–1234

Siddiqui KS, Cavicchioli R (2006) Cold-adapted enzymes. Annu Rev Biochem 75:403–433

Singh P, Mohanta TK, Sinha AK (2015) Unraveling the intricate nexus of molecular mechanisms governing rice root development: OsMPK3/6 and auxin-cytokinin interplay. PLoS One 10(4): e0123620

Singh M, Nara U, Kumar A, Choudhary A, Singh H, Thapa S (2021) Salinity tolerance mechanisms and their breeding implications. J Genet Eng Biotechnol 19(1):1–18

Soma F, Takahashi F, Suzuki T, Shinozaki K, Yamaguchi-Shinozaki K (2020) Plant Raf-like kinases regulate the mRNA population upstream of ABA-unresponsive SnRK2 kinases under drought stress. Nat Commun 11(1):1373

Stief A, Altmann S, Hoffmann K, Pant BD, Scheible WR, Bäurle I (2014) Arabidopsis miR156 regulates tolerance to recurring environmental stress through SPL transcription factors. Plant Cell 26(4):1792–1807

Su CF, Wang YC, Hsieh TH, Lu CA, Tseng TH, Yu SM (2010) A novel MYBS3-dependent pathway confers cold tolerance in rice. Plant Physiol 153(1):145–158

Sunkar R, Zhu JK (2004) Novel and stress-regulated microRNAs and other small RNAs from Arabidopsis. Plant Cell 16(8):2001–2019

Suri SS, Dhindsa RS (2008) A heat-activated MAP kinase (HAMK) as a mediator of heat shock response in tobacco cells. Plant Cell Environ 31(2):218–226

Suzuki N, Sejima H, Tam R, Schlauch K, Mittler R (2011) Identification of the MBF1 heat-response regulon of Arabidopsis thaliana. Plant J 66(5):844–851

Takahashi F, Mizoguchi T, Yoshida R, Ichimura K, Shinozaki K (2011) Calmodulin-dependent activation of MAP kinase for ROS homeostasis in Arabidopsis. Mol Cell 41(6):649–660

Takahashi F, Suzuki T, Osakabe Y, Betsuyaku S, Kondo Y, Dohmae N, Fukuda H, Yamaguchi-Shinozaki K, Shinozaki K (2018) A small peptide modulates stomatal control via abscisic acid in long-distance signalling. Nature 556(7700):235–238

Tan T, Cai J, Zhan E, Yang Y, Zhao J, Guo Y, Zhou H (2016) Stability and localization of 14-3-3 proteins are involved in salt tolerance in Arabidopsis. Plant Mol Biol 92:391–400

Teige M, Scheikl E, Eulgem T, Doczi R, Ichimura K, Shinozaki K, Dangl JL, Hirt H (2004) The MKK2 pathway mediates cold and salt stress signaling in Arabidopsis. Mol Cell 15(1):141–152

Thalmann M, Pazmino D, Seung D, Horrer D, Nigro A, Meier T, Kölling K, Pfeifhofer HW, Zeeman SC, Santelia D (2016) Regulation of leaf starch degradation by abscisic acid is important for osmotic stress tolerance in plants. Plant Cell 28(8):1860–1878

Thomine S, Schroeder JI (2013) Plant metal transporters with homology to proteins of the NRAMP family. In: Madame Curie bioscience database [Internet]. Landes Bioscience

To TK, Nakaminami K, Kim JM, Morosawa T, Ishida J, Tanaka M, Yokoyama S, Shinozaki K, Seki M (2011) Arabidopsis HDA6 is required for freezing tolerance. Biochem Biophys Res Commun 406(3):414–419

Tunc-Ozdemir M, Tang C, Ishka MR, Brown E, Groves NR, Myers CT, Rato C, Poulsen LR, McDowell S, Miller G, Mittler R (2013) A cyclic nucleotide-gated channel (CNGC16) in pollen is critical for stress tolerance in pollen reproductive development. Plant Physiol 161(2): 1010–1020

Uozumi N, Kim EJ, Rubio F, Yamaguchi T, Muto S, Tsuboi A, Bakker EP, Nakamura T, Schroeder JI (2000) The Arabidopsis HKT1 gene homolog mediates inward Na+ currents in Xenopus laevis oocytes and Na+ uptake in Saccharomyces cerevisiae. Plant Physiol 122(4):1249–1260

Vainonen JP, Jaspers P, Wrzaczek M, Lamminmäki A, Reddy RA, Vaahtera L, Brosche M, Kangasjärvi J (2012) RCD1–DREB2A interaction in leaf senescence and stress responses in Arabidopsis thaliana. Biochem J 442(3):573–581

Verbruggen N, Hermans C, Schat H (2009) Mechanisms to cope with arsenic or cadmium excess in plants. Curr Opin Plant Biol 12(3):364–372

Vlachonasios KE, Thomashow MF, Triezenberg SJ (2003) Disruption mutations of ADA2b and GCN5 transcriptional adaptor genes dramatically affect Arabidopsis growth, development, and gene expression. Plant Cell 15(3):626–638

Volkov RA, Panchuk II, Mullineaux PM, Schöffl F (2006) Heat stress-induced H_2O_2 is required for effective expression of heat shock genes in Arabidopsis. Plant Mol Biol *61*:733–746

Wahid A, Gelani S, Ashraf M, Foolad MR (2007) Heat tolerance in plants: an overview. Environ Exp Bot 61(3):199–223

Wang L, Guo Y, Jia L, Chu H, Zhou S, Chen K, Wu D, Zhao L (2014) Hydrogen peroxide acts upstream of nitric oxide in the heat shock pathway in Arabidopsis seedlings. Plant Physiol 164(4):2184–2196

Wang P, Zhao Y, Li Z, Hsu CC, Liu X, Fu L, Hou YJ, Du Y, Xie S, Zhang C, Gao J (2018) Reciprocal regulation of the TOR kinase and ABA receptor balances plant growth and stress response. Mol Cell 69(1):100–112

Wang P, Hsu CC, Du Y, Zhu P, Zhao C, Fu X, Zhang C, Paez JS, Macho AP, Tao WA, Zhu JK (2020) Mapping proteome-wide targets of protein kinases in plant stress responses. Proc Natl Acad Sci 117(6):3270–3280

Waseem M, Liu P, Aslam MM (2023) Salinity and drought stress in plants: understanding physiological, biochemical and molecular responses. Front Plant Sci 14:1277859

Wei Z, Yuan T, Tarkowská DE, Kim J, Nam HG, Novák O, He K, Gou X, Li J (2017) Brassinosteroid biosynthesis is modulated via a transcription factor cascade of COG1, PIF4, and PIF5. Plant Physiol 174(2):1260–1273

Wu A, Allu AD, Garapati P, Siddiqui H, Dortay H, Zanor MI, Asensi-Fabado MA, Munne-Bosch S, Antonio C, Tohge T, Fernie AR (2012) JUNGBRUNNEN1, a reactive oxygen species–responsive NAC transcription factor, regulates longevity in Arabidopsis. Plant Cell 24(2):482–506

Wu F, Chi Y, Jiang Z, Xu Y, Xie L, Huang F, Wan DI, Ni J, Yuan F, Wu X, Zhang Y (2020) Hydrogen peroxide sensor HPCA1 is an LRR receptor kinase in Arabidopsis. Nature 578(7796): 577–581

Wurzinger B, Mair A, Pfister B, Teige M (2011) Cross-talk of calcium-dependent protein kinase and MAP kinase signaling. Plant Signal Behav 6(1):8–12

Xie K, Chen J, Wang Q, Yang Y (2014) Direct phosphorylation and activation of a mitogen-activated protein kinase by a calcium-dependent protein kinase in rice. The Plant Cell 26(7): 3077–3089

Xu J, Li Y, Wang Y, Liu H, Lei L, Yang H, Liu G, Ren D (2008) Activation of MAPK kinase 9 induces ethylene and camalexin biosynthesis and enhances sensitivity to salt stress in Arabidopsis. J Biol Chem 283(40):26996–27006

Xu W, Miao Y, Kong J, Lindsey K, Zhang X, Min L (2024) ROS signaling and its involvement in abiotic stress with emphasis on heat stress-driven anther sterility in plants. Crop Environ 3(2): 65–74

Xue M, Zhang H, Zhao F, Zhao T, Li H, Jiang D (2021) The INO80 chromatin remodeling complex promotes thermomorphogenesis by connecting H2A. Z eviction and active transcription in Arabidopsis. Mol Plant 14(11):1799–1813

Yadav SK (2010) Heavy metals toxicity in plants: an overview on the role of glutathione and phytochelatins in heavy metal stress tolerance of plants. S Afr J Bot 76(2):167–179

Yamada K, Fukao Y, Hayashi M, Fukazawa M, Suzuki I, Nishimura M (2007) Cytosolic HSP90 regulates the heat shock response that is responsible for heat acclimation in Arabidopsis thaliana. J Biol Chem 282(52):37794–37804

Yamaguchi-Shinozaki K, Shinozaki K (2006) Transcriptional regulatory networks in cellular responses and tolerance to dehydration and cold stresses. Annu Rev Plant Biol 57:781–803

Yang Z, Wang C, Xue Y, Liu X, Chen S, Song C, Yang Y, Guo Y (2019) Calcium-activated 14-3-3 proteins as a molecular switch in salt stress tolerance. Nat Commun 10(1):1199

Ye L, Li L, Wang L, Wang S, Li S, Du J, Zhang S, Shou H (2015) MPK3/MPK6 are involved in iron deficiency-induced ethylene production in Arabidopsis. Front Plant Sci 6:953

Yoo SD, Sheen J (2008) MAPK signaling in plant hormone ethylene signal transduction. Plant Signal Behav 3(10):848–849

Yu F, Qian L, Nibau C, Duan Q, Kita D, Levasseur K, Li X, Lu C, Li H, Hou C, Li L (2012) FERONIA receptor kinase pathway suppresses abscisic acid signaling in Arabidopsis by activating ABI2 phosphatase. Proc Natl Acad Sci 109(36):14693–14698

Yuan F, Yang H, Xue Y, Kong D, Ye R, Li C, Zhang J, Theprungsirikul L, Shrift T, Krichilsky B, Johnson DM (2014) OSCA1 mediates osmotic-stress-evoked Ca^{2+} increases vital for osmosensing in Arabidopsis. Nature 514(7522):367–371

Zhang W, Zhou RG, Gao YJ, Zheng SZ, Xu P, Zhang SQ, Sun DY (2009) Molecular and genetic evidence for the key role of AtCaM3 in heat-shock signal transduction in Arabidopsis. Plant Physiol 149(4):1773–1784

Zhang Z, Li J, Li F, Liu H, Yang W, Chong K, Xu Y (2017) OsMAPK3 phosphorylates OsbHLH002/OsICE1 and inhibits its ubiquitination to activate OsTPP1 and enhances rice chilling tolerance. Dev Cell 43(6):731–743

Zhang M, Liang X, Wang L, Cao Y, Song W, Shi J, Lai J, Jiang C (2019) A HAK family Na+ transporter confers natural variation of salt tolerance in maize. Nature Plants 5(12):1297–1308

Zhang H, Liu D, Yang B, Liu WZ, Mu B, Song H, Chen B, Li Y, Ren D, Deng H, Jiang YQ (2020) Arabidopsis CPK6 positively regulates ABA signaling and drought tolerance through phosphorylating ABA-responsive element-binding factors. J Exp Bot 71(1):188–203

Zhang H, Zhu J, Gong Z, Zhu JK (2022) Abiotic stress responses in plants. Nat Rev Genet 23(2): 104–119

Zhao MG, Chen L, Zhang LL, Zhang WH (2009) Nitric reductase-dependent nitric oxide production is involved in cold acclimation and freezing tolerance in Arabidopsis. Plant Physiol 151(2): 755–767

Zhao FY, Wang K, Zhang SY, Ren J, Liu T, Wang X (2014) Crosstalk between ABA, auxin, MAPK signaling, and the cell cycle in cadmium-stressed rice seedlings. Acta Physiol Plant 36: 1879–1892

Zhao C, Wang P, Si T, Hsu CC, Wang L, Zayed O, Yu Z, Zhu Y, Dong J, Tao WA, Zhu JK (2017) MAP kinase cascades regulate the cold response by modulating ICE1 protein stability. Dev Cell 43(5):618–629

Zhou H, Lin H, Chen S, Becker K, Yang Y, Zhao J, Kudla J, Schumaker KS, Guo Y (2014) Inhibition of the Arabidopsis salt overly sensitive pathway by 14-3-3 proteins. Plant Cell 26(3): 1166–1182

Zhu JK (2016) Abiotic stress signaling and responses in plants. Cell 167(2):313–324
Zhu J, Jeong JC, Zhu Y, Sokolchik I, Miyazaki S, Zhu JK, Hasegawa PM, Bohnert HJ, Shi H, Yun DJ, Bressan RA (2008) Involvement of Arabidopsis HOS15 in histone deacetylation and cold tolerance. Proc Natl Acad Sci 105(12):4945–4950
Zhu JY, Oh E, Wang T, Wang ZY (2016) TOC1–PIF4 interaction mediates the circadian gating of thermoresponsive growth in Arabidopsis. Nat Commun 7(1):13692
Zhu JY, Li Y, Cao DM, Yang H, Oh E, Bi Y, Zhu S, Wang ZY (2017) The F-box protein KIB1 mediates brassinosteroid-induced inactivation and degradation of GSK3-like kinases in Arabidopsis. Mol Cell 66(5):648–657

Chapter 2
Functional Genomics Approaches for Abiotic Stress Tolerance Mechanism in Crop Plants

Mohammad Faizan, Anjuman Hussain, and Pravej Alam

Abstract The primary factors limiting agricultural productivity include a variety of abiotic stressors, including heat, salinity, drought, flooding, ion toxicity, and radiation. Plant functional genomics has completely changed not only the methods for identifying and understanding the roles of important genes, but also the development of defense mechanisms against abiotic stressors. Functional genomics has the potential to greatly aid crop improvement initiatives by providing an understanding of the gene regulatory networks functioning under stress, especially with the advent of novel high throughput technologies in this field. This chapter presents recent discoveries in the field of genome-editing technologies for abiotic stress tolerance for the identification and validation of candidate genes, including use of functional genomics in plant physiology and growth under abiotic stress. We now have a better grasp of how plants respond to abiotic stressors and the various components that are used in signaling and response pathways thanks to the identification and isolation of abiotic stress-tolerant genes and transformation technologies.

Keywords Abiotic stress · Functional genomics · Genome editing · Plant physiology

M. Faizan
Botany Section, School of Sciences, Maulana Azad National Urdu University, Hyderabad, India

A. Hussain
Department of Botany, Faculty of Life Sciences, Aligarh Muslim University, Aligarh, India

P. Alam (✉)
Department of Biology, College of Science and Humanities, Prince Sattam Bin Abdulaziz University, Alkharj, Saudi Arabia

S. K. Panda, M. A. Khan (eds.), *Plant Functional Genomics for Abiotic Stress Resilience*, Plant in Challenging Environments 7,
https://doi.org/10.1007/978-3-032-01704-8_2

2.1 Introduction

By the end of 2050, it is predicted that there will be up to 10 billion people on the planet. According to this scenario, in order to ensure global food security, food crop production must increase by 60% in the upcoming years (Jaganathan et al. 2018; Bhatta et al. 2021). However, some of the main obstacles that severely reduce crop plant productivity and yield include abiotic factors such salinity, drought, and extremely high or low temperatures (Bhat et al. 2021; Biswas et al. 2021). According to Kaur and Pati (2019), they cause over 50% of crop plant yield loss, which results in \$14–19 million in annual economic losses globally. Because of the climate uncertainties in the near future, it is anticipated that the effects of these pressures may worsen even more (Bhat et al. 2021; Vaughan et al. 2018). Thus, one of the main goals of research to ensure sustainable agriculture is the regulation of abiotic stressors in various crops. Researchers have frequently attempted to devise diverse methods to help plants develop resistance to abiotic stress (Anwar and Kim 2020; Kerchev et al. 2020). The most common method is to cross two parental plants to create a progeny plant with the desired traits. But because the characteristics linked to abiotic stress tolerance are so complicated, this strategy has only had limited success (Zafar et al. 2017). Plant stress tolerance has increased significantly in the last 10 years owing to advances in genomics and molecular biology and the advent of genetic engineering techniques (Jan et al. 2013; Lal 2020). By mobilising genes or their regulatory elements into the genome of interest, this method creates crops that are resistant to abiotic stress (Wang et al. 2016). However, due to several health and environmental issues, the use of genetically modified crops is banned globally (Scott et al. 2018). Therefore, it is essential to develop cutting-edge and effective tactics that can aid in the creation of abiotic stress-resilient crops with improved growth and yield in order to meet the expanding global population's need for food. In this case, plant science and agriculture have undergone a radical transformation due to the advent of genome editing (Jaganathan et al. 2018; Gao 2021).

For thousands of years, genome change has been a natural occurrence on Earth. Through evolution, certain genetic differences in plants have long been picked for over others. A significant amount of genetic variety has historically been present in almost all crops farmed now (Holmes et al. 2016; Sedeek et al. 2019). One notable example is the modern maize, farmed all over the world, which differs greatly from its wild predecessor, teosinte. To increase the many crop qualities, however, our ancestors relied on naturally occurring genome variations/mutations (Sedeek et al. 2019). Researchers used chemical mutagens and radiation to alter the genome once it was determined later in the twentieth century that DNA is the genetic basis of life and determine an organism's phenotype. In the 1970s Green Revolution, mutational breeding proved to be a highly successful method of breeding thanks to the resourcefulness of this sort of mutagenesis (Sedeek et al. 2019). But the limitations of random mutagenesis and strict mutant selection protocols have made it imperative to create novel strategies that can specifically modify the DNA sequence (Pacher and Puchta 2017). According to Kaur et al. (2019), genome editing provides an integration-free method for precisely modifying the target genome in this situation. In the 1980s, Capecchi came up with the idea of genome editing (Sedeek et al. 2019). This method allows for the addition, deletion, or modification of genetic material at particular genomic loci (Chen et al. 2019). Sequence-specific endonucleases (SSEs) are used in this process to produce double-stranded (ds) DNA breaks (DSB) in a sequence-specific way. SSEs are transiently expressed within the cell as either a protein that is not passed down to subsequent generations or as mRNA that rapidly breaks down after its function is completed (Yin et al. 2015; Ilardi and Tavazza 2015). The host DNA repair mechanism subsequently uses homologous recombination (HR) and nonhomologous end-joining (NHEJ) to close the gaps created during this process. Ku-dependent and Ku-independent pathways are the two ways that the NHEJ system functions (Deriano and Roth 2013; Osakabe and Osakabe 2015). The Ku proteins KU70/80 attach to the ends of ds breaks at the target site during the Ku-dependent pathway. When this binding occurs, the ligase IV enzyme and its cofactors are drawn in as part of the repair machinery, which closes the gaps (Deriano and Roth 2013). On the other hand, the broken ends of the DNA sequence are resected to produce a single-stranded overhang in the Ku-independent route. Then, with the aid of 5–25 bp length micro homologous sequences, these overhangs are aligned. DNA polymerase is then used to fill up the resulting gaps, and DNA ligase is used to ligate nicks (Crespan et al. 2012). The target genome may experience deletions or insertions as a result of the error-prone NHEJ repair pathway. As a result, HR causes insertions or replacements in the target genome. In contrast, the HR repair mechanism is error-free because it uses a DNA template to precisely replace the nucleotides (Osakabe and Osakabe 2015). Since nearly 10 years ago, genome editing has been utilised to improve a variety of plant properties. However, due to the advent of more accessible genome editing tools, the field has seen incredible success in recent years. While several previous research studies have addressed the amelioration of abiotic stress in plants by CRISPR/Cas mediated systems, there is a dearth of review articles that comprehensively examine the potential of various genomes editing technologies to address the issue of abiotic stress in plants. This study provides an update on the several kinds of genome editing tools that are being utilised to improve crops, as well as the latest developments regarding their ability to confer resistance to abiotic stress. Furthermore, it draws attention to the various obstacles that stand in the way of fully using the genome editing technologies currently accessible to enhance plants' resistance to abiotic stress.

2.2 Abiotic Stress Effects in Plants

Any environmental condition that has the potential to restrict plant development and productivity is referred to as abiotic stress. Abiotic stress has historically drawn the attention of plant scientists and agronomists in the last 50 years because of the significant increase in human population (which has jeopardised world food security). It became necessary to look into the often-ignored causes of decreased crop productivity due to rising food demand. It was at this time that scientists began to estimate the amount of crop loss due to environmental restrictions and looked into the causes. Boyer (1982) calculated that environmental conditions might cut crop productivity by up to 70% in the early 1980s. Since then, a number of studies have been released that discuss crop loss as a result of different abiotic stresses (Wu et al. 2007). A few of these stresses damage agricultural soils permanently, which eventually affects crop quality and productivity. A rough calculation based on FAO reports indicates that approximately 96.5% of worldwide rural land is influenced by abiotic pressures, although it is difficult to determine the impacts of abiotic stresses on agricultural areas (Cramer et al. 2011). Many unfavourable elements, including heat, cold, drought, and salinity (among others), can have an impact on agricultural land and crop productivity in a constantly shifting environment. Here, we will consider a few of these key variables along with how they affect plant production and yield.

2.2.1 Heat Stress

Since greenhouse gas accumulation is being greatly accelerated by gaseous emissions and other related human activities, plant scientists are increasingly concerned about global warming. The global ambient temperature is expected to rise gradually due to greenhouse gas emissions, specifically carbon dioxide, methane, nitrous oxide, and chlorofluorocarbons, according to a number of global circulation models. According to a report by the Inter-Governmental Panel on Climatic Change (IPCC), the average global temperature would increase by 0.3 °C every 10 years (Jones et al. 1999). By 2025, the global temperature is expected to rise by roughly 1 °C at this rate (Wahid et al. 2007), which, considering that, by 2050, we will likely need to feed roughly nine billion people, might be fairly significant (Stratonovitch and Semenov 2015). One of the key elements in maintaining cellular metabolism is temperature, and plants interpret any increase in temperature over optimal levels as heat stress. Heat stress, according to Wahid et al. (2007), is an increase in temperature that lasts long enough to harm plants' health irreversibly. Generally speaking, "heat shock" refers to a brief increase in temperature (10–15 °C). On the other hand, heat stress is a result of a number of intricate processes, such as the length and severity of the temperature increase over optimal ranges. The frequency and duration of heat stress during the day and night determine the severity of heat stress in a given climatic zone. Heat is a severe hazard to agricultural productivity worldwide because it can disrupt cellular homeostasis, which can ultimately result in growth retardation or even death (Kotak et al. 2007). (Hall 2018). Because of variations in the ideal temperature for crop maturity, etc., rising temperate temperatures may cause major agriculturally significant crops like rice, wheat, and maize to be geographically redistributed and experience altered growing seasons. This will have an indirect impact on humans as well as agriculture and livestock. For instance, Porter (2005) found that increasing temperatures might help pasture grasses to achieve their ideal growth-starting temperature earlier, thus extending the growing season. According to reports, most cereals' life cycles are shortened by rising temperatures, which causes senescence and shortens the growth season. The temperature range that staple crops (such as corn, wheat, and rice) can withstand is limited. If this range is surpassed during the floral transition, it can harm seed production, which lowers crop yield.

2.2.2 *Cold Stress*

Another environmental issue that restricts many plant species' growing seasons is cold (Guy 1990). Since most agriculturally significant crops that are consumed by humans, like maize, rice, tomatoes, and fruits like bananas, mangoes, and papayas, are from tropical or subtropical regions and are susceptible to chilling, it is crucial to understand the effects that cold stress has on crops (Solanke and Sharma 2008; Eremina et al. 2016). Freezing and chilling stresses are two common distinctions made between cold stresses. The degree of stress and length of exposure determine how plants respond to cold stress. The cold primarily affects the plant when it is at the seedling stage. According to Solanke and Sharma (2008), prevalent symptoms include the development of the surface lesion, discolouration, desiccation, induced senescence, decreased shelf life, and ethylene production. Chilling causes significant physiological and biochemical alterations in addition to significant phenotypic changes, which have been thoroughly examined (Cramer et al. 2011; Ruelland et al. 2009). In summary, these alterations comprise elevated levels of calcium in cells, a build-up of reactive oxygen species (ROS) (Cramer et al. 2011), a reduction in membrane fluidity as a result of fatty acid unsaturation, and modifications in the ratio of lipids to proteins in the membrane. Dehydration can also result from low temperatures, primarily from the body's incapacity to absorb water. Similar to this, freezing can harm membranes because it results in a sharp drop in water content and the creation of ice. Cells and tissues may physically be disrupted when ice forms in intracellular areas. According to Uemura et al. (1995), freezing can also result in solute precipitation and denaturation of proteins. Both freezing and chilling cause the creation of ROS, which upsets the redox balance of cells, just like other abiotic stresses do.

2.2.3 *Drought Stress*

One of the main environmental factors that significantly lower crop productivity is drought (Fahad et al. 2017). Conditions of a water deficit may worsen due to climate change's impact on rainfall variability, rising water demand, and declining water supply (Lobell et al. 2011). All species will be impacted, but plants that lack the ability to move when food and water become scarce will be especially hard hit, which will have an effect on crop productivity. Since drought stress is influenced by a number of variables, including rainfall patterns, the soil's ability to hold water, and water losses from evapotranspiration, its severity is difficult to assess (Fahad et al. 2017). Drastic stress's effects on crop species linked to physiology and growth have been well studied (Daryanto et al. 2016). According to a study citing studies of data released globally between 1980 and 2015, drought-related yield decreases in wheat and maize were estimated to be 21% and 40%, respectively (Daryanto et al. 2016). Poor germination (Farooq et al. 2009), decreased seedling growth (Okçu et al. 2005), decreased nutrient availability, photosynthesis (Manickavelu et al. 2006), the quantity and size of individual leaves, the fresh and dry weight of the plants, etc. are some of the main effects of drought. Furthermore, drought exposure may cause pearl millet (Pennisetum glaucum L.) to become completely sterile; this is most likely the result of disrupted assimilate transport to the developing ear (Yadav et al. 2004). Similarly, up to 80% less coffee might be produced when dehydrated (DaMatta and Ramalho 2006). According to a research by Franchini et al. (2009), a 45-day drought in some parts of the Brazilian state of Parana in 2008–2009 reduced the harvest of soybeans by 80%. Drought induces a variety of physiological responses in plants, including stomatal closure (Fahad et al. 2017) and decreased transpiration (Turner et al. 2001). Furthermore, plants alter the expression of certain genes to hasten their tolerance to drought stress.

2.2.4 Salinity Stress

Another abiotic factor that can restrict agricultural productivity, particularly in dry or semi-arid environments, is salinity in the soil or water (Shannon 1997). An estimate from 2011 states that high salinity affects more than 800 million hectares of land worldwide (Isayenkov 2012). Similarly, by 2050, salinity will damage roughly 17 million hectares of agricultural land, according to the same study. The low osmotic potential of the soil, a nutritional imbalance, the specific ion action, or a combination of these factors are linked to the detrimental effects of salinity on plants (Marschner 2011). At different levels, these variables have detrimental effects on the growth and development of plants (Levitt 2012). Reduced growth rate, fewer and smaller leaves, and shorter roots are some of the general impacts (Munns and Termaat 1986). According to Shannon and Grieve (1998), the osmotic effect of salinity influences root/shoot ratio, maturity rate, and colour variations in leaves. The effects of salt on plants and how they react to it are dependent upon the length and intensity of the stress, just like with other abiotic stresses. According to Munns and Termaat 1986; Munns and Passioura 1984, salinity generally has two types of effects: short-term (ion-independent growth reduction) that occurs minutes, hours, or days after the stimuli are perceived; and long-term (which can occur over days or even weeks) effects that build up cytotoxic ion levels, slow down metabolic activities, cause early senescence and ultimately cell death (Munns and Tester 2008; Roy et al. 2014). Many physio-molecular processes, osmotic and ionic tolerance, and tissue tolerance allow a plant to tolerate salinity (Roy et al. 2014; Rajendran et al. 2009). Among these, osmotic tolerance uses speedy signalling between roots and shoots to store water in response to a rapid decrease in stomatal conductance (Roy et al. 2014; Ismail et al. 2007; Maischak et al. 2010). According to Isayenkov and Maathuis (2019), the activation of several signalling cascades that limit net Na^+ influx and decrease net Na+ translocation results in ionic tolerance. Toxic ions are translocated to the vacuole in order to prevent negative effects on cytoplasmic-based metabolic processes, hence achieving tissue-specific tolerance (Isayenkov and Maathuis 2019). In this instance, the key to directing harmful ions to the vacuole is the salt excessively sensitive (SOS) pathway, which is made up of many SOS genes (Ji et al. 2013).

2.2.5 Heavy Metal Stress

Heavy metals (HMs) like manganese, copper, nickel, cobalt, chromium, zinc, and mercury have accumulated in soils as a result of various human activities like fertiliser use, inappropriate industrial waste disposal, uncontrolled sewage disposal, or pococurante disposal of vehicle effluents (Ghori et al. 2019). They either accumulate on the soil's surface or are leached to groundwater from soil (Dağhan et al. 2015). The fact that HM stress is non-biodegradable—that is, it cannot be eliminated from the environment naturally is one of its severe characteristics. Enzymes are negatively impacted by HM stress by denaturation or inactivation. According to reports, HMs disrupt the process of substituting necessary metallic ions with biomolecules, which compromises membrane integrity and alters respiration, photosynthetic capacity, and other processes (Hossain et al. 2012). Additionally, HMs cause oxidative stress by promoting the generation of superoxide radicals (O^{-2}) and hydroxyl radicals (OH), as well as hydrogen peroxide (H_2O_2) (Ghori et al. 2019; Rascio and Navari-Izzo 2011). The build-up of heavy metals (HMs), especially in the roots where they are trapped by root cell walls or blocked by Casparian strips, affects plant physio-morphological functions and eventually reduces agricultural productivity (Shahid et al. 2015). To deal with heavy metals, plants use both physiological and molecular defences. Plants use cellular and root exudates to decrease metal uptake. This prevents metals from entering the cell by boosting efflux, which is a stress-avoidance strategy (Ghori et al. 2019). In order to prevent delicate cellular components from coming into contact with these metals, many plants have developed specialised systems to deal with a specific kind of heavy metal (HM) by sequestering it in cellular compartments (Ghori et al. 2019). Glutathione, phytochelatins, metallothioneins, amino acids, and specific enzymes like superoxide dismutase (SOD) are among the substances involved in this process (Hossain et al. 2012). Remarkably, plants also make use of their symbiotic association with mycorrhizal fungi by tricking them into accumulating heavy metals (HMs) in the rhizosphere, hence blocking plant availability (Ghori et al. 2019). A plant can also deal with HMs in a number of other ways, such as by altering the pH of the rhizosphere, which causes HMs to precipitate; moving metallic ions from symplastic to aplastic space via metal transport and carrier proteins; secreting root exudates like oxalate and malate (Pinto et al. 2008); releasing nitric oxide (NO), which triggers the antioxidant system and aids in the reduction of HM stress (Nabi et al. 2019); using phytochelatins to move metallic ions via xylem and phloem (Hossain et al. 2012), among other strategies. We shall, however, concentrate more on the chemical reactions of plants to heavy metal stress. Compared to the non-accumulator A. thaliana, transcriptome analyses have revealed over thirty potential genes that were substantially expressed in the metal hyper-accumulator *Arabidopsis helleri* (Becher et al. 2004). According to reports, *Noccaea caerulescens*, a hyperaccumulator in comparison to the non-accumulator *N. arvense*, has an overexpression of the Zn^{2+} transporter AnT1 (Pence et al. 2000). Significant potential genes are those that encode metal pumps associated with the heavy-metal ATPase (HMA) transporter family. They are in charge of the translocation of metals across biological membranes, as their name suggests. In A. thaliana, for instance, HMA2 and HMA4 cause metal efflux out of the cell (Eren and Arguello 2004). In a similar vein, hyper accumulating *A. helleri* plants had higher expression levels of MTP1, a member of the cation diffusion facilitator (CDF) protein family, which functions as a vacuolar membrane transporter, than non-accumulator plants (Dräger et al. 2004). Plants respond to HM stress by activating signalling cascades, just like they do to other stresses. According to Ghori et al. (2019), the most prevalent of these are the Ca-Calmodulin route, the ROS signalling system, and the mitogen-activated protein kinase (MAPK) cascade.

2.3 Genome-Editing Technologies for Abiotic Stress Tolerance

It is essential to increase food crop production enormously in order to feed the world's growing population. However, the vulnerability of plants to an array of abiotic stressors is a key hurdle to accomplishing this goal (Dhankher and Foyer 2018). Previously, conventional marker-assisted breeding was used to generate a large number of crop plants that are tolerant of abiotic stress. However, because of the rigorous screening and backcrossing processes involved, this technique takes nearly 10 years to successfully generate crops resistant to abiotic stress (Hoang et al. 2016). Although the stress-tolerant plants created by genetic engineering have demonstrated favourable results, a number of obstacles stand in the way of their widespread commercialization. Therefore, in this case, genome editing seems like a sophisticated way to create crops that are resistant to abiotic stress in the future since it allows for precise manipulation of different gene loci in a shorter amount of time, which lowers the cost of crop development programs (Schaart et al. 2016) (Fig. 2.1). Plant characteristics such as yield and nutritional value have previously been

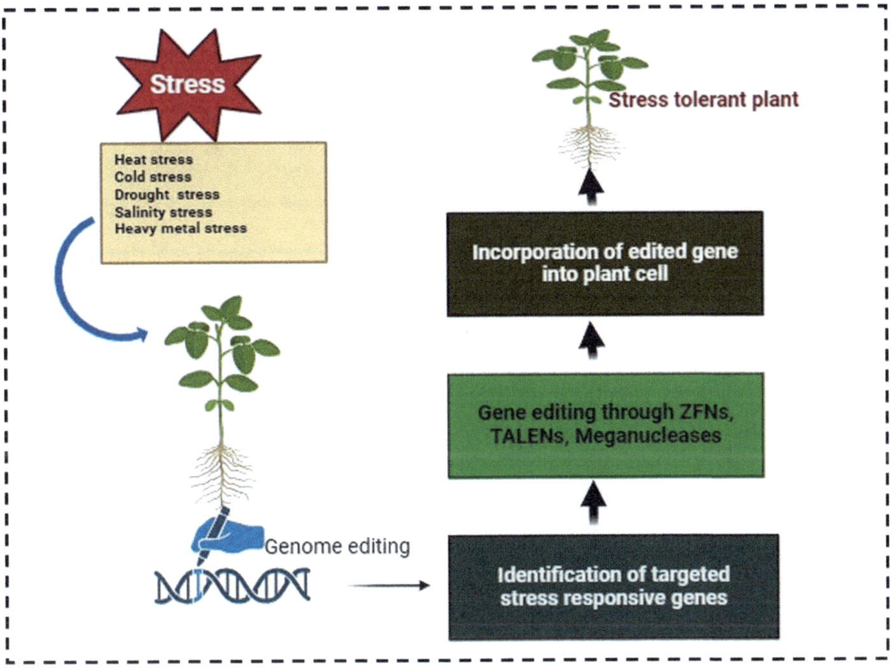

Fig. 2.1 Genome editing in crops against abiotic stresses

successfully improved through genome editing (Sedeek et al. 2019). Conversely, there aren't many studies on its application for developing plants resistant to abiotic stress. However, there are many potential to increase the abiotic stress resilience of plants now that extremely effective genome editing techniques like CRISPR/Cas have been developed. Additionally, it can be applied to comprehend how different stress signalling cascades contribute to abiotic stress adaption (Jain 2015).

One of the most promising methods for improving crop plants and understanding the genome is genome editing. The identification of target genomic loci and binding of effector DNA-binding domain (DBD), double-stranded breaks (DSBs) in target DNA caused by restriction endonucleases (FokI and Cas), and repair of DSBs by homology-directed recombination (HDR) or non-homologous end joining are the basic mechanisms involved in genetic modification by programmable nucleases (NHEJ) (Zhang and Showalter 2020). Nucleotides are either deleted or inserted by the well-organised but error-prone NHEJ, whereas nucleotides are replaced by the less effective but more precise HDR. ZFN, TALEN, and CRISPR/Cas are examples of genome-editing techniques that are being used to add desired traits and eliminate unwanted ones. Site-specific nuclease (SSN) or site-specific recombinase (SSR) systems can be used in a variety of ways to modify a genome. A known sequence needs to be found by both systems. The SSN system triggers endogenous DNA repair processes and breaks single or double strands of DNA. SSR technology, such as Cre/loxP- and Flp/FRT-mediated systems, can either knock down or knock in genes in the eukaryotic genome surrounding the target region, depending on how the sites (loxP, FLP, etc.) are orientated (Gao 2021). The four main categories of plant genome editing methods are based on onsite-specific endonucleases. They are CRISPR-Cas9, TALENs, ZFNs, and meganucleases. Each of these methods is covered in more depth below.

2.3.1 Zinc-Finger Nucleases

ZFNs are collections of DNA recognition modules that are based on the FokI restriction enzyme's DNA cleavage domain and zinc fingers. They can be used to introduce a range of genetic modifications, including point mutations, insertions, inversions, translocations, and deletions, into the target genome (Carroll 2011). There are two domains present in them, namely the nuclease domain and the DNA-binding domain. Nucleotide sequences ranging in length from 9 to 18 bases can be recognized by the 3- to 6-zinc finger repeats of the DNA-binding domain. According to Urnov et al. (2010), the restriction enzyme Flavobacterium okeanokoites I (FokI) is the component of the second domain and is essential for DNA cleavage.ZFN-1, ZFN-2, and ZFN-3 are the three artificial restriction enzymes used in this procedure (Bibikova et al. 2001). ZFN-1: ZFN has now been inserted into the plant genome without requiring a repair template. When it reaches the plant genome, it causes double-strand breaks (DSB) in the host DNA, which results in non-homologous end joining (NHEJ) of DNA (Puchta 2005). This can result in a modest deletion or insertion or in arbitrary changes at specific sites. ZFN-2: Unlike ZFN-I, this gene is inserted into the crop genome adjacent to the ZFN enzyme and is repaired using a homology-directed repair (HDR) in conjunction with a brief repair template (Lusser et al. 2011). The target DNA, which binds to a particular sequence and causes a double-stranded break, is homologous to the template DNA. During homologous recombination (HR), the template starts to repair itself via an endogenous repair mechanism that targets point mutations specific to the spot. ZFN-3: The ZFN transcribing gene is designated ZFN3 as soon as it is inserted into the plant genome adjacent to the big repair template (Lusser et al. 2011; Araki et al. 2014). ZFN has been successfully incorporated into tobacco, soybeans, maize, and Arabidopsis (Shukla et al. 2009; Curtin et al. 2013). One application of ZFNs in crop breeding was the disruption of the endogenous ZmIPK1 gene in maize through the insertion of PAT gene cassettes, which changed the inositol phosphate profile of developing maize seeds and enhanced their resistance to herbicides (Shukla et al. 2009).ZFNs can be designed to target almost any specific DNA stretch using a variety of protein-engineering methods (Petolino 2015). ZFNs with increased activity and specificity have been created to cause gain-of-function changes and knockouts, which impair the function of the gene (Khalil 2020).

2.3.2 Meganucleases

Meganucleases are endonucleases that can specifically recognise and cleave longer DNA sequences (greater than 12 bp). Numerous organisms, including bacteria, fungus, algae, yeast, and archaebacteria, have been found to use this strategy. Mild polymorphisms can be maintained by meganucleases at the target site (Arnould et al. 2011). Meganucleases have been categorised into five classes according to structural characteristics and sequence. These include GIY-YIG, PD-(D/E) XK, PD-(Cys box), LAGLIDADG, and HNH (Silva et al. 2011; Belfort and Bonocora 2014).The majority of genome editing has employed LAGLIDADG meganuclease (LMN) family members. The name of this protein family, according to Silva et al. (2011), is derived from the sequence of the primary motif present in its structure. In unicellular eukaryotes, LMNs are usually expressed in the mitochondria and chloroplast. Most of these endonucleases are dimeric proteins with two distinct roles: as RNA maturases, they splice their own introns, and as specialised endonucleases, they cleave exon sequences (Sultan et al. 2016). The Saccharomyces cerevisiae mitochondrial DNA's rRNA gene can be used for editing the genomes of I-SceI and I-CreI. The location of the I-SceI gene is found in the 21S. It was discovered that the unicellular alga Chlamydomonas reinhardtii contains I-CreI, which is located in the 23S rRNA gene, in its chloroplast. However, meganucleases are not very useful for genome editing since it is hard to reengineer them to target certain regions of DNA (Kaur et al. 2022).

2.3.3 Transcription Activator-Like Effector Nucleases (TALENs)

Transcription activator-like effector nucleases, or TALENs, are restriction enzymes that are designed to cut specific DNA sequences. A nuclease that can cleave DNA in cells and a TALE domain that mimics naturally occurring transcription activator-like effector proteins make up TALENs. Transcription activator-like effector nucleases, or TALENs, are being studied by numerous scientists in a variety of organisms. A free designable DNA-binding domain and a nuclease comprise TALENs (Joung and Sander 2013). Lately, TALENs have gained recognition as a cutting-edge method for genome editing in various species and cell types. It has been discovered that several plants are vulnerable to TALEN-induced genome change, including Arabidopsis, Nicotiana, Brachypodium, barley, potatoes, tomatoes, sugarcane, flax, rapeseed, soybean, rice, maize, and wheat (Malzahn et al. 2017). According to a study, rice was the first crop to gain from the use of TALENs technology. According to Li et al. (2012), the main cause of blight disease, Xanthomonas oryzae, significantly reduces global rice production each year. By altering the fatty acid desaturase (FAD) genes, soybeans with high oleic acid and low linoleic acid levels were produced, improving the heat stability and shelf life of soybean oil (Demorest et al. 2016). TALENs are type III effector proteins that are naturally produced by Xanthomonas species and modify the expression of genes in host plants. The TALENs proteins consist of a nuclear localization signal, a core DNA-binding domain, and a transcriptional activation domain (Boch and Bonas 2010). Bogdanove et al. (2010) state that the nuclear localization signal helps TALENs penetrate the nucleus, whereas the activation domain initiates the transcriptional machinery to start synthesizing genes.

2.3.4 Clustered Regularly Interspaced Short Palindromic Repeats (CRISPR)/CRISPR-Associated Protein 9 (Cas9)

Small, repeated genetic alterations called CRISPR/Cas9, or clustered regularly interspaced short palindromic repeats, are present in most bacterial and archaeal species. Because of CRISPR/Cas9 and associated proteins, plants have an incredibly strong defensive mechanism that keeps them safe from outside threats including bacteria, viruses, and other elements. The first application of CRISPR/Cas9 in an adaptive immune system was documented in a 2007 study (Barrangou et al. 2007). The CRISPR/Cas9 gene-editing system has revolutionized research in animal and plant biology since its use in genome editing was first demonstrated in mammalian cells in 2012 (Jinek et al. 2012). According to Rathore et al. (2021), first-generation CRISPR/Cas9 genome editing entails fundamental manipulation and cloning techniques that may be applied to a variety of guide RNAs in order to alter different locations in the targeted organism's genome. With the application of CRISPR/Cas, crop species can be precisely modified, leading to the development of advantageous germplasm and novel, more sustainable agricultural systems. Agriculture is expanding more quickly now that precise and targeted genetic crop modification is achievable thanks to recent developments in CRISPR/Cas9 technology (Chen et al. 2019). Only a few species have benefited from this approach thus far (Zhou et al. 2020). The quality, yield, disease resistance, and environmental adaptability of monocots and dicots have all improved attributed to the CRISPR/Cas9 system (Ma and Liu 2016). The genomes of cereal crops including wheat, maize, rice, and cotton, as well as fruits and vegetables like potatoes and tomatoes, have all had their genomes altered using the CRISPR/Cas9 technique (Ricroch et al. 2017). Makarova et al. (2011) have distinguished between three types of CRISPR/Cas systems: type I, type II, and type III. Type I CRISPR/Cas systems are found in bacteria and archaea, based on the specific CAS protein signature. Through its endonucleases activity, the Cas3 protein binds itself to the DNA sequence (Makarova et al. 2011). In bacteria, the type II CRISPR/Cas system has been demonstrated. The four protein pairings Cas1, Cas2, Cas4/Csn2 proteins paired with Cas9 make up the simplest system. Sometimes, the type III CRISPR/Cas system looks for DNA and RNA in bacteria and archaea. Cas6, Cas10, and repeat associated mysterious proteins (RAMP) are markers for its presence. Cleaving DNA is the ultimate aim of the Cas10 protein's processing of crRNA (Makarova et al. 2011). Streptococcus pyogenes (SpCas9) is the source of the type II CRISPR system, which mostly targets genes that negatively affect other genes (Zhang et al. 2019).

In contrast to the first two generations of genome-editing technologies, the CRISPR/Cas method is straightforward, stable, and adaptable. These features allowed CRISPR/Cas to quickly replace the well-known genome-editing methods TALEN and ZFN. The technique was adapted from a bacterial defensive mechanism. The CRISPR/Cas system is used by many bacterial and archaeal species as a defence mechanism against invasive viruses (Bortesi and Fischer 2015). Many studies are being conducted these days to improve the CRISPR/Cas system and increase the tool's capacity for genomic targeting. For instance, non-canonical NGA and NG PAM sites in plants can be found using xCas9, SpCas9-VRQR, and Cas9-NG variants (Ming et al. 2020). The orthologs of SpCas9 have been found in Brevibacillus laterosporus (BlatCas9), Staphylococcus aureus (SaCas9), Streptococcus thermophiles (St1Cas9), and Streptococcus canis (ScCas9).Tan et al. (2020) report that they have been demonstrated to alter plant genomic loci that contain PAM sequences of NNGRRT, NNG, NNAG AAW, and NNNCND, in that order. Moreover, the AT-rich PAM requirements for the type V Cas12a and Cas12b, which are generated from different bacterial systems, have been determined. These reagents have been employed in the plant genome editing process (Wang et al. 2020).

The CRISPR/Cas9 gene-editing method has increased yields while reduced biotic and abiotic stress in over 20 crop species to date (Hamdan et al. 2022). Agricultural disease resistance and tolerance to severe abiotic conditions, like salt and drought, have been enhanced by the use of CRISPR/Cas9-based genome-editing techniques. Three rice genes phytoene desaturase (OsPDS), betaine aldehyde dehydrogenase (OsBADH2), and mitogen-activated protein kinase (OsMPK2)—that are crucial for regulating responses to various abiotic stress stimuli have been subjected to sequence-specific CRISPR/Cas9-mediated genome modification. Using CRISPR/Cas9 technology, Shan et al. (2014) successfully introduced the TaMLO gene (mildew resistance locus O) into wheat protoplasts. Furthermore, it was discovered that Blumeria graminis f. sp. Tritici, the causative agent of powdery mildew illness, is resistant to the CRISPR TaMLO knockdown (Btg). Wheat ethylene responsive factor 3 (TaERF3) and wheat dehydration response element binding protein 2 (TaDREB2) are two abiotic stress-related genes that were targeted by the CRISPR/Cas9 genome-editing technique in wheat protoplasts, according to Kim et al. (2018). The CRISPR/Cas9 technology can be used in conjunction with current and upcoming breeding tactics, such as speed breeding and omics-assisted breeding, to boost agricultural production and ensure food security.

2.4 Use of Functional Genomics in Plant Physiology and Growth Under Abiotic Stress

For millions of years, wild plants have endured a changing climate due to biotic and abiotic processes selectively pressing against them. Natural selection has resulted in the accumulation of genes that enable plants to withstand, endure, or steer clear of harsh weather conditions, pests, and illnesses. Some of the plants became our current remarkably productive crops with minimal genetic variety, but many of those now significant features and associated genetic material were lost during future domestication. The remaining genetic resources were abandoned and primarily handled like weeds. Perspectives obtained from the genome sequencing initiatives of several crops (Appels et al. 2018; Springer et al. 2018) highlighted our modern crops' susceptibilities to climate change and their limited germplasm. All contemporary crop plants, however, are domesticated from their crop wild relatives (CWRs), which persist in the wild and offer a rich reservoir of genetic material that is frequently left out of current breeding initiatives (Brozynska et al. 2016). Elite farmed crops, including wheat, maize, rice, and barley, frequently rely on resources provided by farmers, such as irrigation for water, fertilizers for nutrition, and pesticides for resistance to biotic stressors. As a result, the superior kinds are now less resilient than their wild relatives. Furthermore, limited diversity and a smaller gene pool were the outcomes of intense artificial selection for a few key qualities in breeding programs. As much as 30% of the increases in crop yields in the late twentieth century can be linked to the use of CWRs in plant breeding projects. CWRs represent an additional source of genetic variety that can be utilised during agricultural improvement initiatives (Brozynska et al. 2016). According to Vincent et al. (2013), more than 1500 CWRs of food crops have been discovered as a possible source of genetic variation for 173 globally significant crops. The creation of CRW reference genomes was made easier by developments in sequencing technologies. These genomes can then be utilised in comparative genomics investigations to find new genes governing important features (Brozynska et al. 2016).

Historically, new genes have been inserted into elite cultivar backgrounds to transfer new genetic material from CWRs to agricultural plants (Dempewolf et al. 2017). Utilising genomic resources, such as the transfer of disease resistance genes in banana, apple, and grape vine, has accelerated the process through marker-assisted selection (Migicovsky and Myles 2017). Though evidently successful in substantial gene transfer, the process is laborious and limited to sexually compatible species. For instance, there has been limited success crossing cultivated barley with *Hordeum bulbosum*, where a chromosome segment from *H. bulbosum* can be transferred to the chromosomes of cultivated barley. *Hordeum vulgare* (cultivated barley) has been crossed extensively with cross-compatible *Hordeum spontaneum* (wild progenitor) (Westerbergh et al. 2018). However, according to Bothmer et al. (1995), there are 32 species in the genus *Hordeum*, including diploid, tetraploid, and hexaploid variants. The great majority of these species are unusable because of crossing barriers. Nevertheless, transgenics and genome editing technologies can be utilised to transfer the desired genetic material between species independent of natural crossing barriers provided the relevant genes have been discovered. Rich CWR genetic resources have been established for many important crop species, such as maize, rice, and soybean, to help with improvement. Due to artificial selection during domestication and further loss as a result of contemporary breeding procedures, Glycine soja, a wild relative of farmed soybean (Glycine max), has been demonstrated to have a substantially more varied gene pool than G. max (Kofsky et al. 2018). Numerous agriculturally significant characteristics, such as determinate growth habit (Tian et al. 2010), pod fragmenting (Dong et al. 2014), and seed size (Kofsky et al. 2018), are different between cultivated and wild soybean. Furthermore, it was demonstrated that neither the landraces nor the cultivars of *G. soja* had any of the half of the resistance-related sequences that had been identified. *G. soja* and G. max are cross-compatible, which makes the transfer of desirable traits easier even in spite of their distinct phenotypic variances.

According to Mammadov et al. (2018), nine species of warm-season perennial grasses in the genus Tripsacum, as well as lowland teosinte (*Zea mays* ssp. parviglumis), highland teosinte (*Zea mays* ssp. mexicana), and maize (*Zea mays*) have all been identified as donors of significant traits that could be utilised for improvement. According to genome-wide studies, introgression from the mexicana genome is present in more than 10% of the maize genome, indicating a role for it in adaptation and improvement (Yang et al. 2017). Within the genus Oryza, which includes around 20 species, rice (*Oryza sativa* L.) is one of them (*Oryza glaberrima* S. and *O. sativa* L. are farmed). There are multiple groups within the species, and not all of them are compatible with one another. *Oryza rufipogon*, a wild species thought to be *O. sativa's* direct ancestor, has been shown in recent analyses to exhibit higher sequence diversity and to carry genes and sequences that are entirely absent from the population of cultivated rice (Zhao et al. 2018), suggesting that it may be useful for improving modern rice. When nine cultivated lines of Brassica oleracea were compared to a CWR Brassica macrocarpa, it was shown that the former contained distinct disease resistance genes that were probably lost during the domestication and enhancement of elite B. oleracea genotypes (Golicz et al. 2016). Future breeding efforts will benefit greatly from the growing amount of genomic materials for CWRs, which will help determine the best crossings and targets for genome editing.

2.5 Conclusion and Future Direction

Genome editing's effectiveness, simplicity, and specificity makes it a useful tool for crop development initiatives. It has created new possibilities for plant breeding and functional genomics. According to preliminary research, genome editing is a viable strategy for creating crops resistant to abiotic stress in the future, which could help end world hunger. There are numerous successful instances of "genome editing" in the modern period. Designed nucleases are powerful instruments that target specific DNA sequences in the genomes of model and crop plants, as well as a variety of other organisms, to precisely modify genes. An analysis of the literature in the fields of phonemics, transcriptomics, biotechnology, and genomics has demonstrated the efficacy of this innovative crop growth strategy. Genome editing based on CRISPR/Cas9 is a truly novel approach. Crops can efficiently incorporate a range of genetic features through the process of genome editing. When these strong and accurate techniques are used to speed up plant breeding, specific results are produced. Plant breeding will improve with the aid of this multidisciplinary approach in order to achieve a second Green Revolution and meet the increasing food demands of a rapidly growing global population under constantly changing environmental conditions. Improved plant genetics is ushered in using genome-editing technology, which overcomes the drawbacks of existing transgenic approaches. Researchers and plant breeders may find this information helpful in their in-depth analysis of the application of different gene-editing techniques to enhance crops by concentrating on the targeted gene. We think that in the next days, CRISPR/Cas9 technology will probably help to heal the rift between society and GMOs.

References

Anwar A, Kim JK (2020) Transgenic breeding approaches for improving abiotic stress tolerance: recent progress and future perspectives. Int J Mol Sci 21(8):2695

Appels R, Eversole K, Feuillet C, Keller B, Rogers J, Stein N, Pozniak CJ, Choulet F, Distelfeld A, Poland J, Ronen G (2018) Shifting the limits in wheat research and breeding using a fully annotated reference genome. Science 361:eaar7191

Araki M, Nojima K, Ishii T (2014) Caution required for handling genome editing technology. Trends Biotechnol 32(5):234–237

Arnould S, Delenda C, Grizot S, Desseaux C, Paques F, Silva GH, Smith J (2011) The I-CreI meganuclease and its engineered derivatives: applications from cell modification to gene therapy. Protein Eng Design Select 24(1–2):27–31

Barrangou R, Fremaux C, Deveau H, Richards M, Boyaval P, Moineau S, Romero DA, Horvath P (2007) CRISPR provides acquired resistance against viruses in prokaryotes. Science 315(5819): 1709–1712

Becher M, Talke IN, Krall L, Krämer U (2004) Cross-species microarray transcript profiling reveals high constitutive expression of metal homeostasis genes in shoots of the zinc hyperaccumulator Arabidopsis halleri. Plant J 37(2):251–268

Belfort M, Bonocora RP (2014) Homing endonucleases: from genetic anomalies to programmable genomic clippers. Methods and Protocols, Homing Endonucleases, pp 1–26

Bhat MA, Mir RA, Kumar V, Shah AA, Zargar SM, Rahman S, Jan AT (2021) Mechanistic insights of CRISPR/Cas-mediated genome editing towards enhancing abiotic stress tolerance in plants. Physiol Plant 172(2):1255–1268

Bhatta M, Sandro P, Smith MR, Delaney O, Voss-Fels KP, Gutierrez L, Hickey LT (2021) Need for speed: manipulating plant growth to accelerate breeding cycles. Curr Opin Plant Biol 60:101986

Bibikova M, Carroll D, Segal DJ, Trautman JK, Smith J, Kim YG, Chandrasegaran S (2001) Stimulation of homologous recombination through targeted cleavage by chimeric nucleases. Mol Cell Biol

Biswas D, Saha SC, Dey A (2021) CRISPR-Cas genome-editing tool in plant abiotic stress-tolerance. Plant Gene 26:100286

Boch J, Bonas U (2010) Xanthomonas AvrBs3 family-type III effectors: discovery and function. Annu Rev Phytopathol 48(1):419–436

Bogdanove AJ, Schornack S, Lahaye T (2010) TAL effectors: finding plant genes for disease and defense. Curr Opin Plant Biol 13(4):394–401
Bortesi L, Fischer R (2015) The CRISPR/Cas9 system for plant genome editing and beyond. Biotechnol Adv 33(1):41–52
Bothmer R von, Sato K, Komatsuda T, Yasuda S, Fischbeck G (1995) Diversity in Barley (Hordeum vulgare). In: von Bothmer R, Hintum H, Knüpffer H, Sato K (eds) Barley Genetic Resources: Diversity and Conservation. Cambridge, UK: Cambridge University Press 3–27
Boyer JS (1982) Plant productivity and environment. Science 218(4571):443–448
Brozynska M, Furtado A, Henry RJ (2016) Genomics of crop wild relatives: expanding the gene pool for crop improvement. Plant Biotechnol J 14(4):1070–1085
Carroll D (2011) Genome engineering with zinc-finger nucleases. Genetics 188(4):773–782
Chen K, Wang Y, Zhang R, Zhang H, Gao C (2019) CRISPR/Cas genome editing and precision plant breeding in agriculture. Annu Rev Plant Biol 70(1):667–697
Cramer GR, Urano K, Delrot S, Pezzotti M, Shinozaki K (2011) Effects of abiotic stress on plants: a systems biology perspective. BMC Plant Biol 11:1–14
Crespan E, Czabany T, Maga G, Hübscher U (2012) Microhomology-mediated DNA strand annealing and elongation by human DNA polymerases λ and β on normal and repetitive DNA sequences. Nucleic Acids Res 40(12):5577–5590
Curtin SJ, Anderson JE, Starker CG, Baltes NJ, Mani D, Voytas DF, Stupar RM (2013) Targeted mutagenesis for functional analysis of gene duplication in legumes. In: Legume genomics: methods and protocols, pp 25–42
Dağhan H, Öztürk M, Hakeem KR, Sabir M, Mermut AR (2015) Soil pollution in Turkey and remediation methods. In: Soil remediation and plants: prospects and challenges, pp 287–312
DaMatta FM, Ramalho JDC (2006) Impacts of drought and temperature stress on coffee physiology and production: a review. Braz J Plant Physiol 18:55–81
Daryanto S, Wang L, Jacinthe PA (2016) Global synthesis of drought effects on maize and wheat production. PLoS One 11(5):e0156362
Demorest ZL, Coffman A, Baltes NJ, Stoddard TJ, Clasen BM, Luo S, Retterath A, Yabandith A, Gamo ME, Bissen J, Mathis L (2016) Direct stacking of sequence-specific nuclease-induced mutations to produce high oleic and low linolenic soybean oil. BMC Plant Biol 16:1–8
Dempewolf H, Baute G, Anderson J, Kilian B, Smith C, Guarino L (2017) Past and future use of wild relatives in crop breeding. Crop Sci 57(3):1070–1082
Deriano L, Roth DB (2013) Modernizing the nonhomologous end-joining repertoire: alternative and classical NHEJ share the stage. Annu Rev Genet 47(1):433–455
Dhankher OP, Foyer CH (2018) Climate resilient crops for improving global food security and safety. Plant Cell Environ 41(5):877–884
Dong Y, Yang X, Liu J, Wang BH, Liu BL, Wang YZ (2014) Pod shattering resistance associated with domestication is mediated by a NAC gene in soybean. Nat Commun 5(1):3352
Dräger DB, Desbrosses-Fonrouge AG, Krach C, Chardonnens AN, Meyer RC, Saumitou-Laprade-P, Krämer U (2004) Two genes encoding Arabidopsis halleri MTP1 metal transport proteins co-segregate with zinc tolerance and account for high MTP1 transcript levels. Plant J 39(3): 425–439
Eremina M, Rozhon W, Poppenberger B (2016) Hormonal control of cold stress responses in plants. Cell Mol Life Sci 73:797–810
Eren E, Arguello JM (2004) Arabidopsis HMA2, a divalent heavy metal-transporting PIB-type ATPase, is involved in cytoplasmic Zn2+ homeostasis. Plant Physiol 136(3):3712–3723
Fahad S, Bajwa AA, Nazir U, Anjum SA, Farooq A, Zohaib A, Sadia S, Nasim W, Adkins S, Saud S, Ihsan MZ (2017) Crop production under drought and heat stress: plant responses and management options. Front Plant Sci 8:1147
Farooq M, Wahid A, Kobayashi NSMA, Fujita DBSMA, Basra SM (2009) Plant drought stress: effects, mechanisms and management. In: Sustainable agriculture, pp 153–188
Franchini JC, Debiasi H, Sacoman A, Nepomuceno AL, Farias JRB (2009) Manejo do solo para redução das perdas de produtividade pela seca

Gao C (2021) Genome engineering for crop improvement and future agriculture. Cell 184(6): 1621–1635
Ghori NH, Ghori T, Hayat MQ, Imadi SR, Gul A, Altay V, Ozturk M (2019) Heavy metal stress and responses in plants. Int J Environ Sci Technol 16:1807–1828
Golicz AA, Bayer PE, Barker GC, Edger PP, Kim H, Martinez PA, Chan CKK, Severn- Ellis A, McCombie WR, Parkin IA, Paterson AH (2016) The pangenome of an agronomically important crop plant Brassica oleracea. Nat Commun 7(1):13390
Guy CL (1990) Cold acclimation and freezing stress tolerance: role of protein metabolism. Annual Review of Plant Physiology and Plant Molecular Biology 41:187–223
Hall AE (2018) Crop responses to environment: adapting to global climate change. CRC Press, New York
Hamdan MF, Karlson CKS, Teoh EY, Lau SE, Tan BC (2022) Genome editing for sustainable crop improvement and mitigation of biotic and abiotic stresses. Plants 11(19):2625
Hoang TML, Tran TN, Nguyen TKT, Williams B, Wurm P, Bellairs S, Mundree S (2016) Improvement of salinity stress tolerance in rice: challenges and opportunities. Agronomy 6(4):54
Holmes MW, Hammond TT, Wogan GO, Walsh RE, LaBarbera K, Wommack EA, Martins FM, Crawford JC, Mack KL, Bloch LM, Nachman MW (2016) Natural history collections as windows on evolutionary processes. Mol Ecol 25(4):864–881
Hossain MA, Piyatida P, da Silva JAT, Fujita M (2012) Molecular mechanism of heavy metal toxicity and tolerance in plants: central role of glutathione in detoxification of reactive oxygen species and methylglyoxal and in heavy metal chelation. J Bot 2012(1):872875
Ilardi V, Tavazza M (2015) Biotechnological strategies and tools for Plum pox virus resistance: trans-, intra-, cis-genesis, and beyond. Front Plant Sci 6:379
Isayenkov SV (2012) Physiological and molecular aspects of salt stress in plants. Cytol Genet 46(5):302–318
Isayenkov SV, Maathuis FJ (2019) Plant salinity stress: many unanswered questions remain. Front Plant Sci 10:80
Ismail AM, Heuer S, Thomson MJ, Wissuwa M (2007) Genetic and genomic approaches to develop rice germplasm for problem soils. Plant Mol Biol 65:547–570
Jaganathan D, Ramasamy K, Sellamuthu G, Jayabalan S, Venkataraman G (2018) CRISPR for crop improvement: an update review. Front Plant Sci 9:985
Jain M (2015) Function genomics of abiotic stress tolerance in plants: a CRISPR approach. Front Plant Sci 6:375
Jan AT, Singhal P, Haq QMR (2013) Plant abiotic stress: deciphering remedial strategies for emerging problem. J Plant Interact 8(2):97–108
Ji H, Pardo JM, Batelli G, Van Oosten MJ, Bressan RA, Li X (2013) The salt overly sensitive (SOS) pathway: established and emerging roles. Mol Plant 6(2):275–286
Jinek M, Chylinski K, Fonfara I, Hauer M, Doudna JA, Charpentier E (2012) A programmable dual-RNA–guided DNA endonuclease in adaptive bacterial immunity. Science 337(6096): 816–821
Jones PD, New M, Parker DE, Martin S, Rigor IG (1999) Surface air temperature and its changes over the past 150 years. Rev Geophys 37(2):173–199
Joung JK, Sander JD (2013) TALENs: a widely applicable technology for targeted genome editing. Nat Rev Mol Cell Biol 14(1):49–55
Kaur N, Pati PK (2019) Harnessing the potential of brassinosteroids in abiotic stress tolerance in plants. In: Brassinosteroids: plant growth and development, pp 407–423
Kaur N, Kaur G, Pati PK (2019) Deciphering strategies for salt stress tolerance in rice in the context of climate change. In: Advances in rice research for abiotic stress tolerance. Woodhead Publishing, New York, pp 113–132
Kaur N, Sharma S, Hasanuzzaman M, Pati PK (2022) Genome editing: a promising approach for achieving abiotic stress tolerance in plants. Int J Genomics 2022(1):5547231

Kerchev P, van der Meer T, Sujeeth N, Verlee A, Stevens CV, Van Breusegem F, Gechev T (2020) Molecular priming as an approach to induce tolerance against abiotic and oxidative stresses in crop plants. Biotechnol Adv 40:107503
Khalil AM (2020) The genome editing revolution. J Genet Eng Biotechnol 18(1):68
Kim D, Alptekin B, Budak H (2018) CRISPR/Cas9 genome editing in wheat. Funct Integr Genomics 18:31–41
Kofsky J, Zhang H, Song BH (2018) The untapped genetic reservoir: the past, current, and future applications of the wild soybean (Glycine soja). Front Plant Sci 9:949
Kotak S, Larkindale J, Lee U, von Koskull-Döring P, Vierling E, Scharf KD (2007) Complexity of the heat stress response in plants. Curr Opin Plant Biol 10(3):310–316
Lal R (2020) Genetic engineering of plants for crop improvement. CRC Press, New York
Levitt J (2012) Chilling, freezing, and high temperature stresses, vol 1. Elsevier, London
Li T, Liu B, Spalding MH, Weeks DP, Yang B (2012) High-efficiency TALEN-based gene editing produces disease-resistant rice. Nat Biotechnol 30(5):390–392
Lobell DB, Schlenker W, Costa-Roberts J (2011) Climate trends and global crop production since 1980. Science 333(6042):616–620
Lusser M, Parisi C, Plan D, Rodríguez-Cerezo E (2011) New plant breeding techniques. State of-the-art and prospects for commercial development. (= JRC Scientific and Technical Reports/ EUR 24760 EN)
Ma X, Liu YG (2016) CRISPR/Cas9-based multiplex genome editing in monocot and dicot plants. Curr Protocols Mol Biol 115(1):31–36
Maischak H, Zimmermann MR, Felle HH, Boland W, Mithöfer A (2010) Alamethicin-induced electrical long distance signaling in plants. Plant Signal Behav 5(8):988–990
Makarova KS, Haft DH, Barrangou R, Brouns SJ, Charpentier E, Horvath P, Moineau S, Mojica FJ, Wolf YI, Yakunin AF, Van Der Oost J (2011) Evolution and classification of the CRISPR–Cas systems. Nat Rev Microbiol 9(6):467–477
Malzahn A, Lowder L, Qi Y (2017) Plant genome editing with TALEN and CRISPR. Cell Biosci 7: 1–18
Mammadov J, Buyyarapu R, Guttikonda SK, Parliament K, Abdurakhmonov IY, Kumpatla SP (2018) Wild relatives of maize, rice, cotton, and soybean: treasure troves for tolerance to biotic and abiotic stresses. Front Plant Sci 9:886
Manickavelu A, Nadarajan N, Ganesh SK, Gnanamalar RP, Chandra Babu R (2006) Drought tolerance in rice: morphological and molecular genetic consideration. Plant Growth Regul 50: 121–138
Marschner H (ed) (2011) Marschner's mineral nutrition of higher plants. Academic, London
Migicovsky Z, Myles S (2017) Exploiting wild relatives for genomics-assisted breeding of perennial crops. Front Plant Sci 8:460
Ming M, Ren Q, Pan C, He Y, Zhang Y, Liu S, Zhong Z, Wang J, Malzahn AA, Wu J, Zheng X (2020) CRISPR–Cas12b enables efficient plant genome engineering. Nature Plants 6(3): 202–208
Munns R, Passioura JB (1984) Hydraulic resistance of plants. III. Effects of NaCl in barley and lupin. Funct Plant Biol 11(5):351–359
Munns R, Termaat A (1986) Whole-plant responses to salinity. Funct Plant Biol 13(1):143–160
Munns R, Tester M (2008) Mechanisms of salinity tolerance. Annu Rev Plant Biol 59(1):651–681
Nabi RBS, Tayade R, Hussain A, Kulkarni KP, Imran QM, Mun BG, Yun BW (2019) Nitric oxide regulates plant responses to drought, salinity, and heavy metal stress. Environ Exp Bot 161:120–133
Okçu G, Kaya MD, Atak M (2005) Effects of salt and drought stresses on germination and seedling growth of pea (Pisum sativum L.). Turk J Agric For 29(4):237–242
Osakabe Y, Osakabe K (2015) Genome editing with engineered nucleases in plants. Plant Cell Physiol 56(3):389–400
Pacher M, Puchta H (2017) From classical mutagenesis to nuclease-based breeding–directing natural DNA repair for a natural end-product. Plant J 90(4):819–833

Pence NS, Larsen PB, Ebbs SD, Letham DL, Lasat MM, Garvin DF, Eide D, Kochian LV (2000) The molecular physiology of heavy metal transport in the Zn/Cd hyperaccumulator Thlaspi caerulescens. Proc Natl Acad Sci 97(9):4956–4960
Petolino JF (2015) Genome editing in plants via designed zinc finger nucleases. In Vitro Cell Dev Biol Plant 51:1–8
Pinto AP, Sim [Otilde] Es I, Mota AM (2008) Cadmium impact on root exudates of sorghum and maize plants: a speciation study. J Plant Nutr 31(10):1746–1755
Porter JR (2005) Rising temperatures are likely to reduce crop yields. Nature 436(7048):174–174
Puchta H (2005) The repair of double-strand breaks in plants: mechanisms and consequences for genome evolution. J Exp Bot 56(409):1–14
Rajendran K, Tester M, Roy SJ (2009) Quantifying the three main components of salinity tolerance in cereals. Plant Cell Environ 32(3):237–249
Rascio N, Navari-Izzo F (2011) Heavy metal hyperaccumulating plants: how and why do they do it? And what makes them so interesting? Plant Sci 180(2):169–181
Ricroch A, Clairand P, Harwood W (2017) Use of CRISPR systems in plant genome editing: toward new opportunities in agriculture. Emerg Topics Life Sci 1(2):169–182
Roy SJ, Negrão S, Tester M (2014) Salt resistant crop plants. Curr Opin Biotechnol 26:115–124
Ruelland E, Vaultier MN, Zachowski A, Hurry V (2009) Cold signalling and cold acclimation in plants. Adv Bot Res 49:35–150
Schaart JG, van de Wiel CC, Lotz LA, Smulders MJ (2016) Opportunities for products of new plant breeding techniques. Trends Plant Sci 21(5):438–449
Scott SE, Inbar Y, Wirz CD, Brossard D, Rozin P (2018) An overview of attitudes toward genetically engineered food. Annu Rev Nutr 38(1):459–479
Sedeek KE, Mahas A, Mahfouz M (2019) Plant genome engineering for targeted improvement of crop traits. Front Plant Sci 10:114
Shahid M, Khalid S, Abbas G, Shahid N, Nadeem M, Sabir M, Aslam M, Dumat C (2015) Heavy metal stress and crop productivity. In: Crop production and global environmental issues, pp 1–25
Shan Q, Wang Y, Li J, Gao C (2014) Genome editing in rice and wheat using the CRISPR/Cas system. Nat Protoc 9(10):2395–2410
Shannon MC (1997) Adaptation of plants to salinity. Adv Agron 60:75–120
Shannon MC, Grieve CM (1998) Tolerance of vegetable crops to salinity. Sci Hortic 78(1–4):5–38
Shukla VK, Doyon Y, Miller JC, DeKelver RC, Moehle EA, Worden SE, Mitchell JC, Arnold NL, Gopalan S, Meng X, Choi VM (2009) Precise genome modification in the crop species Zea mays using zinc-finger nucleases. Nature 459(7245):437–441
Silva G, Poirot L, Galetto R, Smith J, Montoya G, Duchateau P, Pâques F (2011) Meganucleases and other tools for targeted genome engineering: perspectives and challenges for gene therapy. Curr Gene Ther 11(1):11–27
Singh Rathore M, Tiwari S, Tripthi N, Tripathi MK, Tiwari S (2021) Status and scenario of genome editing device CRISPR-Cas9 in crop advancement. Curr J Appl Sci Technol 40(48):8–20
Solanke AU, Sharma AK (2008) Signal transduction during cold stress in plants. Physiol Mol Biol Plants 14:69–79
Springer NM, Anderson SN, Andorf CM, Ahern KR, Bai F, Barad O, Barbazuk WB, Bass HW, Baruch K, Ben-Zvi G, Buckler ES (2018) The maize W22 genome provides a foundation for functional genomics and transposon biology. Nat Genet 50(9):1282–1288
Stratonovitch P, Semenov MA (2015) Heat tolerance around flowering in wheat identified as a key trait for increased yield potential in Europe under climate change. J Exp Bot 66(12):3599–3609
Sultan LD, Mileshina D, Grewe F, Rolle K, Abudraham S, Głodowicz P, Niazi AK, Keren I, Shevtsov S, Klipcan L, Barciszewski J (2016) The reverse transcriptase/RNA maturase protein MatR is required for the splicing of various group II introns in Brassicaceae mitochondria. Plant Cell 28(11):2805–2829

Tan J, Zhao Y, Wang B, Hao Y, Wang Y, Li Y, Luo W, Zong W, Li G, Chen S, Ma K (2020) Efficient CRISPR/Cas9-based plant genomic fragment deletions by microhomology-mediated end joining. Plant Biotechnol J 18(11):2161

Tian Z, Wang X, Lee R, Li Y, Specht JE, Nelson RL, McClean PE, Qiu L, Ma J (2010) Artificial selection for determinate growth habit in soybean. Proc Natl Acad Sci 107(19):8563–8568

Turner NC, Wright GC, Siddique KHM (2001) Adaptation of grain legumes (pulses) to water-limited environments

Uemura M, Joseph RA, Steponkus PL (1995) Cold acclimation of Arabidopsis thaliana (effect on plasma membrane lipid composition and freeze-induced lesions). Plant Physiol 109(1):15–30

Urnov FD, Rebar EJ, Holmes MC, Zhang HS, Gregory PD (2010) Genome editing with engineered zinc finger nucleases. Nat Rev Genet 11(9):636–646

Vaughan MM, Block A, Christensen SA, Allen LH, Schmelz EA (2018) The effects of climate change associated abiotic stresses on maize phytochemical defenses. Phytochem Rev 17:37–49

Vincent H, Wiersema J, Kell S, Fielder H, Dobbie S, Castañeda-Álvarez NP, Guarino L, Eastwood R, León B, Maxted N (2013) A prioritized crop wild relative inventory to help underpin global food security. Biol Conserv 167:265–275

Wahid A, Gelani S, Ashraf M, Foolad MR (2007) Heat tolerance in plants: an overview. Environ Exp Bot 61(3):199–223

Wang H, Wang H, Shao H, Tang X (2016) Recent advances in utilizing transcription factors to improve plant abiotic stress tolerance by transgenic technology. Front Plant Sci 7:67

Wang M, Xu Z, Gosavi G, Ren B, Cao Y, Kuang Y, Zhou C, Spetz C, Yan F, Zhou X, Zhou H (2020) Targeted base editing in rice with CRISPR/ScCas9 system. Plant Biotechnol J 18(8): 1645

Westerbergh A, Lerceteau-Köhler E, Sameri M, Bedada G, Lundquist PO (2018) Towards the development of perennial barley for cold temperate climates—evaluation of wild barley relatives as genetic resources. Sustainability 10(6):1969

Wu G, Zhang C, Chu LY, Shao HB (2007) Responses of higher plants to abiotic stresses and agricultural sustainable development. J Plant Interact 2(3):135–147

Yadav RS, Hash CT, Bidinger FR, Devos KM, Howarth CJ (2004) Genomic regions associated with grain yield and aspects of post-flowering drought tolerance in pearl millet across stress environments and tester background. Euphytica 136:265–277

Yang N, Xu XW, Wang RR, Peng WL, Cai L, Song JM, Li W, Luo X, Niu L, Wang Y, Jin M (2017) Contributions of Zea mays subspecies mexicana haplotypes to modern maize. Nat Commun 8(1):1874

Yin K, Han T, Liu G, Chen T, Wang Y, Yu AYL, Liu Y (2015) A geminivirus-based guide RNA delivery system for CRISPR/Cas9 mediated plant genome editing. Sci Rep 5(1):14926

Zafar ZU, Manzoor H, Rasul S, Noreen S, Ali Q, Iqbal M, Javed M, Gul HS, Ahmad Z, Shahzad F, Ogbaga CC (2017) Strategies to improve crop salt and drought tolerance: Success and limitations. In: Quality and quantum improvement in field crops. Agribios, New Delhi, pp 265–298

Zhang Y, Showalter AM (2020) CRISPR/Cas9 genome editing technology: a valuable tool for understanding plant cell wall biosynthesis and function. Front Plant Sci 11:589517

Zhang HX, Zhang Y, Yin H (2019) Genome editing with mRNA encoding ZFN, TALEN, and Cas9. Mol Ther 27(4):735–746

Zhao Q, Feng Q, Lu H, Li Y, Wang A, Tian Q, Zhan Q, Lu Y, Zhang L, Huang T, Wang Y (2018) Pan-genome analysis highlights the extent of genomic variation in cultivated and wild rice. Nat Genet 50(2):278–284

Zhou J, Li D, Wang G, Wang F, Kunjal M, Joldersma D, Liu Z (2020) Application and future perspective of CRISPR/Cas9 genome editing in fruit crops. J Integr Plant Biol 62(3):269–286

Chapter 3
Regulatory functions of long non-coding RNA for Stress Tolerance in Plants

Temesgen Assefa Gelaw and Neeti Sanan-Mishra

Abstract Being sessile in nature, plants have evolved sophisticated gene regulatory mechanisms to drive their growth and development as well as cope with environmental cues. This may involve expression and silencing of specific genes. A large family of non-coding RNAs (ncRNAs) are involved in silencing the gene expression at both transcriptional and post- transcriptional levels. The role of small ncRNAs such as microRNAs and small interfering RNAs has been extensively studied in response to various stresses. The increasing insights on role of lncRNA and their functional validations have unveiled their potential as genetic regulators in plants. In this chapter, we present the detailed biology, classification, mode of action and gene regulatory function of lncRNA in plant growth, development and biotic and abiotic stress responses. We also present an update on the methods and databases for lncRNA identification and characterization in plants.

Keywords Non-coding RNAs · lncRNA · Plants · Stress

3.1 Introduction

Being sessile in nature, plants frequently face various stresses that affect their entire life cycle and ultimately the yield (Jia et al. 2024). Among the stresses, drought, salinity, light, temperature, heavy metal and biotic stress have emerged as major challenges (Jha et al. 2020; Patra et al. 2023). In the current scenario of global

T. A. Gelaw
Plant RNAi Biology Group, International Centre for Genetic Engineering and Biotechnology, New Delhi, India

Department of Biotechnology, College of Agriculture and Natural Resource Sciences, Debre Birhan University, Debre Birhan, Ethiopia

N. Sanan-Mishra (✉)
Plant RNAi Biology Group, International Centre for Genetic Engineering and Biotechnology, New Delhi, India
e-mail: neeti@icgeb.res.in

S. K. Panda, M. A. Khan (eds.), *Plant Functional Genomics for Abiotic Stress Resilience*, Plant in Challenging Environments 7,
https://doi.org/10.1007/978-3-032-01704-8_3

climate change, the frequency and duration of these stresses on the plants has increased (Zandalinas et al. 2021). Evolutionarily, plants have adopted complex mechanisms for intricately regulating a large set of physiological, biochemical and molecular networks to guide their growth, development and response to stress. It was observed that about 90% of the plant genome could be transcribed into messenger RNAs (mRNAs) and from this pool, approximately 2% of the transcribed RNAs were translated to functional proteins (Yu et al. 2019). The remaining noncoding RNAs (ncRNAs) were considered as transcriptional noise or junk (Kuska 1998). But since the late 1990s, the existence of regulatory ncRNAs has driven exciting findings and this research field has expanded exponentially.

We now understand that the ncRNAs constitute vital regulatory hubs of gene function (Kung et al. 2013). They are involved in the transcriptional and post-transcriptional regulation of gene expression, as well as regulation of RNA stability and translation (Jin et al. 2023). The ncRNAs can be classified based on their size; with 18–30 nucleotides (nt) range comprising the small ncRNAs, 31–200 nt classifying as medium-sized ncRNAs and >200 nt categorizing as long ncRNAs (lncRNA). The functional roles of small ncRNAs, particularly microRNAs (miRNA) and small interfering RNAs (siRNA), have been explained in several plants in association with diverse biological processes such as growth, development, hormone signalling and stress responses (Tang and Chu 2017; Yu et al. 2019). Several recent studies have now shed light on the essential regulatory roles of lncRNA in plants (Jia et al. 2022, 2024; Mansuri et al. 2022). In this chapter, we present the concise update on the progresses of plant lncRNA research, with an emphasis on their biogenesis, mode of action including interaction with other small RNAs, methods for study and their role in various stress response.

3.2 Long Non-coding RNA

LncRNA have emerged as a pivot of genome regulation, adding a new layer of epigenetic control (Espinosa 2017; Wu et al. 2017; Zampetaki et al. 2018). This particular group of ncRNAs lacks polypeptide-coding potential and possesses significant biochemical versatility (Hirose and Nakagawa 2016; Gloss and Dinger 2016; Wu et al. 2017). Innovations in High- throughput RNA-seq have primarily contributed to the identification of lncRNA. Numerous studies have been performed in animals (Derrien et al. 2012; Bai et al. 2015; Kazemzadeh et al. 2015); however, in plants, only a few molecules have been well characterized (Liu et al. 2015a, b; Wang and Chekanova 2017; Wu et al. 2017).

3.2.1 Biogenesis and Classification

The process of transcription, in eukaryotes, involves the generation and processing of the nascent RNA to form the messenger (mRNA). The biogenesis of a fully mature mRNA requires the coordinated action of enzymes that perform 7-methyl Guanosine (m7G) capping, splicing, polyadenylation, maturation, export and surveillance (Moore and Proudfoot 2009; Bentley 2014). In a similar manner, primary transcripts of lncRNA, which are produced by RNA Polymerases (Pol) II or III and IV or V, undergo polyadenylation, m7G capping and splicing (Wierzbicki et al. 2008; Yu et al. 2019). Most of the plant lncRNA are polyadenylated, but in mammals and yeast, some non-polyadenylated lncRNA have been found (Andersson et al. 2014). Therefore, the presence of non-polyadenylated functional lncRNA in plants cannot be completely ruled out (Wilusz et al. 2008). The non-polyadenylated lncRNA are processed by Ribonuclease P to generate free ends that are stabilized by the capping with small nucleolar RNAs (snoRNA). In some cases, the snoRNA have been found at both the 3′ and 5′ ends of the lncRNA. Reports have also shown that Ribonuclease P processed free ends can also self-ligate to form circular structures (Zhang et al. 2014a, b; Sibley et al. 2016; Wu et al. 2017; Zampetaki et al. 2018).

The classification methods of lncRNA are accordingly diverse (Kashi et al. 2016). On the basis of their location i.e. the relative position between lncRNA coding sequence and protein- coding gene, they could be classified as (Smith et al. 2014);

(a) sense lncRNA: when lncRNA sequence overlaps with the sense strand of the protein-coding gene.
(b) antisense lncRNA: when lncRNA sequence overlaps with the anti-sense strand of the protein-coding gene.
(c) bidirectional lncRNA: when lncRNA sequences are located in the opposite chain of a protein-coding gene.
(d) intergenic lncRNA: when lncRNA sequence is not positioned nearby any other protein-coding genes.
(e) intronic lncRNA: when the lncRNA sequence is completely emerged from the intron of another transcript.

On the basis of their biogenesis with respect to the location, the lncRNAs can be classified into several subgroups (Lucero et al. 2021) as described below.

1. Long intergenic ncRNAs (lincRNA) are derived from the intergenic region and are 200–2000 nt in length. They are also referred to as large/long intervening ncRNAs (Yu et al. 2019; Jha et al. 2020). LincRNA are polyadenylated, weakly spliced, exhibit tissue-specific expression and possess a *trans*-regulatory role (Wu et al. 2017; Statello et al. 2021). These transcripts are characterized by rapid turnover rates, which pose a challenge to fully understand their function (Fan et al. 2015; Wu et al. 2017). According to one report, lincRNA are localized in gene deserts, at least 5 kb away from protein-coding regions (de Santa et al. 2010). Based on their association with specific regions, (Mattick and Rinn 2015;

St. Laurent et al. 2015), lincRNA have been further sub-classified in to different groups such as enhancer RNAs (eRNA) (Andersson et al. 2014), upstream antisense RNA (uaRNA), promoter-associated long RNA (PALR) or promoter upstream transcripts (PROMPT) (Preker et al. 2011) and telomeric repeat-containing RNA (TERRA). The PROMPT and eRNA are short-lived lincRNA identified mostly in humans.

2. Transposable element (TE)-derived lncRNA (TE-lncRNA) are generated from the TE (Jha et al. 2020), also called jumping genes, which were first discovered in maize (McClintock 1951). TE have the ability to copy/cut and paste themselves into other genome regions (Bourgeois and Boissinot 2019). In the maize genome, more than 85% of DNA is derived from TE (Schnable et al. 2009; Bourgeois and Boissinot 2019), so a large number of TE-lncRNA have been predicted in maize (Lv et al. 2019; Noshay et al. 2019). In some cases, the TE-lncRNA act as precursors to miRNA and siRNA (Li et al. 2014; Burgess et al. 2020; Ali et al. 2021). In *Arabidopsis*, rice, maize and tomato, TE-lncRNA were reported to play a role in abiotic stresses response (Wang et al. 2016; Cho 2018; Zhang et al. 2018a, b, c; Chen et al. 2020a, b).
3. Intron-derived lncRNA (incRNA) originate mainly from the introns (Ma et al. 2013) and may include totally intronic or partially intronic RNAs. They are regulated by various transcription activation pathways (Prensner et al. 2011) and were considered to act as progenitors of the small ncRNAs like miRNA and siRNA (Kim and Kim 2007; Zhou and Lin 2008; Louro et al. 2009). However, recently their independent existence was proven in experiments, which also showed these transcripts are usually stable, contain polyadenylated modifications but may not be highly conserved across different plant species (Derrien et al. 2012; Tani et al. 2015).
4. Natural antisense transcripts (NAT) are amongst the widespread lncRNA and originate from coding regions. They regulate gene expression by silencing via both *cis*- and *trans*- action (Wight and Werner 2013; Yu et al. 2013). *Cis*-NAT are transcribed from the antisense strand to regulate the corresponding sense transcript (Wight and Werner 2013). *Trans*-NAT arise from a locus away the target gene (Lavorgna et al. 2004; Wight and Werner 2013) and their binding to complementary strand, triggers the production of specific siRNA, which also exhibit a predominantly *trans* mode of action.
5. Circular lncRNA (circncRNA) are highly conserved but low in abundance. Since they cannot be degraded easily, they are more stable than linear lncRNA (Li et al. 2018). circncRNA were first characterized as non-polyadenylated circular RNA, in plant viroids, (Sanger et al. 1976). It was assumed that during canonical splicing of the exon, failure of lariat debranching might result in the biogenesis of circncRNAs (Zhang et al. 2013; Vicens and Westhof 2014; Liu et al. 2020; Qin et al. 2020). circncRNA mostly arise in the nucleus (Zhang et al. 2013; Wu et al. 2017; Chu et al. 2018; Li et al. 2018), while some arise in the cytoplasm. Mostly, circncRNA consist of one or more extra exons and they are called extra- exon circular ncRNA (eecircncRNA). However, some may originate from the intron region of the parent gene and are called circular intronic RNAs (circincRNA) or

intron retained circular ncRNA. Some others might arise from the overlapping regions and are classified as exon-intron circncRNAs (eicincRNA) (Meng et al. 2017; Liu et al. 2020). Studies on their functions indicated that circncRNA act as endogenous target mimics (eTM) of miRNA, miRNA sponges (Liu et al. 2015a; Qin et al. 2020), protein scaffolds or templates for protein translation.

3.3 Mode of Action of lncRNA in Plants

Numerous lncRNA are transcribed within cells, yet their underlying molecular mechanisms remain largely elusive (Weidong et al. 2020; Qin et al. 2020). Primarily, they serve as regulators, facilitating gene silencing to modulate transcriptional regulation and genome imprinting in response to cues from environment, hormones, photosynthesis, metabolism and so on (Hu et al. 2022). These regulatory functions extend to diverse biological processes, including root organogenesis (Song et al. 2016), photo-morphogenesis (Wang et al. 2014a, b, c), control of flowering time, reproduction and nutrient homeostasis (Yuan et al. 2016), among others (Chen et al. 2020a, b; Statello et al. 2021). The gene regulatory functions of lncRNA, their associated transcripts and the small ncRNA show multigenic interactions.

It is widely accepted that there is lot of similarity in the transcription and processing of lncRNA and mRNA, but expression levels of lncRNA are lower as compared to the mRNA (Yuan et al. 2023). Also, their expression levels exhibit significant variations across different tissues and developmental stages. lncRNA could act as signal molecules, precursors and target mimics for small ncRNA, regulators of pre-mRNA splicing and modulators of chromatin (Geisler and Coller 2013; Bardou et al. 2014; Wang et al. 2017; Cho 2018). The formation of lncRNA-RNA duplexes is linked to post-transcriptional gene regulation (Zhang et al. 2023). lncRNA can serve as precursors of miRNA and siRNA (Burgess et al. 2020). In case of regulatory interplay with miRNA, they also act as decoys or sponges to sequester miRNA (Hansen et al. 2011), or disrupt the precise cleavage process of pri-miRNA, thereby influencing the expression of target mRNAs (Wang et al. 2014a, b, c; Qin et al. 2020). The lncRNA transcribed by Pol V aid in recruiting silencing machinery to gene promoters for repressing gene expression (Zheng et al. 2012) and preventing gene read- through (McKinlay et al. 2018).

3.3.1 *lncRNA as Precursors of Small RNAs*

Some lncRNA can be processed to small ncRNA for functioning (Hirsch et al. 2006), so they can also function as precursors for siRNA. The lncRNA can be transcribed by RNA Pol II as in case of primary miRNA (pri-miRNA) or may be derived from Pol IV/Pol V- dependent pathways as in case of double stranded RNA (dsRNA) precursors of siRNA (Kim and Sung 2012; Liu et al. 2015a, b).

Genome-wide computational analysis of full-length cDNA databases in *Arabidopsis* led to the identification of 76 lncRNA including *npc34, npc351, npc375, npc520* and *npc523*. Numerous 24-nt siRNA were matched to those lncRNA, suggesting that these npcRNA are precursors of siRNA (Amor et al. 2009). Moreover, mapping study identified some of the lncRNA as potential miRNA precursors (pre-miRNA) such as, *npc83* and *npc521* were identified as precursors of miR869a and miR160c, respectively (Hirsch et al. 2006).

In another study, about 110 lincRNA and 46 genic lncRNA were predicted as precursors for *mediator of paramutation 1* (*MOP1*) sensitive siRNA in maize (Boerner and McGinnis 2012). In addition, 26 lincRNA and 97 genic lncRNA were predicted as precursors for short hairpin RNA (shRNA), while one lincRNA and two genic lncRNA were predicted as pre- miRNA. In tomato, *lncRNAZ081* was reported as the pre-miRNA of miR6027 while *lncRNAZ114* as precursor of miR1919b and miR1919c (Wang et al. 2018a, b). In *S. lycopersicum, TCONS_00061773* acts as the precursor of ptc-miR1448 and ptc-miR482a, and is involved in response to nitrogen deficiency. Using small RNA-seq, 43 regions having high complementarity to long-stem transcripts were identified in *Arabidopsis* (Ma et al. 2014), that could be possibly recognized by Dicer-like 1 (DCL1), suggesting their potential function as precursors for siRNA.

3.3.2 lncRNA as Mimic for miRNA Targets

Studies have shown that some of the lncRNA can regulate transcription by mimicking as targets for the miRNA (Jha et al. 2020; Waseem et al. 2021) and have important roles in plant growth and development (Todesco et al. 2010). The miRNA binding mimic- sites possessed by the lncRNA are non-cleavable and can quench the miRNA thereby preventing translational inhibition. The concept of target mimicry was introduced after the discovery of *IPS1* (*induced by phosphate starvation*) and *ASCO* (*alternative splicing competitor*), genes in *Arabidopsis* (Borah et al. 2018). These transcripts contain miRNA complementary sequences, through which they can act as decoys or sponges or eTM, by competing for miRNA binding sites (Franco-Zorrilla et al. 2007; Wu et al. 2013; Fan et al. 2015). *IPS1* contains a region complementary to miR399, though it has a loop at miR399 cutting site, which prevents its cleavage. Thus, *IPS1*:miR399 forms a stable pair and quenches the miRNA activity (Franco-Zorrilla et al. 2007). The expression of both *IPS1* and miR399 is induced upon low phosphate abundance indicating that *IPS1* is required to fine-tune the miR399 activity during phosphate starvation. Moreover, *ASCO*-lncRNA binds to *AtNSRa* and *AtNSRb* to regulate lateral root development (Bardou et al. 2014).

As other examples, *lncRNA23468* functions as a decoy for miR482b to compete with transcripts of *NBS-LRR* genes during the interaction between tomato and *Phytophthora infestans* (Jiang et al. 2019). Similarly, *lncRNA16397* targets *slGRX22* (*glutaredoxin* gene) to induce *GRX21* expression and reduce reactive oxygen species (ROS) (Cui et al. 2017). The dual regulators act to modulate gene expression during *P. infestans* infection in tomatoes.

Several other studies have reported the interaction between lncRNA and miRNA (Shuai et al. 2014). A computational study in maize revealed about 466 lncRNA being targeted by 165 miRNA and 86 lncRNA as decoys for 58 miRNA (Fan et al. 2015). Also, in *Populus* 51 lncRNA were found to be putative miRNA targets and other 20 lncRNA as target mimics of the known miRNA during drought stress (Shuai et al. 2014). Additionally, in *Cleistogenes* 52 lncRNA were identified as target mimics for miRNA using RNA-seq analysis (Yan et al. 2019). These reports showed the association between lncRNA and miRNA in regulating gene expression in plants.

3.3.3 lncRNA in DNA Modification

Trans-acting lncRNA can regulate various processes such as transcription, translation, or epigenetic modification of target genes by binding with diverse molecules such as Transcription factors (TFs), chromatin modifiers, or other regulatory factors (Wang et al. 2014a, b, c). The *cis*-NATs can occasionally serve as enhancers of translation for their corresponding mRNAs. A point in case is the differential expression of *cis-NATPHO1;2*, which impacts the synthesis of the PHO1;2 protein without altering the sense *PHO1;2* transcript (Jabnoune et al. 2013). Some NATs can increase the expression of their cognate sense strand. For instance, a NAT-lncRNA, *MAS* promotes the expression of *MAF4* (*MADS Affecting Flowering 4*) genes (Zhao et al. 2018).

Gene silencing via histone modification and DNA methylation is a vital hub for regulation of gene expression. It holds particular importance in the regulation of transposons, repetitive sequences, centromeric regions and imprinting (Liu et al. 2015a, b; Cho 2018; Zhou et al. 2018a, b). Studies have shown that plant lncRNA also participate in the modification of DNA at different developmental stages (Chen et al. 2018a, b; Urquiaga et al. 2021). The classical example is provided by *Cold Assisted Intronic noncoding RNA* (*COLDAIR*). In *Arabidopsis, COLDAIR* inhibits *Flowering Locus C* (*FLC*) during vernalization to regulate the flowering time (Csorba et al. 2014). It is achieved by enrichment of Polycomb repressive complex 2 (PRC2) and Histone H3 Lysine 27 (H3K27me3) trimethylation at the *FLC* locus (Heo and Sung 2011).

The gene silencing role of lncRNA could also be extended through RNA directed DNA methylation (RdDM) (Zhang et al. 2019a, b, c). Studies in *Arabidopsis* showed that some antisense lncRNA could influence histone modification and regulate sense transcripts (Wang et al. 2014a, b, c). In rice, the lncRNA were shown to regulate photoperiod-sensitive male sterility by increasing methylation in the promoter region (Ding et al. 2012). In another study, it was shown that *Twisted Leaf*, a NAT lncRNA derived from the antisense strand of *MYB60* locus, inhibits expression of *MYB60* by increasing the expression of H3K27me3 in the promoter region of *MYB60* locus (Deng et al. 2018).

In maize, the RNA directed DNA methylation (RdDM) function is supported by MOP1-1, an ortholog of At-RDR2 (Nobuta et al. 2008). Plant-specific RNA Pol IV and Pol V play a crucial role in RdDM and heterochromatin processes to regulate repetitive intergenic elements (Gao et al. 2010; Haag and Pikaard 2011; Huang and Zhu 2014; Sanan-Mishra et al. 2021). Pol IV, in conjunction with CLASSY chromatin remodeling factors (CCRFs) and homeodomain transcription factors like DTF1/SHH1, transcribes transposons and repetitive sequences. These transcripts are then transformed into dsRNA through the activity of RNA-dependent RNA polymerase 2 (RDR2), which are subsequently processed into small ncRNA, specifically siRNA duplexes, by the DCL3 enzyme (Huang and Zhu 2014; Sanan-Mishra et al. 2021). These siRNA are incorporated into Argonaute 4 (AGO4) containing complexes to facilitate RdDM. In an alternative pathway, siRNA are produced through Pol I–RDR6 transcription and are then loaded into the AGO6 complex.

At loci where Pol V is actively transcribing, the siRNA-guided AGO4, 6 complexes interact with the larger subunit of RNA Pol V (Wierzbicki et al. 2009; Zhong et al. 2012) and the interaction is stabilized by KTF1 (a yeast transcription elongation factor and homolog of SPT5). This facilitates establishment of DNA methylation through methyltransferase 2 (DRM2) (Gao et al. 2010; Zhong et al. 2012). The methylated state of DNA is then preserved across cell divisions via pathways mediated by methyltransferase1 (MET1) or chromomethylase3 (CMT3). MOP1 copies the RNA Pol IV transcript for processing into siRNA (Haag and Pikaard 2011). Therefore, RdDM is a complex pathway that has also been implicated to play a role in plant growth and stress response/memory (Urquiaga et al. 2021).

3.3.4 lncRNA as Precursors of Small Peptides

In some recent studies, the lncRNA were shown to produce functional peptides (<100 amino acids), which are involved in a variety of cellular processes. For instance, soybean *ENOD40* can produce two peptides, which can bind to sucrose synthase in nitrogen-fixing nodules (Röhrig et al. 2002). More recently the pri-miRNA encoded peptides have been identified. In *Arabidopsis*, *pri-miR858a* encodes miPEP858a and controls flavonoid biosynthesis (Sharma et al. 2020). Also, *pri-miR165a* of *Arabidopsis* and the *pri-miR171b* of *Medicago truncatula* contain short open reading frame sequences that code for miPEP165a and miPEP171b peptides, respectively (Lauressergues et al. 2015). In *Vitis vinifera*, *pri-miR171d* encodes miPEP171d1 to regulate adventitious root formation (Chen et al. 2020a, b). The association and universality of lncRNA and their encoded peptides is gaining much attention but the understanding of their functional regulatory pathways needs deeper understanding.

3.4 Role of lncRNA in Plant Growth and Development

The lncRNA exhibit low sequence conservation, but they are widely expressed and perform various functional roles in plant growth and development (Chen et al. 2021). These regulatory roles are primarily manifested through auxin transport and signal transduction pathways (Wan et al. 2020). For instance, in *Arabidopsis, RNAAPOLO* lincRNA is transcribed by RNA Pol II and Pol V in response to auxin and affects root organogenesis (Ariel et al. 2014).

The regulatory mechanism involving lncRNA as an eTM for a miRNA is preserved across various plant species and biological pathways. A study by Das et al. (2019) showed that *Cc_lncRNA-2830* sequesters miR160h to induce *auxin-responsive factor* (*ARF*) 18 and thus regulate seed and pod development. In sea buckthorn, the eTM provide an alternative pathway to regulate anthocyanin biosynthesis. The lncRNA, *LNC1* and *LNC2*, act as eTM of miR156a and miR828a to down regulate *SPL9* and up regulate *MYB114*, respectively (Zhang et al. 2018a, b, c). During withering process in tea plant, the *LTCONS_00026271*-novel_miR44-*LOX* and *LTCONS_00020084*-miR169d-5p_1-*ACX* regulatory modules induce the accumulation of jasmonic acid to play a significant role in increasing the terpenoid levels (Zhu et al. 2019).

Moreover, lncRNA are involved in regulating flowering and reproductive development in plants. In *Arabidopsis, MAS* is induced by low temperatures from the *MAF4* locus to regulate flowering. It is also involved in activating the *MAF4* transcription for suppression of precocious flowering. To activate *MAF4*, *MAS* interact with WDR5a to enhance histone 3 lysine 4 trimethylation (H3K4me3) (Zhao et al. 2018). Furthermore during vernalization, lncRNA, *COOLAIR* and *COLDAIR,* regulate epigenetic silencing of *FLC* in *Arabidopsis* (Heo and Sung 2011).

Following *turnip crinkle virus* (TCV) infection in *Arabidopsis*, the up regulation of an intergenic lincRNA, *LINC-AP2* and the simultaneous down regulation of its neighbouring gene, *AP2* (*Apetala2*) was observed (Gao et al. 2016). The notable increase in *LINC-AP2* expression is associated with structural anomalies in flowers (Gao et al. 2016). In *Arabidopsis*, an intronic lncRNA called *AG-incRNA4* is expressed in leaves and it binds to the PRC2 complex component *CLF* (*Curly Leaf*), for the deposition of the H3K27me3 histone mark at the *agamous* (*AG*) loci, contributing to the repression of *AG* expression in leaves. Knocking down *AG-incRNA4* leads to *AG* activation in leaves by diminishing H3K27me3 levels at *AG* sites. Consequently, the mutants display phenotypes characterized by ectopic AG expression (Wu et al. 2018).

In rice, the lncRNA, *MISSEN*, takes over the indispensable role in early endosperm development. *MISSEN* operates by hijacking a helicase family protein to control tubulin activity during endosperm nucleus division and cellularization, leading to irregular cytoskeletal formation (Zhou et al. 2021a, b). The cytoskeletal disruption negatively regulates development of endosperm and leads to a protruding dent and bulge formation in rice seed.

The lncRNA, *LDMAR* regulates photoperiod-sensitive male sterility in rice (Ding et al. 2012). The *BcMF11*, an lncRNA specific to pollen in *Brassica campestris* L. plays a crucial role in both pollen development and male fertility (Song et al. 2013). Reduced expression of BcMF11 led to decreased pollen germination and delayed extension of pollen tubes (Song et al. 2013). Throughout cabbage pollen development, pollination and fertilization, about 15 lncRNA were identified as potential regulators of 13 miRNA, by acting as eTM. Among these, *bra-eTM160-1* and *bra-eTM160-2* were specifically recognized for their role in modulating cabbage miRNA160 activity. This miRNA is crucial in pollen development, influencing the expression of target genes belonging to the ARF family (Huang et al. 2018).

Functional investigations of lncRNA in *Arabidopsis* revealed that overexpression of *npc48* (983-nt-llncRNA expressed in flowers, leaves, roots and stems) alters miR164 expression, resulting in serrated leaves and delayed flowering (Amor et al. 2009). Loss of function of the miR164 leads to prominent leaf serrations, indicative of impaired proliferation of leaf primordial margin cells, along with aberrant vegetative growth characterized by increased rosette leaf diameter and notably delayed flowering time. This observed phenotype bears resemblance to that of *Serrate* (*SE*) and *AGO1* mutants. SE governs miRNA levels and influences miRNA processing, while AGO1 serves as the target cleavage enzyme for most miRNA. In rice, lncRNA-*Twisted Leaf* was identified as a regulator of leaf flatness through modulation of the R2R3-MYB gene expression (Liu et al. 2018).

Conversely, overexpression of *npc536* in *Arabidopsis,* enhanced root growth particularly under salt stress conditions (Amor et al. 2009). In *Arabidopsis*, *ASCO*-lncRNA influences root development by generating alternative splicing variants of various mRNAs (Bardou et al. 2014). The auxin-inducible nuclear speckle RNA binding proteins (NSRs), NSRa and NSRb, are essential for constitutive splicing (Schindler et al. 2008). Treatment with auxin promotes lateral root formation in wild-type seedlings, however, double mutants lacking NSRa and NSRb produce few lateral roots and show insensitivity to auxin. *ASCO-lncRNA* (*npc351*) competes with precursor mRNAs for binding to *NSRs* and *ASCOlncRNA*- overexpression lines exhibit increased accumulation of alternatively spliced transcripts (Bardou et al. 2014).

During growth and development, plants utilize various molecules such as nitrate. The *Arabidopsis* lncRNA, *T5120,* has been identified as a regulator of nitrate assimilation and plant growth, thereby improving nitrogen utilization efficiency (Liu et al. 2019). Overexpression of *T5120* enhanced responsiveness to nitrate, as characterized by improved nitrate assimilation, better biomass and greater root development. Notably, *T5120* is co- regulated by the nitrate transcription factor (NLP7) and the nitrate sensor (NRT1.1) to modulate nitrate signal transduction pathway (Liu et al. 2019). In soybean, lncRNA *GmENOD40* is associated with the formation of symbiotic nitrogen-fixing nodules. Its homologs, *OsENOD40* and *MtENOD40* (Kouchi et al. 1999) were later identified in rice and alfalfa, respectively. *OsENOD40* is specifically expressed in rice stems and functions in organ differentiation and vascular tissue development (Wang et al. 2023).

In wheat, miR9678 targets the lncRNA, *WSGAR*, triggering the production of phasiRNA through cleavage, thus delaying seed germination (Guo et al. 2018). In rice, lncRNA called *LAIR* (*LRK Antisense Intergenic RNA*), which is transcribed from the minus strand adjacent to the leucine-rich repeat receptor kinase (LRK) gene cluster, regulates rice yield. Overexpression of LAIR enhances grain yield and boosts the expression of several LRK genes. Furthermore, it was confirmed by chromatin immunoprecipitation that the activated LRK1 genomic locus was significantly enriched with H3K4me3 and H4K16ac. LAIR interacts with histone modification proteins, OsMOF and OsWDR5 in rice cells (Wang et al. 2018a, b). These results emphasize the importance of the functional study of lncRNA to fully understand the key biological processes of plant development (Yuan et al. 2023).

3.5 Role of lncRNA in Abiotic Stress Tolerance

Several reports have shown the role of lncRNA in regulating abiotic stress responses in plants (Fig. 3.1).

3.5.1 Drought Stress

Numerous drought-responsive lncRNA have been identified across various plant species like maize (Zhang et al. 2014a, b), rice (Chung et al. 2016), foxtail millet

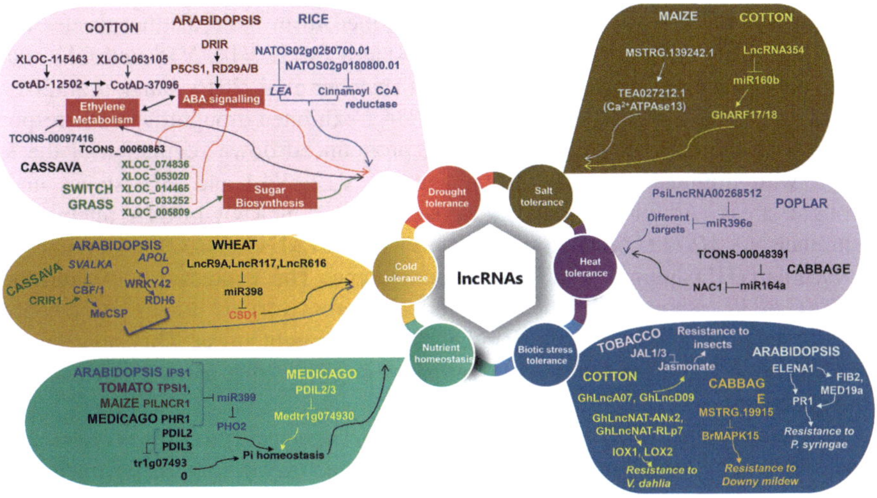

Fig. 3.1 A diagrammatic illustration to show the role of lncRNA in plant response to different abiotic and biotic stress

(Qi et al. 2013), switchgrass (Zhang et al. 2018a, b, c) and cassava (Ding et al. 2019) through genome-wide transcriptome analyses (Gelaw and Sanan-Mishra 2021). A total of 9247, 20 and 28 lncRNA have been identified as *cis*-acting with respect to ethylene, auxin, gibberellin and cytokinin, respectively. Primarily, these lncRNA regulate the drought response by modulating genes involved in ethylene and ABA synthesis, hormone and calcium signalling, starch and sucrose synthesis. Notably, *XLOC-115463* and *XLOC-063105* lncRNA were found to regulate *CotAD-12502* and *CotAD-37096* (*cis*-coding genes) of ethylene metabolic pathway in cotton plants. Additionally, approximately 407 lncRNA were found to correlate with ubiquitin for protein degradation, during stress signal transduction. Moreover, *trans*- activity was observed for 167 lncRNA-target pairs associated with ethylene, 3086 pairs with indole-3-acetic acid, 239 pairs with gibberellic acid, 334 pairs with cytokinin and 2621 pairs with drought stress (Lu et al. 2016).

The potential role of lncRNA as positive regulators of drought stress tolerance was highlighted with the discovery of a novel nucleus-localized 755 nt long *drought-induced lincRNA* (*DRIR*) in *Arabidopsis* (Dong et al. 2022). Overexpression of *DRIR* conferred higher drought tolerance in *Arabidopsis* lines compared to wild-type seedlings (Qin et al. 2017). This lincRNA regulates the transcription of several drought stress-responsive genes, including those involved in ABA signalling (*ABI5, P5CS1, RD29A* and *RD29B*), aquaporin activity (*NIP1* and *TIP4*), annexin function (*ANNAT7*), fucosyltransferase4 (*FUT4*) and transcription factors (*NAC3* and *WARKY8*) (Qin et al. 2017).

In maize, a total of 664 drought-responsive lncRNA have been recognized. Out of these, 567 lncRNA comprising 189 intragenic, 21 antisense, 344 intergenic and 13 overlapping lncRNA showed up regulation, while 97 lncRNA including 16 intragenic, 2 antisense, 77 intergenic and 2 overlapping lncRNA displayed down regulation in the leaves of maize under drought stress conditions (Zhang et al. 2014a, b). A multitude of lncRNA in maize were unearthed from three distinct tissues (ear, tassel and leaf) across four developmental stages (V12, V14, V18 and R1) under both well-watered and drought conditions (Thatcher et al. 2016; Pang et al. 2019). Among the total 3488 lncRNA, *MSTRG.2834.1* exhibited drought- specific responsiveness in the ear, showcasing the most pronounced down regulation at the R1 stage, with minimal changes observed at the V12, V14 and V18 stages under drought stress. Additionally, they discovered 1177 lncRNA-mRNA interactions, with lncRNA *MSTRG.6838.1* targeting *vpp4* (GRMZM2G028432), which encodes a vacuolar (H^+)- pumping ATPase subunit gene. The expressions of *vpp4* and *MSTRG.6838.1* showed significant correlation across various tissues and developmental stages, both being down regulated under drought stress. This suggests that *MSTRG.6838.1* and *vpp4* may represent promising *cis*-acting pair with potential for co-regulation.

In a related study by Zhang et al. (2019a, b, c), large-scale profiling revealed 2174 and 1354 high- confidence circRNA in maize and *Arabidopsis*, respectively. The team identified two circRNA associated with the *GRMZM2G032852* and *GRMZM5G817173* genes encoding *calcium-dependent protein kinase (CPK)* and *cytokinin oxidase/dehydrogenase (CKX)*, respectively. These circRNA, named *circCPK* and *circCKX*, were generated through back- splicing between exons, resulting in the insertion and deletion of alleles in the maize panel. Interestingly, the expression levels of circCPK and the survival rate of maize plants with insertion alleles were notably higher than those with deletion alleles under both normal and drought conditions. This finding suggests a potential link between circRNA expression and plant drought tolerance.

In rice, 98 drought-responsive NAT-lncRNA were identified through RNA-seq analysis. Among these, *NAT Os02g0250700-01* and *NAT Os02g0180800-01* target the *Late Embryogenesis Abundant (LEA) protein* and *Cinnamoyl CoA Reductase genes*, respectively (Chung et al. 2016). In tomato, Eom et al. (2019) identified and characterized a total of 521 drought- responsive lncRNA. These lncRNA are primarily involved in regulating genes associated with stress tolerance, signalling and transporter activity. Drought stress led to the down regulation of Sly-miR169c, while inducing the expression of its target, *lncRNA-tomato-535*. Conversely, Sly-miR396b was up regulated under drought stress, while *lncRNA-tomato-503* was down regulated. During drought conditions, *lncRNA-tomato-146* and *lncRNA-tomato- 178,* acted as eTM for controlling the expression of Sly-miR169c and Sly-miR396b, respectively.

Research on cassava identified 51 drought-specific differentially expressed lncRNA. The expression analysis of selected molecules in response to drought stress indicated the up regulation of *lincRNA101, lincRNA391* and *lincRNA356*, while down regulation of *lincRNA64, lincRNA350, lincRNA182* and *lincRNA392*. Additionally, *TCONS_00060863* and *TCONS_00097416* lncRNA were found to regulate ABA and ethylene signalling pathways, respectively, during drought stress (Ding et al. 2019). In switchgrass, drought stress led to up regulation of the lncRNA, *XLOC_053020, XLOC_014465 and XLOC_033252*, which control ABA synthesis, *XLOC_074836* which regulates ethylene signalling and XLOC_005809 that controls trehalose phosphate synthase (Zhang et al. 2018a, b, c).

In *Brassica juncea*, approximately 21 heat and drought-responsive lncRNA were found to co-express with 10 TFs (Bhatia et al. 2020). Among these pairs, MIKC-MADS exhibits the highest proportion of co-expression with lncRNA, followed by the NAC families (Castelán- Muñoz et al. 2019) and the MYB family (Li et al. 2017a, b). Transcriptome analysis of *Brassica napus* revealed that an lncRNA-mRNA co-expression network is involved in phytohormone signal transduction during drought stress (Tan et al. 2020).

3.5.2 *Salinity Stress*

Transcriptome sequencing resulted in the discovery of more than 7000 lncRNA in different stresses including salinity (Wang et al. 2023). Transcriptome analysis of the salt-tolerant maize variety ‘Lluteño’, identified a total of 48,345 lncRNA after salt stress treatment and among these 41.9% originated from the ‘Lluteño’ maize. The study also identified a cohort of 848 stress-responsive potential *trans*-NATs, which were predicted to regulate genes involved in transcriptional regulation, abiotic stimulus response, stress response and nicotianamine metabolic processes (Huanca-Mamani et al. 2018)

In sorghum, 126 and 133 lncRNA were identified in the M-81E (salt-tolerant) and Roma (salt-sensitive) strains, respectively (Sun et al. 2020a, b). Salt stress triggered the expression of three novel lncRNA in M-81E while suppressing two novel lncRNA in Roma. Among these *lncRNA13472, lncRNA11310, lncRNA2846, lncRNA26929* and *lncRNA14798* act as likely competitive endogenous RNAs. They regulate the expression of target genes associated with ion transport, protein modification, transcriptional regulation and material synthesis and transport.

In cotton, lincRNA, *lncRNA354,* functions as a miRNA sponge by interacting with miR160b. Under salt stress conditions, *lncRNA354* is down regulated, leading to increased accumulation of miR160b that down regulates the expression of GhARF17/18 to enhance resistance to salinity (Zhang et al. 2021a, b). It was also seen that overexpression of nucleus localized *lncRNA-973* increased tolerance to salinity by modulating a series of salt-responsive genes (Zhang et al. 2019a, b, c). In maize, the lncRNA, *MSTRG.139242.1* possibly interacts with *TEA027212.1* (Ca^{2+-} ATPase 13) to play a stress responsive role to high salinity. The study added that 12 lncRNA were predicted to be eTM of 17 known mature miRNA including miR156.

One hundred and seventy two differentially expressed lncRNA responsive to salt stress were identified in *Camellia sinensis* using high-throughput sequencing (Wan et al. 2020). These differentially expressed lncRNA are involved in pathways such as the galactinol synthase (GOLS), calcium signalling and interaction with TFs. Similarly, transcriptome analysis of *Medicago truncatula* under salt stress, identified the presence of large number of differentially expressed lncRNA in roots than leaves, while reverse abundance was observed under osmotic stress (Wang et al. 2015a, b).

In *Populus trichocarpa*, a cytoplasm localized lincRNA, *Ptlinc-NAC72* contains a stem ring with five tandem repeats of "CTTTTT" motif, complementary to salt responsive element "GAAAA" repeats in the 5′ UTR (untranslated region) of target genes. Over expression of the *Ptlinc-NAC72* could augment tolerance to salinity in *Arabidopsis* seedlings (Ye et al. 2022).

3.5.3 Heat Stress

Number of studies have shown the regulatory roles of lncRNA in regulating the heat stress response of plants (Jin et al. 2023). Through a comparative analysis of phenotype and transcriptome, 993 heat regulated lncRNA were identified in the heat-resistant elite maize inbred line. These lncRNA were predicted to regulate nearly a thousand target genes through *cis*-regulation (Hu et al. 2022).

In wheat, 77 lncRNA were characterized during heat stress and among them, *TahlnRNA27* was identified as potential pri-miRNA candidate (Xin et al. 2011). Moreover 5482 differentially expressed lncRNA were identified in response to heat stress (Babaei et al. 2024). They have the potential to target genes coding for TFs like Myc-type, MADS-box etc. and heat-shock proteins like Alpha crystallin/Hsp20 domain. These lncRNA can act both in *cis*- and *trans*- orientation by coupling lncRNA-miRNA-mRNA regulatory networks.

In stress-tolerant poplar, stable interactions between *PsiLNCRNA00268512* and miR396e suggest that the lncRNA may act as an eTM for regulating miR396e levels (Song et al. 2016). Similarly, in Chinese cabbage, *TCONS_00048391* regulates NAC1 expression by acting as an eTM of bra-miR164a during heat stress (Wang et al. 2019a, b).

3.5.4 Cold Stress

Under low temperature or cold stress, the C-Repeat (CRT)-Binding Factor/Dehydration- Responsive Element (DRE)-Binding Factor (CBF/DREB)- regulatory cascade plays a major role in plants by regulating the transcription of a set of cold responsive (*COR*) genes (Shi et al. 2015). RNA-seq analysis in *Arabidopsis* revealed 135 lncRNA exhibiting cold- dependent differential expression and/or alternative splicing of lncRNA. It was verified that the alternative splicing of lncRNA displays sensitivity to even minor temperature variations (Calixto et al. 2019). Among these, the cryptic lncRNA *asCBF1*, which overlaps with CBF1 on the antisense strand, is generated through RNA Pol II read-through transcription of the lncRNA, *SVALKA*. It functions to down regulate CBF1 expression and facilitate cold acclimation. The cascade involving *SVALKA/asCBF1*, establishes a negative feedback loop, enabling plants to adapt to low temperatures by enhancing cold tolerance through precise control of CBF1 expression (Kindgren et al. 2018). In *Arabidopsis*, the lncRNA, *Auxin- regulated Promoter Loop* (APOLO) interacts with WRKY42 to activate RHD6 binding to its promoter. This interaction subsequently modulates the local chromatin 3D conformation, ultimately promoting root hair cell elongation in response to low temperatures (Moison et al. 2021).

During cold stress, 487 lncRNA were significantly up regulated, while 326 were down regulated in grapes. Among these, 203 differentially expressed lncRNA were predicted to regulate 326 target genes such as TFs like *CBF* and *WRKY*, *LEA* and peroxisome biogenesis-related genes (Wang et al. 2019a, b).

In Cassava, a total of 316 cold-related lncRNA were identified. Functional analysis indicated that the *trans*-acting lncRNA, *CRIR1*, enhanced cold stress tolerance via CBF-independent pathway, probably through auxin signalling (Li et al. 2022). *CRIR1* interacted with a cold shock domain protein (MeCSP), thereby improving translation efficiency at low temperatures. Many cold-inducible genes such as *MeNAC, MeNF-YA* and *MeGOLS* were up regulated in both wild-type (WT) and *CRIR1*-overexpressing plants.

In wheat, about 9970 potential lncRNAs were identified in cultivar Dn1 lncRNA libraries. Among theses 7435 lncRNAs were co-expressed with 7191 mRNAs, including those responsive to cold stress (Lu et al. 2020a, b). In another study, it was reported that Tae-miR398 interacted with three ceRNAs, *lncR9A, lncR117* and *lncR616* to regulate *Copper/Zinc Superoxide Dismutase* 1 (*CSD*1) and improve cold tolerance (Lu et al. 2020a, b).

3.5.5 Light Stress

Light serves as a primary external signal that regulates both plant growth and development (Patra et al. 2023). In plants, lncRNA, *HID1* (*Hidden Treasure 1*) and *FLORE* (*CDF5 Long Non-Coding RNA*) are involved in regulating photomorphogenesis, seedling greening and circadian rhythm (Wang et al. 2014a, b, c; Henriques et al. 2017). In *Arabidopsis* seedlings, the involvement of lncRNA in blue light signalling and mannitol stress was explored. Sun et al. (2020a, b) observed differential expression of 481 lncRNA, along with 4197 mRNAs and 375 miRNA, under blue light for 2 h. Similarly, 545 lncRNA, along with 5207 mRNAs and 286 miRNA, exhibited differential expression under 8-h blue light treatment. Among these, a blue light-induced lncRNA (*BLIL1 or NONATHG000085.1*) sequesters miR167 to regulate blue light-directed plant photomorphogenesis, specifically inhibition of hypocotyl elongation, as well as responding to mannitol stress.

Li et al. (2021a, b) utilized RNA-seq to investigate the impact of various far-red light treatments on *Dendrobium officinale*. The analysis revealed differential expression of 147 lncRNA among experimental and control groups. Many of these lncRNA targeted genes involved in hormone signalling, metabolic pathways, membrane function and O-methyltransferase activity. Detailed physiological and biochemical assessments indicated that accumulation of carotenoids, alkaloids, flavonoids and polysaccharides are stimulated upon exposure to far- red light in *D. officinale*. The observed effects may be associated to cell membrane dynamics and Ca^{2+} signalling. Notably, lncRNA, *MSTRG.38867.1, MSTRG.69319.1* and *MSTRG.66273.1* were identified as potential regulators of the far-red light signalling network.

Intense light exposure can boost anthocyanin synthesis and coloration in apple fruits. A study showed that lncRNA, *MdLNC610* plays a role in anthocyanin accumulation under strong light conditions. *MdLNC610* is located 81 kb downstream of the ethylene biosynthesis gene *MdACO1* which regulates ethylene release and anthocyanin accumulation (Yu et al. 2022). Another, photo-induced lncRNA *MdLNC499* was found to play a role in anthocyanin accumulation in apple (Ma et al. 2021). Manipulation of MdLNC499 expression demonstrated its influence on both anthocyanin accumulation and the MdERF109 transcription level. This suggests that MdLNC499 positively regulates the expression of MdERF109, thereby promoting light-induced anthocyanin biosynthesis (Ma et al. 2021).

3.5.6 Nutrient Deficiency

Obtaining nutrients from soil is a fundamental physiological process, which is crucial for optimal plant growth and development (Islam 2019). Within the realm of major nutrients, phosphorus holds a pivotal role as a fundamentally important element for plants (Adnan et al. 2018). In this perspective, the involvement of miRNA and lncRNA in regulating phosphate availability is evident in *Arabidopsis* (Franco-Zorrilla et al. 2007; Yuan et al. 2016), rice (Jabnoune et al. 2013) and Medicago (Wang et al. 2017).

In *Arabidopsis*, the lncRNA, *IPS1* functions as a competing endogenous RNA or decoy for miR399 to regulate *PHO2* expression for maintaining phosphate homeostasis (Franco- Zorrilla et al. 2007). When phosphate levels are abundant, *IPS1* binds to miR399, preventing it from targeting the *PHO2* gene. This action disrupts the function of the phosphate transporters by targeting them for ubiquitination, thereby limiting the excessive uptake of phosphate by roots (Bari et al. 2006). In tomato, its homologue, *TPSI1* (*phosphate starvation induced 1 lncRNA*) is activated under phosphorus stress conditions. The homologue identified in maize is known as *PILNCR1* (*Pi-deficiency-induced lncRNA1*). The *TPSI1* and *PILNCR1* have been shown to compete with miR399 for binding to *PHO2* (Franco-Zorrilla et al. 2007).

Rice *PHO1;2* encodes a protein that load phosphoric acid into the xylem. Its complementary strand yields a related *cis-lncNAT*. Both *PHO1;2* and the *cis-lncNAT* are under the control of active promoters and are expressed in vascular tissues. However, phosphorus starvation induction occurs solely through the *cis-lncNAT* promoter. During phosphorus starvation, both *PHO1;2* and *cis-lncNAT* usually get accumulated, yet *PHO1;2* mRNA levels remain steady. The *cis-lncNAT* RNAi lines show reduced *PHO1;2* levels and diminished seed yield. Conversely, constitutive overexpression of cis-lncNAT markedly enhances *PHO1;2* expression, even under phosphorus deficiency conditions (Jabnoune et al. 2013).

In *Medicago truncatula*, the *PHR1-miR399-PHO2* phosphate regulatory cascade in conjunction with phosphate deficiency-expressed lncRNA *PDIL1* (paralog of Mt4), has also been elucidated (Wang et al. 2017). The lncRNA, *PDIL2* and *PDIL3,* exert a negative regulatory function in controlling the expression of the phosphate transporter gene, *Medtr1g074930.*

Nitrogen is another vital nutrient crucial for the growth and development of plants (Hawkesford and Griffiths 2019). In rice, *cis-NATAMT1.1* (Os04g0509500) and *cis-NATAMT1.2* (Os02g0620533) were reported as nitrogen responsive lncRNA (Shin et al. 2018). *cis-NATAMT1* transcripts and their corresponding *AMT1* genes exhibited distinct expression: *cis-NATAMT1.2* and *AMT1.2* showed elevated expression in rice roots under nitrogen starvation conditions, while *cis-NATAMT1.1* and *AMT1.1* displayed tissue- dependent expression discrepancies. The *AMT1.1* gene was predominantly expressed in roots, whereas *cis-NATAMT1.1* exhibited higher abundance in shoots but minimal expression in roots (Shin et al. 2018).

Similar studies were performed to investigate the role of lncRNA within the Arabidopsis genome using different nutrient-deficient conditions. These studies identified the significance of *trans-acting siRNA3* (*TAS3*) transcript as a crucial lincRNA. *TAS3* targets the *nitrate transporter 2* gene in nitrogen-starved environments (Fukuda et al. 2019).

Another comprehensive investigation was conducted to study the genome-wide regulation of lncRNA in response to boron deficiency in *Poncirus trifoliate.* Strand-specific deep transcriptome sequencing and subsequent analysis revealed 2101 unique lncRNA. Among them, 729 were up regulated and 721 down regulated under boron deficient conditions (Zhou et al. 2019).

These studies have provided a solid groundwork for performing further comprehensive investigations into the regulatory functions of diverse lncRNA governing nutrient deficiency responses in plants.

3.5.7 Heavy Metal Stress

Plants use lncRNA to control the uptake of heavy metals and these constitute a part of stress response mechanisms (Ali et al. 2019). In *Brassica napus,* RNA-seq revealed 301 cadmium- responsive lncRNA, among which 67 acted as eTM for 36 miRNA. Molecular validations of *TCONS_00091906, TCONS_00033487* and *TCONS_00097191* lncRNA under cadmium stress demonstrated their significant roles as target mimics for *EL628609, TC182597* and *TC203372* mRNAs, which are involved in cadmium uptake and detoxification (Feng et al. 2016). In rice, genome-wide analysis of response to cadmium stress showed the up- and down-regulation of lncRNA. Analysis of differentially expressed lncRNA shed light to their association with cysteine and methionine metabolism during stress (Chen et al. 2018a, b).

Tang et al. (2019) identified arsenic responsive small RNA and lncRNA in rice. During arsenic stress, 4762 and 18,152 small RNAs were up regulated, while 14,603 and 8308 small RNAs were down regulated in roots and shoots, respectively. Among them, 252 AGO1-enriched small RNAs were identified and 56 lncRNA were discovered, which were potential targets of nine AGO1-enriched small ncRNA.

A more recent study in rice identified about 171 differently expressed lncRNA under iron deficiency using strand-specific RNA-seq. Several lncRNA were predicted to be miRNA precursors or eTM and were linked with iron transport and photosynthesis. Additionally, 76 lncRNA were identified as potential targets of *OsbHLH156*, a key transcriptional regulator to iron deficiency (Wang et al. 2021).

3.6 Role of lncRNA in Biotic Stress Tolerance

Plants are attacked by a variety of viruses, fungi, bacteria and pests (Patra et al. 2023; Wang et al. 2023), so they have developed suitable strategies for survival, which also employ the lncRNA (Li et al. 2021a, b; Patra et al. 2023). In wild tobacco, *Nicotiana attenuate*, certain lncRNA have been implicated in providing insect resistance by modulating the jasmonate hormone-signalling pathway. Upon infestation by phytophagous insects, the silencing of two lincRNA (*JAL1* and *JAL3*) results in reduced levels of jasmonate thereby compromising host resistance to insects. Notably, the jasmonate signalling pathway could also regulate expression of specific late-response lincRNA (Li et al. 2021a, b).

In another study, whitefly infestation responsive lncRNA were profiled in cotton (Zhang et al. 2022). Briefly, a total of 6651 lncRNA transcripts and 606 differentially expressed lncRNA were identified using RNA-seq. Co-expression network analysis showed that *lncA07* and *lncD09* act as a potential hub in defense against insect infestation. Further experiments using CRISPR/Cas9 knock-out of *lncD09* and *lncA07* showed that the mutants exhibited down regulation of jasmonic acid, which lead to decreased tolerance to insect infestation.

The role of lncRNA in mediating bacterial and fungal stress response has also been explored (Patra et al. 2023). Strand-specific RNA-seq analysis in potato (*Solanum tuberosum*) identified 1113 lincRNA from stem tissues. In a time-course RNA-seq analysis comparing a tolerant and a susceptible potato cultivar, 559 lincRNA were found to respond to the challenge posed by *Pectobacterium carotovorum* subsp. *brasiliense*, in contrast to mock- inoculated controls. Co-expression analysis identified 17 lincRNA as closely linked with 12 potato defense-related genes (Kwenda et al. 2016).

In rice, 567 lncRNA were identified from *Xanthomonas oryzae* infected leaves. Interaction analysis between lncRNA and jasmonate pathway enriched targets revealed that about 39 jasmonate-related protein-coding genes interact with 73 lncRNA. This further confirmed the role of lncRNA in plant defense through regulation of jasmonate signalling pathway (Yu et al. 2020). In *Arabidopsis*, overexpression of the lncRNA, *ELENA1* (*ELF18-Induced Long noncoding RNA 1*) up regulates the expression of *pathogenesis-related gene 1 (PR1)* for boosting resistance to *Pseudomonas syringae* (Seo et al. 2017). It was implicated that *ELENA1* had association with FIB2 and MED19a, as dissociation of FIB2 from MED19a tends to activate *PR1* (Seo et al. 2017). In tomato, *lncRNA33732* enhanced resistance to *Phytophthora infestans* through expression of respiratory burst oxidase gene and increasing H_2O_2 accumulation (Cui et al. 2019).

In banana, 5294 lncRNA responsive to *Fusarium oxysporum* infection were identified (Li et al. 2017a, b). Zhang et al. (2018a, b, c) identified *Verticillium dahlia* responsive lncRNA; *GhlncNAT- ANX2* (lncRNA *L2*) and *GhlncNAT-RLP7* (lncRNA L3) from cotton. Silencing the function of *L2* and *L3* could increase *LOX1* and *LOX2* expression to augment resistance to *V. dahlia*. In Chinese cabbage *MSTRG.19915*, a NATs of the MAPK gene *BrMAPK15*, has been linked to susceptibility to downy mildew (*Hyaloperonospora brassicae*). *BrMAPK15* plays a crucial role of resistance against downy mildew, which was further confirmed by silencing *MSTRG.19915*. The loss in expression of *MSTRG.19915* resulted in increased resistance to downy mildew in the seedlings, due to the up-regulation of BrMAPK15 (Zhang et al. 2021a, b).

Furthermore, strand-specific paired-end RNA-seq in tomato plants infected with *Tomato yellow leaf curl virus* (*TYLCV*) led to the identification of about 1565 lncRNA (lincRNA and NAT). Several of these lncRNA acted as eTM for tomato miRNA responsive to TYLCV infection (Wang et al. 2015a, b).

3.7 Methods for Studying Plant lncRNA

The family of ncRNA has emerged as a key regulator of genome and phenotypic diversity in plants (Wang et al. 2014a, b, c). The advancement of next-generation sequencing technologies and computational analysis has been instrumental in advancing our understanding of the functional role of lncRNA in plants (Heo and Sung 2011). While recent discoveries have shed light on their functions and mechanisms of action, plant lncRNA still hold mysteries yet to be fully unraveled (Van Dijk et al. 2011). Common techniques utilized for the discovery of lncRNA encompass RNA- seq, serial analysis of gene expression, transcriptome deep sequencing, cap analysis of gene expression and low-abundance transcript detection/single-cell sequencing. In Table 3.1 we present some of the databases for lncRNA identification used in plants.

The validation of high-throughput analysis involves techniques for molecular characterization like RT-qPCR, northern blotting and fluorescence in situ hybridization (Kashi et al. 2016; Wang et al. 2023). This is followed by gene manipulation or editing to generate gain- and- loss-of-function mutants for understanding their biological function and mechanism of action target (Mattick 2009; Zhu et al. 2013).

3.8 Conclusions and Prospects

Numerous regulatory RNA, including small ncRNA and lncRNA, are implicated in regulating a variety of responses in plants, yet their mechanistic details remain largely elusive (Fig. 3.1). The interactions between small ncRNA, lncRNA and the target transcripts form carefully regulated genetic loops that modulate various aspects of plant biology, such as pathogen or stress perception, signal transduction and other downstream responses. The identification of ncRNA, characterization of their targets, detecting the subcellular localization and potential mode of action are essential for understanding their functions. Although research on plant lncRNA is in its early stages, their roles in growth, development, stress responses and immunity are increasingly recognized. The advancements in experimental techniques and bioinformatics tools will allow the identification and detection of expression zones of the lncRNA. Cutting-edge experimental tools such as CRISPR/Cas, RNAi will help in exploration of their function and mechanisms. The comprehensive understanding of lncRNA holds immense promise for improvements in plant biology.

Acknowledgement This work was financially supported by grants from ICGEB. TAG acknowledges the Arturo Falaschi ICGEB Fellowships Programme.

Table 3.1 Databases for lncRNA identification and study in plants

S. No	Name of the lncRNA database	Link access	Description	References
1	PlantNATsDB	http://bis.zju.edu.cn/pnatdb/	It is the most comprehensive resource of NATs and it contains approx. 2 million NAT pairs in 69 plant species	Chen et al. (2012)
2	PNRD	http://structuralbiology.cau.edu.cn/PNRD	28,214 entries from 166 species, including 16 new plant species	Yi et al. (2015)
3	PLNlncRbase	http://bioinformatics.ahau.edu.cn/PLNlncRbase	Based on manually collected data from nearly 200 published literatures, comprising 1187 plant lncRNAs in 43 plant species	Xuan et al. (2015)
4	CANTATAdb	http://cantata.amu.edu.pl or http://yeti.amu.edu.pl/CANTATA/	A database of lncRNAs in 108 plant species and computationally identified 571,688 lncRNAs. It also contains 112,980 lncRNAs expressed specifically in embryo	Szcześniak et al. (2016)
5	GREENC	http://greenc.sciencedesigners.com/	A database of 37 plant species and six algae, containing 120, 000 annotated lncRNAs	Gallart et al. (2016)
6	PLncPRO	http://ccbb.jnu.ac.in/plncpro/	It predicts lncRNAs using transcriptome data, it is based on machine learning and uses random forest algorithm	Singh et al. (2017)
7	Araportll	https://www.araport.org/	Based on Arabidopsis Col-0 version 11 (Araport11) and includes additional coding and noncoding annotations compared to TAIR10, such as lincRNAs, NATs and other ncRNAs	Cheng et al. (2017)
8	EVLncRNAs	http://biophy.dzu.edu.cn/EVLncRNAs/	Contains all experimentally confirmed LncRNAs from various species, such as mammals, plants and	Zhou et al. (2018a, b)

(continued)

Table 3.1 (continued)

S. No	Name of the lncRNA database	Link access	Description	References
			microorganisms. It also contains disease-associated lncRNAs, helps to study interaction between lncRNAs and other macromolecules	
	EVlncRNAs 2.0	https://www.sdklab-biophysics-dzu.net/EVLncRNAs2/	A comprehensive, manually curated, high-quality and freely accessible resource of integrated sequence, structure, functional and phenotypic information of experimentally validated lncRNAs from all species	Zhou et al. (2021a, b)
9	PlaNC-TE	http://planc-te.cp.utfpr.edu.br	Contains 14,350 overlaps between ncRNAs and TEs in 40 plant genomes	Pedro et al. (2018)
10	CRISPRlnc	http://www.crisprlnc.org or http://crisprlnc.xtbg.ac.cn	It is a manually curated database of validated CRISPR/Cas9 sgRNAs for lncRNAs from all species	Chen et al. (2019)
11	ItLnc-BXE	https://github.com/BioMedicalBigDataMiningLab/ItLnc-BXE	Used to classify plant lncRNAs using extensive characteristics and ensemble learning mechanism	Zhang et al. (2020)
12	AlnC	http://14.139.61.8/AlnC/	Contains 10,855,598 lncRNAs, annotated in 682 flowering plants using machine learning approach	Singh et al. (2021)
13	LncMachine	https://github.com/hbusra/lncMachine	Provides crop-specific and alignment-free coding potential prediction method	Cagirici et al. (2021)
14	NONCODEV6	http://www.noncode.org/	It is an updated database dedicated to annotate lncRNAs in both animals and plants	Zhao et al. (2021)

(continued)

Table 3.1 (continued)

S. No	Name of the lncRNA database	Link access	Description	References
15	lncRNADetector 1.0	https://lncrnapipe.cimap.res.in	Enables identification of novel lncRNAs, especially from medicinal and aromatic plants and contains 88,459 lncRNAs from 21 species. It is also developed to identify lncRNAs from NGS data of different organisms for downstream studies	Shukla et al. (2021)
16	PLncDB V2.0	http://plncdb.tobaccodb.org	Contains 1,246,372 lncRNAs and 13,834 RNA-Seq datasets for 80 species; it is the updated version of PLncDB	Jin et al. (2021)

References

Adnan M, Shah Z, Fahad S, Arif M, Alam M, Khan IA, Mian IA, Basir A, Ullah H, Arshad M et al (2018) Phosphate-solubilizing bacteria nullify the antagonistic effect of soil calcification on bioavailability of phosphorus in alkaline soils. Sci Rep 8(1):4339. https://doi.org/10.1038/s41598-018-22653-7

Ali H, Khan E, Ilahi I (2019) Environmental chemistry and ecotoxicology of hazardous heavy metals: environmental persistence, toxicity, and bioaccumulation. J Chem 2019:6730305. https://doi.org/10.1155/2019/6730305

Ali A, Han K, Liang P (2021) Role of transposable elements in gene regulation in the human genome. Life 11(2):1–23. https://doi.org/10.3390/life11020118

Amor BB, Wirth S, Merchan F, Laporte P, D'Aubenton-Carafa Y, Hirsch J, Maizel A, Mallory A, Lucas A et al (2009) Novel long non-protein coding RNAs involved in *Arabidopsis* differentiation and stress responses. Genome Res 19(1):57–69. https://doi.org/10.1101/gr.080275.108

Andersson R, Gebhard C, Miguel-Escalada I, Hoof I, Bornholdt J, Boyd M, Chen Y, Zhao X, Schmidl C, Suzuki T et al (2014) An atlas of active enhancers across human cell types and tissues. Nature 507(7493):455–461. https://doi.org/10.1038/nature12787

Ariel F, Jegu T, Latrasse D, Romero-Barrios N, Christ A, Benhamed M, Crespi M (2014) Noncoding transcription by alternative RNA polymerases dynamically regulates an auxin-driven chromatin loop. Mol Cell 55(3):383–396. https://doi.org/10.1016/j.molcel.2014.06.011

Babaei S, Bhalla PL, Singh MB (2024) Identifying long non-coding RNAs involved in heat stress response during wheat pollen development. Front Plant Sci 15:1–17. https://doi.org/10.3389/fpls.2024.1344928

Bai Y, Dai X, Harrison AP, Chen M (2015) RNA regulatory networks in animals and plants: a long noncoding RNA perspective. Brief Funct Genomics 14(2):91–101. https://doi.org/10.1093/bfgp/elu017

Bardou F, Ariel F, Simpson CG, Romero-Barrios N, Laporte P, Balzergue S, Brown JWS, Crespi M (2014) Long noncoding RNA modulates alternative splicing regulators in *Arabidopsis*. Dev Cell 30(2):166–176. https://doi.org/10.1016/j.devcel.2014.06.017

Bari R, Pant BD, Stitt M (2006) PHO2, MicroRNA399, and PHR1 define a phosphate-signaling pathway in plants. Plant Physiol 141:988–999. https://doi.org/10.1104/pp.106.079707.988

Bentley DL (2014) Coupling mRNA processing with transcription in time and space. Nat Rev Genet 15(3):163–175. https://doi.org/10.1038/nrg3662

Bhatia G, Singh A, Verma D, Sharma S, Singh K (2020) Genome-wide investigation of regulatory roles of lncRNAs in response to heat and drought stress in *Brassica juncea* (Indian mustard). Environ Exp Bot 171:103922. https://doi.org/10.1016/j.envexpbot.2019.103922

Boerner S, McGinnis KM (2012) Computational identification and functional predictions of long noncoding RNA in *Zea mays*. PLoS One 7(8):e43047. https://doi.org/10.1371/journal.pone.0043047

Borah P, Das A, Milner MJ, Ali A, Bentley AR, Pandey R (2018) Long non- coding RNAs as endogenous target mimics and exploration of their role in low nutrient stress tolerance in plants. Genes 9(9):1–17. https://doi.org/10.3390/genes9090459

Bourgeois Y, Boissinot S (2019) On the population dynamics of junk: a review on the population genomics of transposable elements. Genes 10(6):419. https://doi.org/10.3390/genes10060419

Burgess D, Li H, Zhao M, Kim SY, Lisch D (2020) Silencing of mutator elements in maize involves distinct populations of small RNAs and distinct patterns of DNA methylation. Genetics 215(2): 379–391. https://doi.org/10.1534/genetics.120.303033

Cagirici HB, Galvez S, Sen TZ, Budak H (2021) LncMachine: a machine learning algorithm for long noncoding RNA annotation in plants. Funct Integr Genomics 21(2):195–204. https://doi.org/10.1007/s10142-021-00769-w

Calixto CPG, Tzioutziou NA, James AB, Hornyik C, Guo W, Zhang R, Nimmo HG, Brown JWS (2019) Cold-dependent expression and alternative splicing of *Arabidopsis* long non-coding RNAs. Front Plant Sci 10:1–16. https://doi.org/10.3389/fpls.2019.00235

Castelán-Muñoz N, Herrera J, Cajero-Sánchez W, Arrizubieta M, Trejo C, García- Ponce B, Sánchez MDLP, Álvarez-Buylla ER, Garay-Arroyo A (2019) MADS-box genes are key components of genetic regulatory networks involved in abiotic stress and plastic developmental responses in plants. Front Plant Sci 10:853. https://doi.org/10.3389/fpls.2019.00853

Chen D, Yuan C, Zhang J, Zhang Z, Bai L, Meng Y, Chen LL, Chen M (2012) PlantNATsDB: a comprehensive database of plant natural antisense transcripts. Nucleic Acids Res 40(D1): 1187–1193. https://doi.org/10.1093/nar/gkr823

Chen L, Shi S, Jiang N, Khanzada H, Wassan GM, Zhu C, Peng X, Xu J, Chen Y, Yu Q, He X et al (2018a) Genome-wide analysis of long non-coding RNAs affecting roots development at an early stage in the rice response to cadmium stress. BMC Genomics 19(1):1–10. https://doi.org/10.1186/s12864-018-4807-6

Chen Y, Li X, Su L, Chen X, Zhang S, Xu X, Zhang Z, Chen Y, Xuhan X, Lin Y, Lai Z (2018b) Genome-wide identification and characterization of long non- coding RNAs involved in the early somatic embryogenesis in *Dimocarpus longan* Lour. BMC Genomics 19(1):1–19. https://doi.org/10.1186/s12864-018-5158-z

Chen W, Zhang G, Li J, Zhang X, Huang S, Xiang S, Hu X, Liu C (2019) CRISPRlnc: a manually curated database of validated sgRNAs for lncRNAs. Nucleic Acids Res 47(D1):D63–D68. https://doi.org/10.1093/nar/gky904

Chen L, Zhu QH, Kaufmann K (2020a) Long non-coding RNAs in plants: emerging modulators of gene activity in development and stress responses. Planta 252(5):1–14. https://doi.org/10.1007/s00425-020-03480-5

Chen QJ, Deng BH, Gao J, Zhao ZY, Chen ZL, Song SR, Wang L, Zhao LP, Xu WP, Zhang CX, Ma C, Wang SP (2020b) A miRNA-encoded small peptide, vvi-miPEP171d1, regulates adventitious root formation. Plant Physiol 183(2):656–670. https://doi.org/10.1104/pp.20.00197

Chen Q, Liu K, Yu R, Zhou B, Huang P, Cao Z, Zhou Y, Wang J (2021) From "Dark Matter" to "Star": insight into the regulation mechanisms of plant functional long non-coding RNAs. Front Plant Sci 12:1–13. https://doi.org/10.3389/fpls.2021.650926

Cheng Y, Krishnakumar V, Chan AP, Thibaud-Nissen F, Schobel S, Town CD (2017) Araport11: a complete reannotation of the *Arabidopsis thaliana* reference genome. Plant J 89(4):789–804. https://doi.org/10.1111/tpj.13415

Cho J (2018) Transposon-derived non-coding RNAs and their function in plants. Front Plant Sci 9: 1–6. https://doi.org/10.3389/fpls.2018.00600

Chu Q, Bai P, Zhu X, Zhang X, Mao L, Zhu QH, Fan L, Ye CY (2018) Characteristics of plant circular RNAs. Brief Bioinform 21(1):135–143. https://doi.org/10.1093/bib/bby111

Chung PJ, Jung H, Jeong DH, Ha SH, Do Choi Y, Kim JK (2016) Transcriptome profiling of drought responsive noncoding RNAs and their target genes in rice. BMC Genomics 17(1):563. https://doi.org/10.1186/s12864-016-2997-3

Csorba T, Questa JI, Sun Q, Dean C (2014) Antisense COOLAIR mediates the coordinated switching of chromatin states at FLC during vernalization. Proc Natl Acad Sci USA 111(45): 16160–16165. https://doi.org/10.1073/pnas.1419030111

Cui J, Luan Y, Jiang N, Bao H, Meng J (2017) Comparative transcriptome analysis between resistant and susceptible tomato allows the identification of lncRNA16397 conferring resistance to *Phytophthora infestans* by co-expressing glutaredoxin. Plant J 89(3):577–589. https://doi.org/10.1111/tpj.13408

Cui J, Jiang N, Meng J, Yang G, Liu W, Zhou X, Ma N, Hou X, Luan Y (2019) LncRNA33732-respiratory burst oxidase module associated with WRKY1 in tomato- Phytophthora infestans interactions. Plant J 97(5):933–946. https://doi.org/10.1111/tpj.14173

Das A, Nigam D, Junaid A, Tribhuvan KU, Kumar K, Durgesh K, Singh NK, Gaikwad K (2019) Expressivity of the key genes associated with seed and pod development is highly regulated via lncRNAs and miRNAs in Pigeonpea. Sci Rep 1-14:18191. https://doi.org/10.1038/s41598-019-54340-6

de Santa F, Barozzi I, Mietton F, Ghisletti S, Polletti S, Tusi BK, Muller H, Ragoussis J, Wei CL, Natoli G (2010) A large fraction of extragenic RNA Pol II transcription sites overlap enhancers. PLoS Biol 8(5):e1000384. https://doi.org/10.1371/journal.pbio.1000384

Deng P, Liu S, Nie X, Weining S, Wu L (2018) Conservation analysis of long non- coding RNAs in plants. Sci China Life Sci 61(2):190–198. https://doi.org/10.1007/s11427-017-9174-9

Derrien T, Johnson R, Bussotti G, Tanzer A, Djebali S, Tilgner H, Guernec G, Martin D, Merkel A, Knowles DG et al (2012) The GENCODE v7 catalog of human long noncoding RNAs: analysis of their gene structure, evolution, and expression. Genome Res 22(9):1775–1789. https://doi.org/10.1101/gr.132159.111

Ding J, Lu Q, Ouyang Y, Mao H, Zhang P, Yao J, Xu C, Li X, Xiao J, Zhang Q (2012) A long noncoding RNA regulates photoperiod-sensitive male sterility an essential component of hybrid rice. Proc Natl Acad Sci USA 109(7):2654–2659. https://doi.org/10.1073/pnas.1121374109

Ding Z, Wu C, Tie W, Yan Y, He G, Hu W (2019) Strand-specific RNA-seq based identification and functional prediction of lncRNAs in response to melatonin and simulated drought stresses in cassava. Plant Physiol Biochem 140:96–104. https://doi.org/10.1016/j.plaphy.2019.05.008

Dong SM, Xiao L, Li ZB, Shen J, Yan HB, Li SX, Liao WB, Peng M (2022) A novel long non-coding RNA, DIR, increases drought tolerance in cassava by modifying stress-related gene expression. J Integr Agric 21(9):2588–2602. https://doi.org/10.1016/j.jia.2022.07.022

Eom SH, Lee HJ, Lee JH, Wi SH, Kim SK, Hyun TK (2019) Identification and functional prediction of drought-responsive long non-coding RNA in tomato. Agronomy 9(10):629. https://doi.org/10.3390/agronomy9100629

Espinosa JM (2017) On the origin of lncRNAs: missing link found. Trends Genet 33(10):660–662. https://doi.org/10.1016/j.tig.2017.07.005

Fan C, Hao Z, Yan J, Li G (2015) Genome-wide identification and functional analysis of lincRNAs acting as miRNA targets or decoys in maize. BMC Genomics 16(1):793. https://doi.org/10.1186/s12864-015-2024-0

Feng SJ, Zhang XD, Liu XS, Tan SK, Chu SS, Meng JG, Meng JG, Zhao KX, Yang ZM (2016) Characterization of long non-coding RNAs involved in cadmium toxic response in *Brassica napus*. RSC Adv 6:82157. https://doi.org/10.1039/C6RA05459E

Franco-Zorrilla JM, Valli A, Todesco M, Mateos I, Puga MI, Rubio-Somoza I, Leyva A, Weigel D, García JA, Paz-Ares J (2007) Target mimicry provides a new mechanism for regulation of microRNA activity. Nat Genet 39(8):1033–1037. https://doi.org/10.1038/ng2079

Fukuda M, Nishida S, Kakei Y, Shimada Y, Fujiwara T (2019) Genome-wide analysis of long intergenic noncoding RNAs responding to low-nutrient conditions in *Arabidopsis thaliana*: possible involvement of trans-acting siRNA3 in response to low nitrogen. Plant Cell Physiol 60(9):1961–1973. https://doi.org/10.1093/pcp/pcz048

Gallart AP, Pulido AH, De Lagrán IAM, Sanseverino W, Cigliano RA (2016) GREENC: a Wiki-based database of plant lncRNAs. Nucleic Acids Res 44(D1):D116–D1166. https://doi.org/10.1093/nar/gkv1215

Gao Z, Liu HL, Daxinger L, Pontes O, He X, Qian W, Lin H, Xie M, Lorkovic ZJ, Zhang S et al (2010) An RNA polymerase II-and AGO4-associated protein acts in RNA-directed DNA methylation. Nature 465(7294):106–109. https://doi.org/10.1038/nature09025

Gao R, Liu P, Irwanto N, Loh DR, Wong SM (2016) Upregulation of LINC-AP2 is negatively correlated with AP2 gene expression with Turnip crinkle virus infection in *Arabidopsis thaliana*. Plant Cell Rep 35(11):2257–2267. https://doi.org/10.1007/s00299-016-2032-9

Geisler S, Coller J (2013) RNA in unexpected places: long non-coding RNA functions in diverse cellular contexts. Nat Rev Mol Cell Biol 14(11):699–712. https://doi.org/10.1038/nrm3679

Gelaw TA, Sanan-Mishra N (2021) Non-coding RNAs in response to drought stress. Int J Mol Sci 22(22):1–37. https://doi.org/10.3390/ijms222212519

Gloss BS, Dinger ME (2016) The specificity of long noncoding RNA expression. Biochim Biophys Acta 1859(1):16–22. https://doi.org/10.1016/j.bbagrm.2015.08.005

Guo G, Liu X, Sun F, Cao J, Huo N, Wuda B, Xin M, Hu Z, Du J, Xia R et al (2018) Wheat miR9678 affects seed germination by generating phased siRNAs and modulating abscisic acid/gibberellin signaling. Plant Cell 30(4):796–814. https://doi.org/10.1105/tpc.17.00842

Haag JR, Pikaard CS (2011) Multisubunit RNA polymerases IV and V: purveyors of non-coding RNA for plant gene silencing. Nat Rev Mol Cell Biol 12(8):483–492. https://doi.org/10.1038/nrm3152

Hansen TB, Wiklund ED, Bramsen JB, Villadsen SB, Statham AL, Clark SJ, Kjems J (2011) MiRNA-dependent gene silencing involving AGO2-mediated cleavage of a circular antisense RNA. EMBO J 30(21):4414–4422. https://doi.org/10.1038/emboj.2011.359

Hawkesford MJ, Griffiths S (2019) Exploiting genetic variation in nitrogen use efficiency for cereal crop improvement. Curr Opin Plant Biol 49:35–42. https://doi.org/10.1016/j.pbi.2019.05.003

Henriques R, Wang H, Liu J, Boix M, Huang F, Chua H (2017) The antiphasic regulatory module comprising CDF5 and its antisense RNA FLORE links the circadian clock to photoperiodic flowering. New Phytol 216(3):854–867. https://doi.org/10.1111/nph.14703

Heo JB, Sung S (2011) Vernalization-mediated epigenetic silencing by a long intronic noncoding RNA. Science 331(6013):76–79. https://doi.org/10.1126/science.1197349

Hirose T, Nakagawa S (2016) Clues to long noncoding RNA taxonomy. Biochim Biophys Acta 1859(1):1–2. https://doi.org/10.1016/j.bbagrm.2015.11.011

Hirsch J, Lefort V, Vankersschaver M, Bouaiem A, Lucas A, Thermes C, D'Aubenton-Carafa Y, Crespi M (2006) Characterization of 43 non-protein-coding mRNA genes in *Arabidopsis*, including the miR162a-derived transcripts. Plant Physiol 140(4):1192–1204. https://doi.org/10.1104/pp.105.073817

Hu X, Wei Q, Wu H, Huang Y, Peng X, Han G, Ma Q, Zhao Y (2022) Identification and characterization of heat-responsive lncRNAs in maize inbred line CM1. BMC Genomics 23(1):1–15. https://doi.org/10.1186/s12864-022-08448-1

Huanca-Mamani W, Arias-Carrasco R, Cárdenas-Ninasivincha S, Rojas-Herrera M, Sepúlveda-Hermosilla G, Caris-Maldonado JC, Bastías E, Maracaja-Coutinho V (2018) Long non-coding

RNAs responsive to salt and boron stress in the hyper-arid lluteño maize from atacama desert. Genes 9(3):170. https://doi.org/10.3390/genes9030170
Huang CF, Zhu JK (2014) RNA splicing factors and RNA-directed DNA methylation. Biology 3(2):243–254. https://doi.org/10.3390/biology3020243
Huang L, Dong H, Zhou D, Li M, Liu Y, Zhang F, Feng Y, Yu D, Lin S, Cao J (2018) Systematic identification of long non-coding RNAs during pollen development and fertilization in *Brassica rapa*. Plant J 96(1):203–222. https://doi.org/10.1111/tpj.14016
Islam MS (2019) Sensing and uptake of nitrogen in rice plant: a molecular view. Rice Sci 26(6): 343–355. https://doi.org/10.1016/j.rsci.2018.12.007
Jabnoune M, Secco D, Lecampion C, Robaglia C, Shu Q, Poirier Y (2013) A rice cis-Natural antisense RNA acts as a translational enhancer for its cognate mRNA and contributes to phosphate homeostasis and plant fitness. Plant Cell 25(10):4166–4182. https://doi.org/10.1105/tpc.113.116251
Jha UC, Nayyar H, Jha R, Khurshid M, Zhou M, Mantri N, Siddique KHM (2020) Long non-coding RNAs: emerging players regulating plant abiotic stress response and adaptation. BMC Plant Biol 20(1):1–20. https://doi.org/10.1186/s12870-020-02595-x
Jia Y, Zhao H, Niu Y, Wang Y (2022) Identification of birch lncRNAs and mRNAs responding to salt stress and characterization of functions of lncRNA. Hortic Res 10(2):uhac277. https://doi.org/10.1093/hr/uhac277
Jia Y, Zhao H, Niu Y, Wang Y (2024) Long noncoding RNA from *Betula platyphylla*, BplncSIR1, confers salt tolerance by regulating BpNAC2 to mediate reactive oxygen species scavenging and stomatal movement. Plant Biotechnol J 22(1):48–65. https://doi.org/10.1111/pbi.14164
Jiang N, Cui J, Shi Y, Yang G, Zhou X, Hou X, Meng J, Luan Y (2019) Tomato lncRNA23468 functions as a competing endogenous RNA to modulate NBS-LRR genes by decoying miR482b in the tomato-*Phytophthora infestans* interaction. Hortic Res 6(1):28. https://doi.org/10.1038/s41438-018-0096-0
Jin J, Lu P, Xu Y, Li Z, Yu S, Liu J, Wang H, Chua NH, Cao P (2021) PLncDB V2.0: a comprehensive encyclopedia of plant long noncoding RNAs. Nucleic Acids Res 49(D1): D1489–D1495. https://doi.org/10.1093/nar/gkaa910
Jin X, Wang Z, Li X, Ai Q, Wong DCJ, Zhang F, Yang J, Zhang N, Si H (2023) Current perspectives of lncRNAs in abiotic and biotic stress tolerance in plants. Front Plant Sci 14:1–11. https://doi.org/10.3389/fpls.2023.1334620
Kashi K, Henderson L, Bonetti A, Carninci P (2016) Discovery and functional analysis of lncRNAs: methodologies to investigate an uncharacterized transcriptome. Biochim Biophys Acta 1859(1):3–15. https://doi.org/10.1016/j.bbagrm.2015.10.010
Kazemzadeh MINA, Safaralizadeh REZA, Orang AV (2015) LncRNAs: emerging players in gene regulation and disease pathogenesis. J Genet 94(4):771–784. https://doi.org/10.1007/s12041-015-0561-6
Kim YK, Kim VN (2007) Processing of intronic microRNAs. EMBO J 26(3):775–783. https://doi.org/10.1038/sj.emboj.7601512
Kim ED, Sung S (2012) Long noncoding RNA: unveiling hidden layer of gene regulatory networks. Trends Plant Sci 17(1):16–21. https://doi.org/10.1016/j.tplants.2011.10.008
Kindgren P, Ard R, Ivanov M, Marquardt S (2018) Transcriptional read-through of the long non-coding RNA SVALKA governs plant cold acclimation. Nat Commun 9(1):4561. https://doi.org/10.1038/s41467-018-07010-6
Kouchi H, Takane KI, So RB, Ladha JK, Reddy PM (1999) Rice ENOD40: isolation and expression analysis in rice and transgenic soybean root nodules. Plant J 18(2):121–129. https://doi.org/10.1046/j.1365-313X.1999.00432.x
Kung JTY, Colognori D, Lee JT (2013) Long noncoding RNAs: past, present, and future. Genetics 193(3):651–669. https://doi.org/10.1534/genetics.112.146704
Kuska B (1998) Should scientists scrap the notion of junk DNA? J Natl Cancer Inst 90(14): 1032–1033. https://doi.org/10.1093/jnci/90.14.1032

Kwenda S, Birch PRJ, Moleleki LN (2016) Genome-wide identification of potato long intergenic noncoding RNAs responsive to *Pectobacterium carotovorum* subspecies brasiliense infection. BMC Genomics 17(1). https://doi.org/10.1186/s12864-016-2967-9

Lauressergues D, Couzigou JM, San Clemente H, Martinez Y, Dunand C, Bécard G, Combier JP (2015) Primary transcripts of microRNAs encode regulatory peptides. Nature 520(7545):90–93. https://doi.org/10.1038/nature14346

Lavorgna G, Dahary D, Lehner B, Sorek R, Sanderson CM, Casari G (2004) In search of antisense. Trends Biochem Sci 29(2):88–94. https://doi.org/10.1016/j.tibs.2003.12.002

Li L, Eichten SR, Shimizu R, Petsch K, Yeh CT, Wu W, Chettoor AM, Givan SA, Cole RA, Fowler JE et al (2014) Genome-wide discovery and characterization of maize long non-coding RNAs. Genome Biol 15(2):1–15. https://doi.org/10.1186/gb-2014-15-2-r40

Li S, Yu X, Lei N, Cheng Z, Zhao P, He Y, Wang W, Peng M (2017a) Genome- wide identification and functional prediction of cold and/or drought-responsive lncRNAs in cassava. Sci Rep 7: 45981. https://doi.org/10.1038/srep45981

Li W, Li C, Li S, Peng M (2017b) Long noncoding RNAs that respond to *Fusarium oxysporum* infection in "Cavendish" banana (*Musa acuminata*). Sci Rep 7(1):1–13. https://doi.org/10.1038/s41598-017-17179-3

Li X, Yang L, Chen LL (2018) The biogenesis, functions, and challenges of circular RNAs. Mol Cell 71(3):428–442. https://doi.org/10.1016/j.molcel.2018.06.034

Li H, Ye W, Wang Y, Chen X, Fang Y, Sun G (2021a) RNA sequencing-based exploration of the effects of far-red light on lncRNAs involved in the shade-avoidance response of *D. officinale*. PeerJ 9:e10769. https://doi.org/10.7717/peerj.10769

Li R, Jin J, Xu J, Wang L, Li J, Lou Y, Baldwin IT (2021b) Long non-coding RNAs associate with jasmonate-mediated plant defence against herbivores. Plant Cell Environ 44(3):982–994. https://doi.org/10.1111/pce.13952

Li S, Cheng Z, Dong S, Li Z, Zou L, Zhao P, Guo X, Bao Y, Wang W, Peng M (2022) Global identification of full-length cassava lncRNAs unveils the role of cold- responsive intergenic lncRNA 1 in cold stress response. Plant Cell Environ 45(2):412–426. https://doi.org/10.1111/pce.14236

Liu J, Wang H, Chua NH (2015a) Long noncoding RNA transcriptome of plants. Plant Biotechnol J 13(3):319–328. https://doi.org/10.1111/pbi.12336

Liu X, Hao L, Li D, Zhu L, Hu S (2015b) Long non-coding RNAs and their biological roles in plants. Genomics Proteomics Bioinf 13(3):137–147. https://doi.org/10.1016/j.gpb.2015.02.003

Liu X, Li D, Zhang D, Yin D, Zhao Y, Ji C, Zhao X, Li X, He Q, Chen R, Hu S, Zhu L (2018) A novel antisense long noncoding RNA, TWISTED LEAF, maintains leaf blade flattening by regulating its associated sense R2R3-MYB gene in rice. New Phytol 218(2):774–788. https://doi.org/10.1111/nph.15023

Liu F, Xu Y, Chang K, Li S, Liu Z, Qi S, Jia J, Zhang M, Crawford NM, Wang Y (2019) The long noncoding RNA T5120 regulates nitrate response and assimilation in *Arabidopsis*. New Phytol 224(1):117–131. https://doi.org/10.1111/nph.16038

Liu X, Wang X, Li J, Hu S, Deng Y, Yin H, Bao X, Zhang QC, Wang G, Wang B, Shi Q, Shan G (2020) Identification of mecciRNAs and their roles in the mitochondrial entry of proteins. Sci China Life Sci 63(10):1429–1449. https://doi.org/10.1007/s11427-020-1631-9

Louro R, Smirnova AS, Verjovski-Almeida S (2009) Long intronic noncoding RNA transcription: expression noise or expression choice? Genomics 93(4):291–298. https://doi.org/10.1016/j.ygeno.2008.11.009

Lu X, Chen X, Mu M, Wang J, Wang X, Wang D, Yin Z, Fan W, Wang S, Guo L, Ye W (2016) Genome-wide analysis of long noncoding RNAs and their responses to drought stress in cotton (*Gossypium hirsutum* L.). PLoS One 11(6):1–18. https://doi.org/10.1371/journal.pone.0156723

Lu Q, Guo F, Xu Q, Cang J (2020a) LncRNA improves cold resistance of winter wheat by interacting with miR398. Funct Plant Biol 47(6):544–557. https://doi.org/10.1071/FP19267

Lu Q, Xu Q, Guo F, Lv Y, Song C, Feng M, Yu J, Zhang D, Cang J (2020b) Identification and characterization of long non-coding RNAs as competing endogenous RNAs in the cold stress response of *Triticum aestivum*. Plant Biol 22(4):635–645. https://doi.org/10.1111/plb.13119

Lucero L, Ferrero L, Fonouni-Farde C, Ariel F (2021) Functional classification of plant long noncoding RNAs: a transcript is known by the company it keeps. New Phytol 229(3): 1251–1260. https://doi.org/10.1111/nph.16903

Lv Y, Hu F, Zhou Y, Wu F, Gaut BS (2019) Maize transposable elements contribute to long non-coding RNAs that are regulatory hubs for abiotic stress response. BMC Genomics 20(1): 1–17. https://doi.org/10.1186/s12864-019-6245-5

Ma L, Bajic VB, Zhang Z (2013) On the classification of long non-coding RNAs. RNA Biol 10(6): 924–933. https://doi.org/10.4161/rna.24604

Ma X, Shao C, Jin Y, Wang H, Meng Y (2014) Long non-coding RNA: a novel endogenous source for the generation of Dicer-like 1-dependent small RNAs in *Arabidopsis thaliana*. RNA Biol 11(4):373–390. https://doi.org/10.4161/rna.28725

Ma H, Yang T, Li Y, Zhang J, Wu T, Song T, Yao Y, Tian J (2021) The long noncoding RNA MdLNC499 bridges MdWRKY1 and MdERF109 function to regulate early-stage light-induced anthocyanin accumulation in apple fruit. Plant Cell 33(10):3309–3330. https://doi.org/10.1093/plcell/koab188

Mansuri RM, Azizi AH, Sadri AH, Shobbar ZS (2022) Long non-coding RNAs as the regulatory hubs in rice response to salt stress. Sci Rep 12(1):1–14. https://doi.org/10.1038/s41598-022-26133-x

Mattick JS (2009) The genetic signatures of noncoding RNAs. PLoS Genet 5(4):e1000459. https://doi.org/10.1371/journal.pgen.1000459

Mattick JS, Rinn JL (2015) Discovery and annotation of long noncoding RNAs. Nat Struct Mol Biol 22(1):5–7. https://doi.org/10.1038/nsmb.2942

McClintock B (1951) Chromosome organization and genic expression. Cold Spring Harb Symp Quant Biol 16:13–47. https://doi.org/10.1101/SQB.1951.016.01.004

McKinlay A, Podicheti R, Wendte JM, Cocklin R, Rusch DB (2018) RNA polymerases IV and V influence the 3′ boundaries of Polymerase II transcription units in *Arabidopsis*. RNA Biol 15(2): 269–279. https://doi.org/10.1080/15476286.2017.1409930

Meng S, Zhou H, Feng Z, Xu Z, Tang Y, Li P, Wu M (2017) CircRNA: functions and properties of a novel potential biomarker for cancer. Mol Cancer 16(1):94. https://doi.org/10.1186/s12943-017-0663-2

Moison M, Pacheco JM, Lucero L, Fonouni-Farde C, Rodríguez-Melo J, Mansilla N, Christ A, Bazin J, Benhamed M, Ibañez F et al (2021) The lncRNA APOLO interacts with the transcription factor WRKY42 to trigger root hair cell expansion in response to cold. Mol Plant 14(6): 937–948. https://doi.org/10.1016/j.molp.2021.03.008

Moore MJ, Proudfoot NJ (2009) Pre-mRNA processing reaches back to transcription and ahead to translation. Cell 136(4):688–700. https://doi.org/10.1016/j.cell.2009.02.001

Nobuta K, Lu C, Shrivastava R, Pillay M, De Paoli E, Accerbi M, Arteaga-Vazquez M, Sidorenko L, Jeong DH, Yen Y et al (2008) Distinct size distribution of endogenous siRNAs in maize: evidence from deep sequencing in the mop1-1 mutant. Proc Natl Acad Sci USA 105(39):14958–14963. https://doi.org/10.1073/pnas.0808066105

Noshay JM, Anderson SN, Zhou P, Ji L, Ricci W, Lu Z, Stitzer MC, Crisp PA, Hirsch CN, Zhang X et al (2019) Monitoring the interplay between transposable element families and DNA methylation in maize. PLoS Genet 15(9):1–25. https://doi.org/10.1371/journal.pgen.1008291

Pang J, Zhang X, Ma X, Zhao J (2019) Spatio-temporal transcriptional dynamics of maize long non-coding RNAs responsive to drought stress. Genes 10(2). https://doi.org/10.3390/genes10020138

Patra GK, Gupta D, Rout GR, Panda SK (2023) Role of long non coding RNA in plants under abiotic and biotic stresses. Plant Physiol Biochem 194:96–110. https://doi.org/10.1016/j.plaphy.2022.10.030

Pedro DLF, Lorenzetti APR, Domingues DS, Paschoal AR (2018) PlaNC-TE: a comprehensive knowledgebase of non-coding RNAs and transposable elements in plants. Database 2018:1–7. https://doi.org/10.1093/database/bay078

Preker P, Almvig K, Christensen MS, Valen E, Mapendano CK, Sandelin A, Jensen TH (2011) PROMoter uPstream Transcripts share characteristics with mRNAs and are produced upstream of all three major types of mammalian promoters. Nucleic Acids Res 39(16):7179–7193. https://doi.org/10.1093/nar/gkr370

Prensner JR, Iyer MK, Balbin OA, Dhanasekaran SM, Cao Q, Brenner JC, Laxman B, Asangani IA, Grasso CS, Kominsky HD et al (2011) Transcriptome sequencing across a prostate cancer cohort identifies PCAT-1, an unannotated lincRNA implicated in disease progression. Nat Biotechnol 29(8):742–749. https://doi.org/10.1038/nbt.1914

Qi X, Xie S, Liu Y, Yi F, Yu J (2013) Genome-wide annotation of genes and noncoding RNAs of foxtail millet in response to simulated drought stress by deep sequencing. Plant Mol Biol *83*(4-5):459–473. https://doi.org/10.1007/s11103-013-0104-6

Qin T, Zhao H, Cui P, Albesher N, Xionga L (2017) A nucleus-localized long non- coding RNA enhances drought and salt stress tolerance. Plant Physiol 175(3):1321–1336. https://doi.org/10.1104/pp.17.00574

Qin T, Li J, Zhang KQ (2020) Structure, regulation, and function of linear and circular long non-coding RNAs. Front Genet 11:150. https://doi.org/10.3389/fgene.2020.00150

Röhrig H, Schmidt J, Miklashevichs E, Schell J, John M (2002) Soybean ENOD40 encodes two peptides that bind to sucrose synthase. Proc Natl Acad Sci USA 99(4):1915–1920. https://doi.org/10.1073/pnas.022664799

Sanan-Mishra N, Abdul Kader Jailani A, Mandal B, Mukherjee SK (2021) Secondary siRNAs in plants: biosynthesis, various functions, and applications in virology. Front Plant Sci 12:1–32. https://doi.org/10.3389/fpls.2021.610283

Sanger HL, Klotz G, Riesner D, Gross HJ, Kleinschmidt AK (1976) Viroids are single stranded covalently closed circular RNA molecules existing as highly base paired rod like structures. Proc Natl Acad Sci USA 73(11):3852–3856. https://doi.org/10.1073/pnas.73.11.3852

Schindler S, Szafranski K, Hiller M, Ali GS, Palusa SG, Backofen R, Platzer M, Reddy ASN (2008) Alternative splicing at NAGNAG acceptors in *Arabidopsis thaliana* SR and SR-related protein-coding genes. BMC Genomics 9:1–11. https://doi.org/10.1186/1471-2164-9-159

Schnable PS, Ware D, Fulton RS, Stein JC, Wei F, Pasternak S, Liang C, Zhang J, Fulton L, Graves TA et al (2009) The B73 maize genome: complexity, diversity, and dynamics. Science 326(5956):1112–1115. https://doi.org/10.1126/science.1178534

Seo JS, Sun HX, Park BS, Huang CH, Yeh SD, Jung C, Chua NH (2017) ELF18-INDUCED LONG-NONCODING RNA associates with mediator to enhance expression of innate immune response genes in *Arabidopsis*. Plant Cell 29(5):1024–1038. https://doi.org/10.1105/tpc.16.00886

Sharma A, Badola PK, Bhatia C, Sharma D, Trivedi PK (2020) Primary transcript of miR858 encodes regulatory peptide and controls flavonoid biosynthesis and development in *Arabidopsis*. Nat Plants 6(10):1262–1274. https://doi.org/10.1038/s41477-020-00769-x

Shi Y, Ding Y, Yang S (2015) Cold signal transduction and its interplay with phytohormones during cold acclimation. Plant Cell Physiol 56(1):7–15. https://doi.org/10.1093/pcp/pcu115

Shin SY, Jeong JS, Lim JY, Kim T, Park JH, Kim JK, Shin C (2018) Transcriptomic analyses of rice (*Oryza sativa*) genes and non-coding RNAs under nitrogen starvation using multiple omics technologies. BMC Genomics 19(1):1–20. https://doi.org/10.1186/s12864-018-4897-1

Shuai P, Liang D, Tang S, Zhang Z, Ye CY, Su Y, Xia X, Yin W (2014) Genome-wide identification and functional prediction of novel and drought-responsive lincRNAs in *Populus trichocarpa*. J Exp Bot 65(17):4975–4983. https://doi.org/10.1093/jxb/eru256

Shukla B, Gupta S, Srivastava G, Sharma A, Shukla AK, Shasany AK (2021) lncRNADetector: a bioinformatics pipeline for long non-coding RNA identification and MAPslnc: a repository of medicinal and aromatic plant lncRNAs. RNA Biol 18(12):2290–2295. https://doi.org/10.1080/15476286.2021.1899673

Sibley CR, Blazquez L, Ule J (2016) Lessons from non-canonical splicing. Nat Rev Genet 17(7): 407–421. https://doi.org/10.1038/nrg.2016.46

Singh U, Khemka N, Rajkumar MS, Garg R, Jain M (2017) PLncPRO for prediction of long non-coding RNAs (lncRNAs) in plants and its application for discovery of abiotic stress-responsive lncRNAs in rice and chickpea. Nucleic Acids Res 45(22):15. https://doi.org/10.1093/nar/gkx866

Singh A, Vivek AT, Kumar S (2021) AlnC: an extensive database of long noncoding RNAs in angiosperms. PLoS One 16(4):1–11. https://doi.org/10.1371/journal.pone.0247215

Smith JE, Alvarez-Dominguez JR, Kline N, Huynh NJ, Geisler S, Hu W, Coller J, Baker KE (2014) Translation of small open reading frames within unannotated RNA transcripts in *Saccharomyces cerevisiae*. Cell Rep 7(6):1858–1866. https://doi.org/10.1016/j.celrep.2014.05.023

Song JH, Cao JS, Wang CG (2013) BcMF11, a novel non-coding RNA gene from Brassica campestris, is required for pollen development and male fertility. Plant Cell Rep 32(1):21–30. https://doi.org/10.1007/s00299-012-1337-6

Song Y, Ci D, Tian M, Zhang D (2016) Stable methylation of a non-coding RNA gene regulates gene expression in response to abiotic stress in *Populus simonii*. J Exp Bot 67(5):1477–1492. https://doi.org/10.1093/jxb/erv543

St. Laurent G, Wahlestedt C, Kapranov P (2015) The Landscape of long noncoding RNA classification. Trends Genet 31(5):239–251. https://doi.org/10.1016/j.tig.2015.03.007

Statello L, Guo CJ, Chen LL, Huarte M (2021) Gene regulation by long non-coding RNAs and its biological functions. Nat Rev Mol Cell Biol 22(2):96–118. https://doi.org/10.1038/s41580-020-00315-9

Sun X, Zheng H, Li J, Liu L, Zhang X, Sui N (2020a) Comparative transcriptome analysis reveals new lncRNAs responding to salt stress in sweet sorghum. Front Bioeng Biotechnol 8:1–14. https://doi.org/10.3389/fbioe.2020.00331

Sun Z, Huang K, Han Z, Wang P, Fang Y (2020b) Genome-wide identification of *Arabidopsis* long noncoding RNAs in response to the blue light. Sci Rep 10(1):1–10. https://doi.org/10.1038/s41598-020-63187-1

Szcześniak MW, Rosikiewicz W, Makałowska I (2016) CANTATAdb: a collection of plant long non-coding RNAs. Plant Cell Physiol 57(1):e8. https://doi.org/10.1093/pcp/pcv201

Tan X, Li S, Hu L, Zhang C (2020) Genome-wide analysis of long non-coding RNAs (lncRNAs) in two contrasting rapeseed (*Brassica napus* L.) genotypes subjected to drought stress and re-watering. BMC Plant Biol 20(1):1–20. https://doi.org/10.1186/s12870-020-2286-9

Tang J, Chu C (2017) MicroRNAs in crop improvement: fine-tuners for complex traits. Nat Plants 3(7):1–11. https://doi.org/10.1038/nplants.2017.77

Tang Z, Xu M, Ito H, Cai J, Ma X, Qin J, Yu D, Meng Y (2019) Deciphering the non-coding RNA-level response to arsenic stress in rice (*Oryza sativa*). Plant Signal Behav 14(9). https://doi.org/10.1080/15592324.2019.1629268

Tani H, Imamachi N, Mizutani R, Imamura K, Kwon Y, Miyazaki S, Maekawa S, Suzuki Y, Akimitsu N (2015) Genome-wide analysis of long noncoding RNA turnover. Methods Mol Biol 1262:305–320. https://doi.org/10.1007/978-1-4939-2253-6_19

Thatcher SR, Danilevskaya ON, Meng X, Beatty M, Zastrow-Hayes G, Harris C, Van Allen B, Habben J, Li B (2016) Genome-wide analysis of alternative splicing during development and drought stress in maize. Plant Physiol 170(1):586–599. https://doi.org/10.1104/pp.15.01267

Todesco M, Rubio-Somoza I, Paz-Ares J, Weigel D (2010) A collection of target mimics for comprehensive analysis of MicroRNA function in *Arabidopsis thaliana*. PLoS Genet 6(7):1–10. https://doi.org/10.1371/journal.pgen.1001031

Urquiaga MCDO, Thiebaut F, Hemerly AS, Ferreira PCG (2021) From trash to luxury: the potential role of plant lncRNA in DNA methylation during abiotic stress. Front Plant Sci 11:1–10. https://doi.org/10.3389/fpls.2020.603246

Van Dijk EL, Chen CL, Daubenton-Carafa Y, Gourvennec S, Kwapisz M, Roche V, Bertrand C, Silvain M, Legoix-Ne P, Loeillet S et al (2011) XUTs are a class of Xrn1-sensitive antisense regulatory non-coding RNA in yeast. Nature 475(7354):114–119. https://doi.org/10.1038/nature10118

Vicens Q, Westhof E (2014) Biogenesis of circular RNAs. Cell 159(1):1–14. https://doi.org/10.1016/j.cell.2014.09.005

Wan S, Zhang Y, Duan M, Huang L, Wang W, Xu Q, Yang Y, Yu Y (2020) Integrated analysis of long non-coding RNAs (lncRNAs) and mRNAs reveals the regulatory role of lncrnas associated with salt resistance in *Camellia sinensis*. Front Plant Sci 11:1–14. https://doi.org/10.3389/fpls.2020.00218

Wang HLV, Chekanova JA (2017) Long noncoding RNAs in plants. Adv Exp Med Biol 1008:133–154. https://doi.org/10.1007/978-981-10-5203-3_5

Wang H, Chung PJ, Liu J, Jang IC, Kean MJ, Xu J, Chua NH (2014a) Genome-wide identification of long noncoding natural antisense transcripts and their responses to light in *Arabidopsis*. Genome Res 24(3):444–453. https://doi.org/10.1101/gr.165555.113

Wang Y, Fan X, Lin F, He G, Terzaghi W, Zhu D, Deng XW (2014b) *Arabidopsis* noncoding RNA mediates control of photomorphogenesis by red light. Proc Natl Acad Sci USA 111(28): 10359–10364. https://doi.org/10.1073/pnas.1409457111

Wang Y, Wang X, Deng W, Fan X, Liu TT, He G, Chen R, Terzaghi W, Zhu D, Wang Deng X (2014c) Genomic features and regulatory roles of intermediate- sized non-coding rnas in *Arabidopsis*. Mol Plant 7(3):514–527. https://doi.org/10.1093/mp/sst177

Wang J, Yu W, Yang Y, Li X, Chen T, Liu T, Ma N, Yang X, Liu R, Zhang B (2015a) Genome-wide analysis of tomato long non-coding RNAs and identification as endogenous target mimic for microRNA in response to TYLCV infection. Sci Rep 5:16946. https://doi.org/10.1038/srep16946

Wang TZ, Liu M, Zhao MG, Chen R, Zhang WH (2015b) Identification and characterization of long non-coding RNAs involved in osmotic and salt stress in *Medicago truncatula* using genome-wide high-throughput sequencing. BMC Plant Biol 15(1):1–13. https://doi.org/10.1186/s12870-015-0530-5

Wang X, Ai G, Zhang C, Cui L, Wang J, Li H, Zhang J, Ye Z (2016) Expression and diversification analysis reveals transposable elements play important roles in the origin of Lycopersicon-specific lncRNAs in tomato. New Phytol 209(4):1442–1455. https://doi.org/10.1111/nph.13718

Wang T, Zhao M, Zhang X, Liu M, Yang C, Chen Y, Chen R, Wen J, Mysore KS, Zhang WH (2017) Novel phosphate deficiency-responsive long non-coding RNAs in the legume model plant *Medicago truncatula*. J Exp Bot 68(21-22):5937–5948. https://doi.org/10.1093/jxb/erx384

Wang Y, Luo X, Sun F, Hu J, Zha X, Su W, Yang J (2018a) Overexpressing lncRNA LAIR increases grain yield and regulates neighbouring gene cluster expression in rice. Nat Commun 9(1):1–9. https://doi.org/10.1038/s41467-018-05829-7

Wang Y, Gao L, Li J, Zhu B, Zhu H, Luo Y, Wang Q, Zuo J (2018b) Analysis of long-non-coding RNAs associated with ethylene in tomato. Gene 674:151–160. https://doi.org/10.1016/j.gene.2018.06.089

Wang A, Hu J, Gao C, Chen G, Wang B, Lin C, Song L, Ding Y, Zhou G (2019a) Genome-wide analysis of long non-coding RNAs unveils the regulatory roles in the heat tolerance of Chinese cabbage (*Brassica rapa* ssp.chinensis). Sci Rep 9(1):1–14. https://doi.org/10.1038/s41598-019-41428-2

Wang P, Dai L, Ai J, Wang Y, Ren F (2019b) Identification and functional prediction of cold-related long non-coding RNA (lncRNA) in grapevine. Sci Rep 9(1):1–15. https://doi.org/10.1038/s41598-019-43269-5

Wang S, Sun S, Guo R, Liao W, Shou H (2021) Transcriptomic profiling of Fe- responsive lncRNAs and their regulatory mechanism in rice. Genes 12(4):567. https://doi.org/10.3390/genes12040567

Wang X, Fan H, Wang B, Yuan F (2023) Research progress on the roles of lncRNAs in plant development and stress responses. Front Plant Sci 14:1–12. https://doi.org/10.3389/fpls.2023.1138901

Waseem M, Liu Y, Xia R (2021) Long non-coding RNAs, the dark matter: an emerging regulatory component in plants. Int J Mol Sci 22(1):1–26. https://doi.org/10.3390/ijms22010086

Weidong Q, Hongping C, Zuozhen Y, Biaolin H, Xiangdong L, Bing A, Yuan L, Yu H, Jiankun X, Fantao Z (2020) Systematic characterization of long non- coding RNAs and their responses to drought stress in Dongxiang wild rice. Rice Sci 27(1):21–31. https://doi.org/10.1016/j.rsci.2019.12.003

Wierzbicki AT, Haag JR, Pikaard CS (2008) Noncoding transcription by RNA polymerase Pol IVb/Pol V mediates transcriptional silencing of overlapping and adjacent genes. Cell 135(4):635–648. https://doi.org/10.1016/j.cell.2008.09.035

Wierzbicki AT, Ream TS, Haag JR, Pikaard CS (2009) RNA polymerase V transcription guides ARGONAUTE4 to chromatin. Nat Genet 41(5):630–634. https://doi.org/10.1038/ng.365

Wight M, Werner A (2013) The functions of natural antisense transcripts. Essays Biochem 54(1):91–101. https://doi.org/10.1042/BSE0540091

Wilusz JE, Freier SM, Spector DL (2008) 3′ end processing of a long nuclear- retained noncoding RNA yields a tRNA-like cytoplasmic RNA. Cell 135(5):919–932. https://doi.org/10.1016/j.cell.2008.10.012

Wu HJ, Wang ZM, Wang M, Wang XJ (2013) Widespread long noncoding RNAs as endogenous target mimics for microRNAs in plants. Plant Physiol 161(4):1875–1884. https://doi.org/10.1104/pp.113.215962

Wu H, Yang L, Chen LL (2017) The diversity of long noncoding RNAs and their generation. Trends Genet 33(8):540–552. https://doi.org/10.1016/j.tig.2017.05.004

Wu HW, Deng S, Xu H, Mao HZ, Liu J, Niu QW, Wang H, Chua NH (2018) A noncoding RNA transcribed from the AGAMOUS (AG) second intron binds to CURLY LEAF and represses AG expression in leaves. New Phytol 219(4):1480–1491. https://doi.org/10.1111/nph.15231

Xin M, Wang Y, Yao Y, Song N, Hu Z, Qin D, Xie C, Peng H, Ni Z, Sun Q (2011) Identification and characterization of wheat long non-protein coding RNAs responsive to powdery mildew infection and heat stress by using microarray analysis and SBS sequencing. BMC Plant Biol 11:61. https://doi.org/10.1186/1471-2229-11-61

Xuan H, Zhang L, Liu X, Han G, Li J, Li X, Liu A, Liao M, Zhang S (2015) PLNlncRbase: a resource for experimentally identified lncRNAs in plants. Gene 573(2):328–332. https://doi.org/10.1016/j.gene.2015.07.069

Yan Q, Wu F, Yan Z, Li J, Ma T, Zhang Y, Zhao Y, Wang Y, Zhang J (2019) Differential co-expression networks of long non-coding RNAs and mRNAs in *Cleistogenes songorica* under water stress and during recovery. BMC Plant Biol 19(1):1–19. https://doi.org/10.1186/s12870-018-1626-5

Ye X, Wang S, Zhao X, Gao N, Wang Y, Yang Y, Wu E, Jiang C, Cheng Y, Wu W, Liu S (2022) Role of lncRNAs in cis- and trans-regulatory responses to salt in *Populus trichocarpa*. Plant J 110(4):978–993. https://doi.org/10.1111/tpj.15714

Yi X, Zhang Z, Ling Y, Xu W, Su Z (2015) PNRD: a plant non-coding RNA database. Nucleic Acids Res 43(D1):D982–D989. https://doi.org/10.1093/nar/gku1162

Yu X, Yang J, Li X, Liu X, Sun C, Wu F, He Y (2013) Global analysis of cis- natural antisense transcripts and their heat-responsive nat-siRNAs in *Brassica rapa*. BMC Plant Biol 13(1):208. https://doi.org/10.1186/1471-2229-13-208

Yu Y, Zhang Y, Chen X, Chen Y (2019) Plant noncoding RNAs: hidden players in development and stress responses. Annu Rev Cell Dev Biol 35:407–431. https://doi.org/10.1146/annurev-cellbio-100818-125218

Yu Y, Zhou F, Feng Z, He H, Lian P, Yang W, Lei Q, Zhang C, Chen Q (2020) Transcriptional landscape of pathogen-responsive lncRNAs in rice unveils the role of ALEX1 in jasmonate pathway and disease resistance. Plant Biotechnol J 18(3):679–690. https://doi.org/10.1111/pbi.13234

Yu J, Qiu K, Sun W, Yang T, Wu T, Song T, Zhang J, Yao Y, Tian J (2022) A long noncoding RNA functions in high-light-induced anthocyanin accumulation in apple by activating ethylene synthesis. Plant Physiol 189(1):66–83. https://doi.org/10.1093/plphys/kiac049

Yuan J, Zhang Y, Dong J, Sun Y, Lim BL, Liu D, Lu ZJ (2016) Systematic characterization of novel lncRNAs responding to phosphate starvation in *Arabidopsis thaliana*. BMC Genomics 17(1):1–16. https://doi.org/10.1186/s12864-016-2929-2

Yuan C, He RR, Zhao WL, Chen YQ, Zhang YC (2023) Insights into the roles of long noncoding RNAs in the communication between plants and the environment. Plant Genome 16(4):1–15. https://doi.org/10.1002/tpg2.20277

Zampetaki A, Albrecht A, Steinhofel K (2018) Long non-coding RNA structure and function: is there a link? Front Physiol 9:1–8. https://doi.org/10.3389/fphys.2018.01201

Zandalinas SI, Fritschi FB, Mittler R (2021) Global warming, climate change, and environmental pollution: recipe for a multifactorial stress combination disaster. Trends Plant Sci 26(6): 588–599. https://doi.org/10.1016/j.tplants.2021.02.011

Zhang Y, Zhang XO, Chen T, Xiang JF, Yin QF, Xing YH, Zhu S, Yang L, Chen LL (2013) Circular intronic long noncoding RNAs. Mol Cell 51(6):792–806. https://doi.org/10.1016/j.molcel.2013.08.017

Zhang W, Han Z, Guo Q, Liu Y, Zheng Y, Wu F, Jin W (2014a) Identification of maize long non-coding RNAs responsive to drought stress. PLoS One 9(6):e98958. https://doi.org/10.1371/journal.pone.0098958

Zhang XO, Yin QF, Wang HB, Zhang Y, Chen T, Zheng P, Lu X, Chen LL, Yang L (2014b) Species-specific alternative splicing leads to unique expression of sno-lncRNAs. BMC Genomics 15(1):1–15. https://doi.org/10.1186/1471-2164-15-287

Zhang C, Tang G, Peng X, Sun F, Liu S, Xi Y (2018a) Long non-coding RNAs of switchgrass (*Panicum virgatum* L.) in multiple dehydration stresses. BMC Plant Biol 18(1):1–15. https://doi.org/10.1186/s12870-018-1288-3

Zhang G, Chen D, Zhang T, Duan A, Zhang J, He C (2018b) Transcriptomic and functional analyses unveil the role of long non-coding RNAs in anthocyanin biosynthesis during sea buckthorn fruit ripening. DNA Res 25(5):465–476. https://doi.org/10.1093/dnares/dsy017

Zhang L, Wang M, Li N, Wang H, Qiu P, Pei L, Xu Z, Wang T, Gao E, Liu J et al (2018c) Long noncoding RNAs involve in resistance to *Verticillium dahliae*, a fungal disease in cotton. Plant Biotechnol J 16(6):1172–1185. https://doi.org/10.1111/pbi.12861

Zhang P, Fan Y, Sun X, Chen L, Terzaghi W, Bucher E, Li L, Dai M (2019a) A large-scale circular RNA profiling reveals universal molecular mechanisms responsive to drought stress in maize and *Arabidopsis*. Plant J 98(4):697–713. https://doi.org/10.1111/tpj.14267

Zhang X, Dong J, Deng F, Wang W, Cheng Y, Song L, Hu M, Shen J, Xu Q, Shen F (2019b) The long non-coding RNA lncRNA973 is involved in cotton response to salt stress. BMC Plant Biol 19(1):459. https://doi.org/10.1186/s12870-019-2088-0

Zhang X, Wang W, Zhu W, Dong J, Cheng Y, Yin Z, Shen F (2019c) Mechanisms and functions of long non-coding RNAs at multiple regulatory levels. Int J Mol Sci 20(22):5573. https://doi.org/10.3390/ijms20225573

Zhang G, Liu Z, Dai J, Yu Z, Liu S, Zhang W (2020) ItLnc-BXE: a bagging- XGBoost-ensemble method with comprehensive sequence features for identification of plant lncRNAs. IEEE Access 8:68811–68819. https://doi.org/10.1109/ACCESS.2020.2985114

Zhang B, Su T, Li P, Xin X, Cao Y, Wang W, Zhao X, Zhang D, Yu Y, Li D, Yu S, Zhang F (2021a) Identification of long noncoding RNAs involved in resistance to downy mildew in Chinese cabbage. Hortic Res 8(1):1–15. https://doi.org/10.1038/s41438-021-00479-1

Zhang X, Shen J, Xu Q, Dong J, Song L, Wang W, Shen F (2021b) Long noncoding RNA lncRNA354 functions as a competing endogenous RNA of miR160b to regulate ARF genes in response to salt stress in upland cotton. Plant Cell Environ 44(10):3302–3321. https://doi.org/10.1111/pce.14133

Zhang J, Li J, Saeed S, Batchelor WD, Alariqi M, Meng Q, Zhu F, Zou J, Xu Z, Si H et al (2022) Identification and functional analysis of lncRNA by CRISPR/Cas9 during the cotton response to sap-sucking insect infestation. Front Plant Sci 13:784511. https://doi.org/10.3389/fpls.2022.784511

Zhang L, Lin T, Zhu G, Wu B, Zhang C, Zhu H (2023) LncRNAs exert indispensable roles in orchestrating the interaction among diverse noncoding RNAs and enrich the regulatory network of plant growth and its adaptive environmental stress response. Hortic Res 10(12):uhad234. https://doi.org/10.1093/hr/uhad234

Zhao X, Li J, Lian B, Gu H, Li Y, Qi Y (2018) Global identification of *Arabidopsis* lncRNAs reveals the regulation of MAF4 by a natural antisense RNA. Nat Commun 9(1):1–12. https://doi.org/10.1038/s41467-018-07500-7

Zhao L, Wang J, Li Y, Song T, Wu Y, Fang S, Bu D, Li H, Sun L, Pei D et al (2021) NONCODEV6: an updated database dedicated to long non-coding RNA annotation in both animals and plants. Nucleic Acids Res 49(D1):D165–D171. https://doi.org/10.1093/nar/gkaa1046

Zheng Q, Rowley MJ, Böhmdorfer G, Sandhu D, Gregory BD, Wierzbicki AT (2012) RNA polymerase V targets transcriptional silencing components to promoters of protein-coding genes. Plant J 73(2):179–189. https://doi.org/10.1111/tpj.12034

Zhong X, Hale CJ, Law JA, Johnson LM, Feng S, Tu A, Jacobsen SE (2012) DDR complex facilitates global association of RNA polymerase V to promoters and evolutionarily young transposons. Nat Struct Mol Biol 19(9):870–875. https://doi.org/10.1038/nsmb.2354

Zhou H, Lin K (2008) Excess of microRNAs in large and very 5′ biased introns. Biochem Biophys Res Commun 368(3):709–715. https://doi.org/10.1016/j.bbrc.2008.01.117

Zhou B, Zhao H, Yu J, Guo C, Dou X, Song F, Hu G, Cao Z, Qu Y, Yang Y, Zhou Y, Wang J (2018a) EVLncRNAs: a manually curated database for long non- coding RNAs validated by low-throughput experiments. Nucleic Acids Res 46(D1):D100–D105. https://doi.org/10.1093/nar/gkx677

Zhou M, Palanca AMS, Law JA (2018b) Locus-specific control of the de novo DNA methylation pathway in *Arabidopsis* by the CLASSY family. Nat Genet 50(6):865–873. https://doi.org/10.1038/s41588-018-0115-y

Zhou GF, Zhang LP, Li BX, Sheng O, Wei QJ, Yao FX, Guan G, Liu GD (2019) Genome-wide identification of long non-coding RNA in trifoliate orange (*Poncirus trifoliata* L. raf) leaves in response to boron deficiency. Int J Mol Sci 20(21):5419. https://doi.org/10.3390/ijms20215419

Zhou B, Ji B, Liu K, Hu G, Wang F, Chen Q, Yu R, Huang P, Ren J, Guo C et al (2021a) EVLncRNAs 2.0: an updated database of manually curated functional long non-coding RNAs validated by low-throughput experiments. Nucleic Acids Res 49(D1):D86–D91. https://doi.org/10.1093/nar/gkaa1076

Zhou YF, Zhang YC, Sun YM, Yu Y, Lei MQ, Yang YW, Lian JP, Feng YZ, Zhang Z, Yang L et al (2021b) The parent-of-origin lncRNA MISSEN regulates rice endosperm development. Nat Commun 12(1):6525. https://doi.org/10.1038/s41467-021-26795-7

Zhu JJ, Fu HJ, Wu YG, Zheng XF (2013) Function of lncRNAs and approaches to lncRNA-protein interactions. Sci China Life Sci 56(10):876–885. https://doi.org/10.1007/s11427-013-4553-6

Zhu C, Zhang S, Fu H, Zhou C, Chen L, Li X, Lin Y, Lai Z, Guo Y (2019) Transcriptome and phytochemical analyses provide new insights into long non- coding RNAs modulating characteristic secondary metabolites of Oolong tea (*Camellia sinensis*) in solar-withering. Front Plant Sci 10:1–25. https://doi.org/10.3389/fpls.2019.01638

Chapter 4
Redox Regulation of the Response to Abiotic Stresses by MicroRNAs and Transcription Factors in Plants

G. Kocsy, Z. Gulyás, K. Kulman, K. Jobbágy, K. Singh, and Sanjib Kumar Panda

Abstract A common consequence of various abiotic stresses is the disorder in the balance of the formation and removal of reactive oxygen species. The efficient activation of the antioxidant system can reset the optimal redox balance in the various plant tissues. In this process, even the reactive oxygen species participate through the induction of the enzymatic and non-enzymatic antioxidants. This process can happen at both transcriptional and post-transcriptional levels. The expression of the various redox-controlled genes can be regulated by the redox-responsive transcription factors and microRNAs. The temporal changes in their level in the various tissues, cells and compartments could be very important in the efficient response to adverse environmental conditions. In this chapter, we would like to introduce the most important results of this research field and highlight the unsolved problems to give perspectives for future research.

G. Kocsy · Z. Gulyás
Agricultural Institute, HUN-REN Centre for Agricultural Research, Martonvásár, Hungary
e-mail: kocsy.gabor@atk.hu

K. Kulman · K. Singh
Agricultural Institute, HUN-REN Centre for Agricultural Research, Martonvásár, Hungary

Doctoral School of Plant Sciences, Hungarian University of Agriculture and Life Sciences, MATE, Gödöllő, Hungary

K. Jobbágy
Agricultural Institute, HUN-REN Centre for Agricultural Research, Martonvásár, Hungary

Doctoral School of Biology and Institute of Biology, ELTE Eötvös Loránd University, Budapest, Hungary

S. K. Panda (✉)
Department of Biochemistry, School of Life Sciences, Central University of Rajasthan, Ajmer, India
e-mail: sanjib.panda@curaj.ac.in

S. K. Panda, M. A. Khan (eds.), *Plant Functional Genomics for Abiotic Stress Resilience*, Plant in Challenging Environments 7,
https://doi.org/10.1007/978-3-032-01704-8_4

Keywords Abiotic stress · ROS · Redox · MiRNAs · Transcription factor · Tolerance

4.1 Introduction

The maintenance of the optimal redox environment is important to ensure the appropriate growth and development of plants both under optimal and stress conditions. The redox system plays an important role in this process. The enzymatic and non-enzymatic antioxidants ensure the necessary adjustment of the level of reactive oxygen species (ROS) to the changing environmental conditions in the various developmental stages of plants (Fichman and Mittler 2021; Mittler et al. 2022; Foyer and Kunert 2024). Extreme temperatures, lack or surplus of water and various heavy metals can induce oxidative stress, which harmful effect can be prevented by the redox signalling-dependent setting of ROS-level and activation of the protective mechanisms. Although a large surplus of ROS damage the vital macromolecules, the moderate increase of ROS levels can initiate such transcriptional and post-transcriptional regulatory processes, which can protect plants from the harmful effect of oxidative stress. The ROS waves participates in the transduction of signals from the stressed leaves to systemic leaves, thus they have an important role to adjust the gene expression to environmental effect at whole plant level (Fichman and Mittler 2021). H_2O_2, the most stable ROS has an important role in the long distance signalling.

The redox control of stress response can occur both at transcriptional and post-transcriptional level. The redox system can affect the transcription through the control of various redox-responsive small RNAs and transcription factors. The different ROS, enzymatic and non-enzymatic antioxidants may influence the thiol-disulfide conversion of the cysteines of proteins. These changes in turn affect their conformation and the activity. The thioredoxin and thiol peroxidases, as well as thioredoxins, glutaredoxins and peroxiredoxin play an important role in this process (Hopkins and Neumann 2019; Chae et al. 2023).

Redox-responsive small RNAs such as miRNAs participating in the control of transcription may also mediate the effect of oxidative stress induced by various abiotic effects on the expression of stress-related genes (Gulyás et al. 2023). miRNAs regulate gene expression via transcriptional gene silencing through DNA methylation or via post-transcriptional gene silencing through cleavage of the target mRNAs and translational inhibition. A cross-talk between ROS and miRNAs occurs since they can mutually regulate each other's level. Redox-controlled miRNAs were identified in H_2O_2-treated wheat and by comparison of miRNA profile in Arabidopsis genotypes deficient in ascorbate or glutathione (Cao et al. 2019; Székely et al. 2023). Among them, miR395 controls the expression of genes encoding enzymes of sulfate reduction, which pathway ensures Cys for the synthesis of the antioxidant glutathione.

In the redox control of transcription, the interaction between the redox and hormonal system also plays an important role (Kocsy et al. 2013; Considine and

Foyer 2014). Thus, ROS control the metabolism of several hormones, such as ethylene, salicylic acid, jasmonate and brassinosteroids. On the other hand, certain hormones (e.g. abscisic acid, salicylic acid or auxin) also induce ROS production. They also affect the redox equilibrium by inducing the transcription of genes encoding antioxidants, e.g. glutathione S-transferase (Laskowski et al. 2002).

4.2 Extreme Temperatures

4.2.1 Cold Stress

In our time, the most challenging and emerging problem of humanity is global warming, which dramatically changes our climate, resulting in erratic weather patterns, such as extremely low and high temperatures (de Larminat 2016; Masson-Delmotte et al. 2021). These extraordinary phenomena cause huge losses in crop yields globally (Jägermeyr et al. 2021). One of the most harmful extreme weather events affecting crop production worldwide is cold, including winter and early spring frost. Cold acclimation is a well-known phenomenon in plants, which helps to adapt to cold conditions by developing freezing tolerance (Thomashow 1999). In this complex process, plants can optimize their growth and development under cold stress by focusing on survival and defence against the stressor using different strategies like redox regulatory mechanisms (Dreyer and Dietz 2018). In the cold acclimation process, the most important transcription factor (TF) family is the CBF (C-repeat binding factor) group that is included in the APETALA2/ETHYLENE RESPONSE FACTOR (AP2/ERF) family (Shi et al. 2018). They can regulate numerous genes downstream such as *COR* (*COLD REGULATED GENES*) genes, which can help to plants surviving the low temperature (Li et al. 2013) (Table 4.2). The overexpression of *CBF* genes increase the cold tolerance of the plants, which was verified in several species (Gilmour et al. 2000; Wisniewski et al. 2011; Soltész et al. 2013; Liang et al. 2022). *CBF* genes can regulate genes maintaining the redox homeostasis in plants. According to (Liang et al. 2022) overexpression of *Malus baccata CBF1* gene upregulated the expression of *CATALASE1* (*CAT1*) and the proline-related *P5CS1* (*DELTA1-PYRROLINE-5-CARBOXYLATE SYNTHASE 1*) genes increasing the cold and salt tolerance in Arabidopsis. Supporting this, the overexpression of peach *CBF1* caused upregulation of genes involving in the redox reactions and detoxification of ROS (Artlip et al. 2019). In watermelon, the CBF-responsive pathway has important role in melatonin improved photosynthesis and redox homeostasis under aerial cold stress (Chang et al. 2020). Moreover, (Wang et al. 2021) suggest a relationship between glutathione peroxidases and CBF-COR signal transduction pathway in *Ricinus communis*. These studies support the regulation of downstream redox system-dependent genes by CBF regulon, but whether the CBFs are directly regulated by redox system remains unclear. However, according to (Lee et al. 2021) the redox protein, thioredoxin h2 (Trx-h2) can interact with CBF1, which reduces the inactive CBF oligomers and monomers resulting active forms of CBFs. In this study, the null mutant *trx-h2* Arabidopsis mutant showed lower cold tolerance, while the overexpression lines were more resistant to cold. The other example for the redox regulation of CBFs is related to the *AIR12* gene (*AUXIN INDUCED IN ROOT CULTURE 12*) in *Medicago falcata*, which can alter the H_2O_2 level in apoplast resulting in the upregulation of CBF cold response pathway (Wang et al. 2021).

In addition to the *CBF* genes, there are several redox-responsive TFs, which have crucial role in the cold stress response (Maryan et al. 2021). In Arabiodopsis, 30 quickly induced, cold sensitive TFs were identified, called as "first-wave cold-inducible TFs", such as ZAT12 (zinc finger of (*Arabidopsis thaliana*) transcription factor 12) , HSFC1 (heat stress transcription factor C-1), RAV1 (ethylene-responsive transcription factor RAV1), MYB73 (myeloblastosis transcription factor MYB73), MYB44 (myeloblastosis transcription factor MYB44), CRF2 (ethylene-responsive transcription factor CRF2), WRKY33 (Trp, Arg, Lys, Tyr amino acid signature sequence-containing transcription factor 33), ERF6 (ethylene-responsive transcription factor 6), CRF3 (ethylene-responsive transcription factor CRF3) and RVE2 (Reveille 2 protein) (Park et al. 2015). The largest TF families (MYB, WRKY, AP2/ERF, and ZAT) are up-regulated by cold-induced oxidative stress (Bittner et al. 2020; Yang et al. 2020; Ritonga et al. 2021; Luo et al. 2023). Nevertheless, Redox Responsive Transcription Factor1 (RRTF1) can be induced by elevated H_2O_2 level, abiotic- and biotic stresses. In the *RRTF1* overexpression lines, many genes are activated such as the CBF regulon related *STZ/ZAT10 and RAP2.6* (Ethylene-responsive transcription factor RAP2-6) TFs, which are involved in the cold stress response (Zhu et al. 2010; Miura and Furumoto 2013; Matsuo et al. 2015).

The small non-coding RNAs, miRNAs has crucial role in the regulation of gene expression under abiotic and biotic stresses via mRNA cleavage or translational repression (Wang et al. 2019). Numerous cold responsive miRNAs, such as miR396, miR397, and miR319, have been discovered in various plant species including wheat, rice and Arabidopsis (Zhou et al. 2008; Barrera-Figueroa et al. 2012; Tang et al. 2012) (Fig. 4.1). The most significant targets of the conserved miRNA396 are the *Growth Regulating Factors* (GRFs) Overexpressing the Sp-miR396a-5p in tobacco resulted in higher tolerance against cold, drought and salt stress (Chen et al. 2015). The expression of this stress-responsive miR396 depends largely on the redox environment. (Shen et al. 2013) observed, that the reducing agent, hydrogen sulfide involved in the regulation of the miR396 expression. Nevertheless, the wheat miR396c was downregulated after the 3 h and 6h H_2O_2 treatment, respectively (Cao et al. 2019). The other two miRNAs, the miR319 and miR397 showed H_2O_2 responsivity similar to the miR396 in rice (Li et al. 2011). The miR397 family in plants consists of multiple conserved members, either 21-nt or 22-nt in length, which primarily target *laccase* (*LAC*) genes involved in lignin synthesis, circadian regulation, flowering and plant development under diverse conditions (Huang et al. 2020). Overexpression of miR397 improved the tolerance against chilling and freezing stresses in Arabidopsis through the regulation of *CBF* genes and downstream *COR* genes (Dong and Pei 2014) (Table 4.1). The third mentioned cold-responsive miRNA, the miR319 regulates *TEOSINTE BRANCHED/CYCLOIDEA/PCF (TCP)*

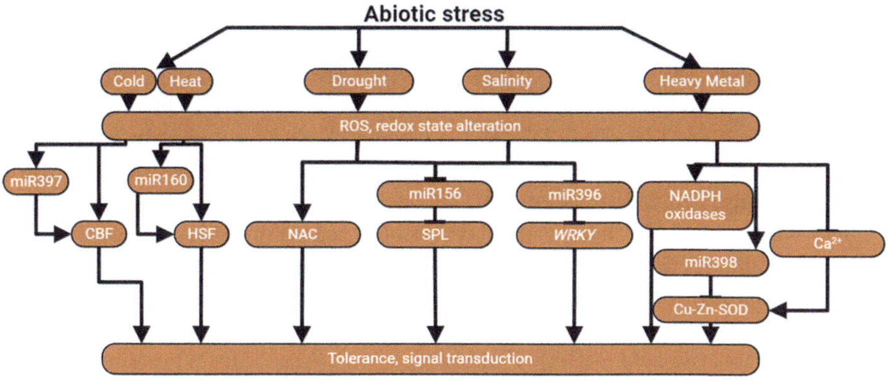

Fig. 4.1 Redox control of the abiotic stress response at transcriptional level. The various abiotic stresses (cold, heat, drought, salinity, heavy metals) induce the accumulation of reactive oxygen species (ROS) which alters the cellular redox state. This change in turn effect the transcription of the redox-responsive microRNAs (miR397, miR160, miR156, miR396, miR398). They modify the level of several transcription factors and the activity of antioxidant enzymes (POD: peroxidases, SOD: superoxide dismutase). miR397 controls C-repeat binding factors (CBF) and cold regulated (COR) genes. miR160 regulates heat shock factors (HSF) and heat shock proteins (HSP). mir156 affects the function of squamosa promoter binding protein-like (SPL) protein and miR396 that of WRKY transcription factors controlling growth and development. The NAC [NAM (no apical meristem), ATAF Arabidopsis thaliana activating factor, CUC (cup-shaped cotyledon)] transcription factors are directly controlled by ROS without the involvement of miRNAs and they can activate various antioxidants such as glutathione (GSH), ascorbate peroxidase (APX), catalase (CAT) and late embryogenesis abundant (LEA) proteins. The ROS can also regulate their own levels through the control of NADPH oxidase producing superoxide radicals and superoxide dismutase removing them

Table 4.1 Most important microRNAs involved in the redox regulation of the stress response

Name of miRNA	Species	Function	References
miR156	*Malus sieversii*	Inhibition of SPL	Feng et al. (2023)
miR160	*Arabidopsis thaliana*	Activation of HSF	Lin et al. (2018)
miR396	*Helianthus annuus*	Inhibition of WRKY	Giacomelli et al. (2012) and Chen et al. (2024)
miR397	*Arabidopsis thaliana*	Activation of CBF	Dong and Pei (2014)
miR398	*Arabidopsis thaliana*	Activation of SOD	Kohli et al. (2017)

SPL squamosa promoter binding protein-like, *HSF* heat shock factor, *WRKY* Trp, Arg, Lys, Tyr amino acid signature sequence-containing transcription factor 33, *CBF* C-repeat binding factor, *SOD* superoxide dismutase

genes, the plant-specific transcription factors containing bHLH motifs (Schommer et al. 2012). Overexpression of miR319 resulted in improved cold tolerance and higher expression level of *DREB1A/B/C (dehydration-responsive element-binding factor 1A/B/C)*, *DREB2A (dehydration-responsive element-binding factor 2A) and TPP1/2 (trehalose-phosphate phosphatase 1)* genes in rice (Wang et al. 2014). In

Marchantia polymorpha, it was described, that miR319 regulated *MpTCP1 (Marchantia polymorpha TCP domain containing protein)* controls cell proliferation and redox processes (Busch et al. 2019). Nevertheless, overexpression of *Solanum habrochaites* miR319d enhanced the expression of *SOD, MYB83, CBF1,* ROS-signalling (*ZAT10, ZAT12*), and *CAT* genes. Based on this, (Shi et al. 2019) suggested that miR319d takes part in regulating the tomato chilling tolerance via ROS signalling, and ROS-scavengers.

It is well-studied that hormones play a central regulatory role in the cold stress response in plants (Eremina et al. 2016). Due to hormonal crosstalk, which creates a complex network, all hormones participate in the cold stress response and adaptation to varying degrees. However, the most relevant and studied hormone related to cold tolerance is abscisic acid (ABA) (Finkelstein 2013). ABA affects proteins involved in photosynthesis (the main source of intracellular ROS), redox homeostasis, and stress signalling (Alvarez et al. 2011, 2013). ABA can induce the expression of genes related to the AsA-GSH cycle (Ghassemian et al. 2008) and ROS accumulation (Li et al. 2020). Additionally, exogenous H_2O_2 boosts ABA catabolism by upregulating the expression of *CYP707A* (*Cytochrome P450 CYP707-1a*) genes (Liu et al. 2010). Based on these findings, a tight relationship is suggested between ABA-related genes and redox regulation at the transcriptional level in optimal and stress conditions.

4.2.2 Heat Stress

Heat stress (HS) is another challenge which changing climate poses to plants. With increasing global temperatures and changing weather patterns, HS causes productivity loss worldwide. For instance, global wheat production is estimated to be limited by HS by more than 6% for every degree (°C) that the temperature rises (Hassan et al. 2021). HS, typically happens at temperatures between 10 and 15 °C above the threshold, and is commonly defined as an elevation in temperature for a pre-determined amount of time that might harm plants irreversibly (Argosubekti 2020). Plants biosynthesise specific proteins - Heat Shock Proteins (HSPs) as a response when challenged with heat stress (Fig. 4.1). These HSPs not only prevent the denaturation of other proteins but also initiate the signal cascade and trigger the expression of genes necessary that are related to heat stress response (Bourgine and Guihur 2021). HS-induced H_2O_2 production creates an oxidised environment which triggers the activation of transcription factors - heat shock factors (HSF). These HSFs bind to a conserved element and activate the transcription of HSP (He et al. 2018) (Table 4.2). Genetically transformed cotton plants expressing the *Agave sisalana* heat shock protein gene, *AsHSP70,* had better tolerance to heat as the transgenic plants exhibited less electrolyte leakage, higher membrane stability, better boll formation along with better performance in terms of chlorophyll, proline and soluble sugars (Batcho et al. 2021). There are studies elaborating on the crosstalk between cellular redox homeostasis and the activation of HSPs. For instance,

Table 4.2 Most important transcription factors involved in the redox regulation of the stress response

Name of TF	Species	Function	References
CBF	*Arabidopsis thaliana*	Activation of Cor genes	Li et al. (2013)
HSF	*Arabidopsis thaliana*	Control of HSP	He et al. (2018)
NAC	*Arabidopsis thaliana*	Activation of CAT, APX	Nguyen et al. (2018)
WRKY	*Larrea tridentata*	Activation of stress defence	Li et al. (2020)
SPL	*Malus domestica*	Activation of SOD, POD	Ma et al. (2021)

CBF C-repeat binding factor, *HSF* heat shock protein, *NAC* no apical meristem, *WRKY* Trp, Arg, Lys, Tyr amino acid signature sequence-containing transcription factor 33, *SPL* squamosa promoter binding protein-like, *COR* cold regulated genes, *HSP* heat shock protein, *CAT* catalase, *APX* ascorbate, *SOD* superoxide dismutase, *POD* peroxidase

(Kumar and Chattopadhyay 2018) elucidated that in Arabidopsis, GSH regulates HSPs through the transcription factors BZIP10 (basic leucine zipper 10) and MYB21. The expression of HSP genes - *BiP3 (luminal binding protein 3)*, *HSP70B*, and *HSP90.1* was positively correlated with GSH level. Moreover, redox signals can be transferred to Hsp70-related signalling pathways when ROS directly induce oxidative cysteine alterations of Hsp70 members, which modify their structure and chaperone activity. These alterations also affect the relationships involving Hsp70 and their clients or cochaperones. However, ROS also cause various redox-related signalling pathways to become active, which in turn affects Hsp70 expression and activity in an indirect manner (Zhang et al. 2022). Another study demonstrates that HsfA1a is activated by heat stress, low/high pH, and hydrogen peroxide treatments, leading to its binding to *HSP18.2* and *HSP70* promoters, with activation and binding correlating with mRNA expression of target genes. This activation can be reversed by reducing agents, indicating a redox-dependent mechanism. Deletion of amino acid residues 48–74 in HsfA1a impairs its ability to sense pH and hydrogen peroxide stress, but not heat stress (Liu et al. 2013).

miRNA regulation of HS is also studied. For instance, miR160 modulates HSP gene expression in Arabidopsis to induce tolerance. miR160 overexpression improved seed germination and seedling survival under HS. The miR160 overexpressing transgenic plants revealed the regulation of *HSP17.6A*, *HSP17.6II*, *HSP21*, and *HSP70B* by targeting ARF10 (auxin response factor 10), ARF16, and ARF17 (Lin et al. 2018) (Table 4.1). In chickpea, miR156, miR166, miR319, miR171 and miR5213 were identified by psRNATarget server to be targets of *sHsps*, *Hsp70*, *Hsfs*, *Hsp90*, and *Hsp100* respectively (Juneja et al. 2023). Interestingly, Hsp70 is phytohormone-responsive as ABA, JA and ethylene are known to show negative effects on *HSP70* expression and the abundance of corresponding proteins. Whereas, the majority of *HSP70s* are upregulated by cytokinin, which also increases HSP70 abundances (Berka et al. 2022). The application of cytokinin was also beneficial in inducing HS tolerance in wheat via S-phase kinase protein1 (SKP1) as it showed positive induction after seed priming (Jaiswal et al. 2022).

The available literature focuses on the signalling cascade, however, the species-specific regulation, their induction and regulation to improve heat tolerance is scarce.

The understanding of the signal transduction and the potential targets in food crops such as wheat and rice is important to secure food for the growing population.

4.3 Drought

Drought is a severe abiotic stress affecting production yield annually. It can reduce grain size and weight, cause spike sterility, and impair cell growth in rice (Khan et al. 2021). Additionally, drought conditions can adversely affect rice morphological traits such as plant height, flag leaf length, flowering, grains per panicle, test weight, and spikelet fertility (Kumar et al. 2020). On a molecular level, drought stress can lead to lipid peroxidation, protein denaturation, nucleic acid damage, and accumulation of reactive oxygen species (ROS), causing oxidative damage (Nahar et al. 2016). To mitigate these effects, plants activate ROS scavenging mechanisms to restore cellular redox balance (Singh et al. 2024), modify their metabolic pathways, and transcription of genes (Gill and Tuteja 2010). This reorientation of gene expression is done by TFs. They either alone or in complex with other molecules can inhibit or activate the interaction of RNA polymerase which ultimately results in modifying the plant morphology to adapt to the altered environment (Rolly et al. 2020). For instance, the induction of bZIP4 (basic leucine zipper 4) TF is observed after drought stress in maize primary roots. Transgenic maize plants overexpressing *ZmbZIP4* exhibited longer primary roots, an adaptation which could help in absorbing more water under water-deficient conditions (Ma et al. 2018). Similarly, GmNAC085 from soybean is identified to be drought-inducible TF. When constitutively overexpressed in Arabidopsis, upregulation of *CSD1 (superoide dismutase [Cu-Zn] 1)*, *CAT1 (catalase 1)*, *sAPX* (stromal ascorbate peroxidase), *NCED3* (9-cis-epoxycarotenoid dioxygenase NCED3), *LEA14 (late embryogenesis abundant protein 14)*, *RD20 (responsive to desiccation 20)*, *RD29B (responsive to desiccation 29B)*, and *ANAC019* (abscisic-acid-responsive NAC domain-containing protein 19) was observed under water-deficient conditions in transgenic plants compared to the wild types (Nguyen et al. 2018). Another study confirmed the role of GmNAC085 (*Glycine max* NAC domain-containing protein 085) in the redox regulation of the cells under drought stress.

GmNAC085 induces drought stress tolerance by scavenging ROS in a GSH-dependent manner and regulating GSH biosynthesis (Fig. 4.1). Transgenic Arabidopsis plants overexpressing *35S:GmNAC085* had higher levels of reduced GSH and GSH/GSSG ratio in the leaves. The study concluded that the overexpression of *GmNAC085* triggered the expression of *GSH1* and *GSH2* genes as well as the activity of GR, imparting stress tolerance to transgenic plants by maintaining cellular redox balance in a GSH-dependent manner and a more reduced state even under stress (Nguyen et al. 2019a). Another NAC TF, SNAC3 in rice is identified to play a positive role in drought and heat stress tolerance by maintaining redox balance through efficient scavenging of ROS. Rice plants overexpressing *SNAC3* were more tolerant to high temperature, drought and oxidative stress whereas, downregulation of it by RNAi caused the plants to be sensitive to the above-mentioned abiotic stresses. The mechanism of action identified was that the overexpression of *SNAC3* led to the upregulation of ROS-scavenging enzymes' genes such as *APX*, *SOD*, and *CAT* as the promoter region of these genes were enriched with cis-elements to which the TF binds. Thus, it could elevate the expression of these genes for the efficient ROS detoxification under stress (Fang et al. 2015). Similarly, *AtbZIP62* was noted to influence the transcriptional control of *AtPYD1 (Arabidopsis thaliana Dihydropyrimidine dehydrogenase (NADP(+)))* while adjusting the redox status of cells and photosynthesis-related metabolism. Furthermore, it has been proposed that *AtbZIP62* favourably regulates the drought stress response in Arabidopsis (Rolly et al. 2020).

There are TFs which regulate hormonal-mediated stress response. For instance, GmNAC109, a NAC TF shows positive crosstalk with ABA signalling and induces drought stress tolerance. The ectopic expression of *GmNAC109* in Arabidopsis caused plants to exhibit higher recovery rates after water deficit conditions than the wild types. Moreover, the transgenic plants showed higher sensitivity to ABA as they had lower germination rates than the wild types at the same ABA concentration when applied exogenously. Along with the ABA signalling, the ectopic expression of the TF also caused a higher activity of the antioxidant enzymes including SOD and CAT under drought stress with lower H_2O_2 accumulation, thus maintaining cellular redox balance (Nguyen et al. 2019b) (Table 4.2). Similar results were obtained by Hoang et al. (2019) as the transgenic Arabidopsis plants had higher survival rates in soil-drying assay. On the contrary, there are TFs which negatively regulate ABA signalling and drought tolerance. For example, AtPPRT1 (*Arabidopsis thaliana Transmembrane Fragile-X-F-associated protein 1*), a putative C3HC4 zinc-finger ubiquitin E3 ligase is induced in an ABA-dependent manner. Plants overexpressing *AtPPRT1* were insensitive to ABA but sensitive to drought while the knockout plants had higher expression of three investigated stress-inducible genes with downregulation of two ABA hydrolysis genes. Thus, the protein (AtPPRT1) encoded by *AT1G68820* negatively regulates ABA signalling and affects drought tolerance (Pei et al. 2019).

The miRNAs are vital regulators of gene expression. They are involved in modulating the expression of various genes under drought stress and ultimately affect stress adaptiveness and tolerance as reviewed by (Riyazuddin et al. 2023). Moreover, these non-coding RNAs also target TFs such as SPL13 *(Squamosa promoter-binding-like protein 13)* and confer drought tolerance as studied in *Malus sieversii*. The transgenic plants overexpressing msi-miR156ab had better growth parameters and higher drought tolerance as the plants exhibited elevated activity of the antioxidant enzymes and proline content under stress. The underlying mechanism was that the miR156ab targets the negative regulator of auxin biosynthesis TF SPL13. The overexpression of miR156ab could induce drought stress tolerance by regulating auxin metabolism and antioxidant activity by inhibiting the TF SPL13 (Feng et al. 2023) (Table 4.1). Similarly, rice plant development and yield were enhanced by overexpressing a miR164b-resistant *OsNAC2* mutant gene, that generates transcripts that are not responsive to miR164b. The transgenic plants had higher ABA levels and drought tolerance than the wild-type in an ABA-dependent manner (Jiang et al. 2019).

There is a gap in our understanding of the spatio-temporal changes at the transcriptomic level that influence the drought stress response. The species- and crop-specific changes mediated by the crosstalk of hormones, metabolites, and TF regulated by miRNAs could be a potential field of study for developing stress-resilient crops and ultimately enhancing the productivity and yield per plant.

4.4 Salt Stress

In every living organism, high salinity causes salt stress, which initiates signal transduction pathways, resulting in complex response mechanisms. There are numerous redox-regulated transcription factors that greatly affect the given response. SQUAMOSA PROMOTER BINDING PROTEIN-LIKE (SPL) transcription factors are one of these main TFs. Due to their metal sulphur clusters, SPLs are sensitive to the redox state and redox-based modifications (Fig. 4.1). Besides salt stress, they also play an important role in other abiotic stresses, embryogenesis, flower morphogenesis, flowering time etc. (Wang and Wang 2015). Their elevated presence increases certain *POD*, *SOD* and *WRKY* expressions (Ma et al. 2021) (Table 4.2). WRKY TFs are also key elements in the alleviation of abiotic stress. They can regulate ABA biosynthesis, redox and ion homeostasis and more specifically, ethylene response pathways (Li et al. 2020) (Table 4.2). The changes in ABA levels lead to decreased activity of the negative regulator NAC089 (NAC domain containing protein 89) as well. NAC098, when active, binds to certain stomatal ascorbate peroxidase sequences, suppressing them. However, during stress, NAC098 is bound to the trans-Golgi network and the ER (endoplasmic reticulum) and only under reducing conditions does it split from it (Klein et al. 2012).

The importance of the aforementioned TFs is clear in transformant, mutant lines, consisting of overexpressing or knockout lines. *SPL13* overexpression mutant apple lines exhibited increased salt tolerance. Compared to non-overexpression lines, they showed higher RWC and chlorophyll content and decreased ROS levels. *WRKY100* overexpression lines also performed better under high salt conditions, with higher RWC and chlorophyll content and lower H_2O_2 and MDA (monodehydroascorbate) levels, than the wild type. *NAC089* knock-down plants grew just as well as the wild types but they were more tolerant to stress, producing more greenish shoots (Yang et al. 2014).

The stress reaction is fine-tuned by miRNAs, affecting the elements of the signal-transduction cascades. SPLs are targeted by the negative regulator miR156 (Cardon et al. 1997). miR156 is generally repressed by abiotic stress, enabling an increased SPL translation, although there are cases, where the opposite trend was observed. This seeming contradiction is because the change is determined by numerous factors, such as; species, age, environment, possible priming, survival strategy etc. WRKY is targeted by miR396. During stress, the miR396 level drops drastically, ending its suppressing effect on WRKY. The upregulated WRKY levels result in better stress response, enabling the plant to utilise defensive measures (Giacomelli et al. 2012; Chen et al. 2024) (Table 4.1). NAC sequences in general are regulated by miR164. During salt conditions, miR164 expression increases, keeping the *NAC089* expressions low, and enabling proper stomatal ascorbate peroxidase expression (Lu et al. 2017).

miR156 overexpression lowers salt resistance in several species, such as *Gossypium hirsutum*, or *Malus domestica*, by abolishing SPL translation (Yin et al. 2012; Ma et al. 2021). During salt stress, overexpression of miR396 leads to enhanced salt sensitivity, resulting in growth inhibition and the disruption of photosynthesis in *Oryza sativa*. On the other hand, in *Glycine max,* gene edited miR396 deficient lines exhibited better branching, greater biomass and grain yield during salinity stress (Chen et al. 2024). Finally, miR164 overexpression results in NAC inactivation, which elevates ethylene and H_2O_2 levels. Meanwhile, miR164 silenced lines exhibit an increased chlorophyll loss (Wang et al. 2023). This indicated, that a careful and balanced level of this miRNA is required for proper stress responses.

It is worth investigating the exact mechanism of all the elements in this intertwined cascade of signalization network. Revealing what aspects make a trait favourable, compared to another, in light of species, environment or even survival strategy could greatly improve our understanding of the complex underlying regulation.

4.5 Heavy Metal Stress

As abiotic stress factors, heavy metals significantly impact plant growth and development, affect crop yields, accumulate in plants and thus can enter the food chain (Berni et al. 2019). Heavy metals are metals and metalloids with an atomic density greater than 4 g/cm^3 or at least five times that of water (Nagajyoti et al. 2010). Certain heavy metals such as copper, zinc, manganese, nickel and iron are essential for plants, as they are involved in photosynthesis or play an important role in plant cells as constituents of enzymes (Berni et al. 2019). Copper plays an important role in photosynthesis as a primary electron donor in photosystem I and as a cofactor for several enzymes involved in the removal of reactive radicals (Nagajyoti et al. 2010). Many enzymes contain zinc, such as superoxide dismutase, alcohol dehydrogenase, or carbon anhydrase. However, zinc also plays a role in the assembly of many transcription factors, such as the RNA polymerase cofactor. Manganese plays an essential role in the reaction of enzymes and is needed for water splitting at photosystem II and for superoxide dismutase. Iron is an essential element in many metabolic processes and is important for all organisms. It is a component of several heme-containing and non-heme, iron-containing proteins. However, in too high concentrations, all heavy metals are toxic to plant physiological processes, inhibiting germination and growth, altering carbohydrate and protein content and interfering with nutrient uptake (Asati et al. 2016).

To reduce stress caused by heavy metals, plants have developed several mechanisms, such as increased extraction of metals from the root, preventing metals from entering the plant, chelating toxic metals to organic compounds and storing them in vacuoles, and binding metals in the cell wall to prevent entry (Mansoor et al. 2023). Low molecular weight thiols have a strong attraction for potentially harmful metals. GSTs are a superfamily of multifunctional, dimeric enzymes that induce the conjugation of the tripeptide glutathione to electrophilic xenobiotics, including heavy metals, and sequence this complex into the vacuole for detoxification (Dubey et al. 2014). In addition, glutathione is essential for heavy metal detoxification and is a precursor to phytochelatin synthesis (Mansoor et al. 2023). Phytochelatins are cysteine-rich polypeptides that bind heavy metals. The phytochelatins form complexes with the harmful metal ions and transport them into the vacuolum. As a result, these compounds protect plants from the harmful effects of heavy metals.

Many transcription factors are expressed under heavy metal stress. For example, cadmium and chrome modulate many WRKY transcription factors. In addition, many MYB family transcription factor-related genes have shown altered expression patterns under heavy metal stress (Dubey et al. 2014). The metal response transcription factor MTF-1 (metal regulatory transcription factor 1) also play an important role in cellular response and tolerance to heavy metal stress by stimulating genes responsible for the uptake, accumulation and detoxification of heavy metals and also stimulating the expression of genes involved in ROS scavenging (Kohli et al. 2017). In addition to these transcription factors, RING-H2 (really interesting new gene H2) finger protein, RING finger, CHY (cysteine, histidine, tyrosine) zinc finger domain-containing protein and bZIP (basic leucine zipper) transcription factors also showed metal-specific expression patterns in response to heavy metal stresses (Dubey et al. 2014).

The first effect of heavy metal stress is the accumulation of reactive oxygen species (ROS) (Mansoor et al. 2023). Heavy metals such as cadmium and copper can affect ROS accumulation in three ways: (1) they stimulate the expression of miR398 which inhibits copper-zinc-superoxide dismutase (Cu-Zn-SOD) and eventually leads to ROS accumulation; (2) excess heavy metal inhibits the regulatory effect of Ca^{2+} whereby copper-zinc-superoxide dismutase activity is inhibited and thus ROS levels increase, and (3) excess heavy metal leads to the activation of NADPH oxidases producing H_2O_2 (Kohli et al. 2017) (Fig. 4.1 and Table 4.1). The increased production of ROS leads to changes in some physiological processes, including the degradation of enzymes, proteins and amino acids and changes in cell structure. However, ROS are well known as secondary messengers in various cellular processes, including the acclimatization of cells to stress conditions. ROS regulate several signalling cascades, including calcium/calmodulin mobilisation and phosphorylation of proteins and certain genes. They also modulate protein kinases, various transcription factors and protein phosphatases. ROS induced by chromium metal stress in *Raphanus sativus* resulted in down-regulation of genes involved in putrescine biosynthesis, up-regulation of other genes involved in spermidine biosynthesis, and increased activity of superoxide dismutase, catalase, and glutathione reductase (Choudhary et al. 2012). In *Brassica napus,* chromium stress-induced

ROS increased the amount of reduced and oxidized glutathione, as well as glutathione reductase activity (Ali et al. 2014). Oxidative damage triggers specific signals that modify cell structure or gene expression in response to metal stress (Kohli et al. 2017). In Dubey et al. (2014) study, rice roots were treated with different concentrations of arsenate, cadmium, lead and chromium. Expression data showed that one of the main detoxification mechanisms is the up-regulation of specific cytochrome P450s under various heavy metal stresses. During chromium stress, a large number of ABC (ATP-binding cassette transporter) transporters were up-regulated, suggesting that they may be involved in the transport of chromium into the cell/vacuole. Some specific peroxidases are upregulated by heavy metal stress. These specific peroxidase family genes may play a key role in the enzymatic defence of plant cells by scavenging ROS under stress conditions.

In the transcriptional control of the response to heavy metal, the interaction between ROS and miRNAs has an important role (Gulyás et al. 2023). Several miRNAs regulate the response to Cu and Cd stress (Yang et al. 2022). Thus, the SPL7 transcription factor-responsive miR397, miR398, miR407, and miR857 control genes encoding Cu/Zn superoxide dismutase, laccase, and phycocyanin, which participate in the response to excess of Cu. Regarding the Cd stress, miR167, miR159, miR192, miR166 and miR268, control the Cd transport, miR156 and miR395 regulate Cd chelation and miR168, miR393, and miR1535 influences the hormone-related signalling pathways. Based on these results, both ROS and miRNAs, and their interactions are important in the reduction of the heavy metal-induced damage.

4.6 Conclusions, Perspectives

Under stress conditions, reactive oxygen species (ROS) have both beneficial and detrimental roles in plants. The initial rise in ROS serves as a primary defence mechanism, triggering signal transduction pathways that regulate the plant's response to stress. Subsequently, key regulators—including transcription factors (TFs), non-coding RNAs (such as miRNAs), and phytohormones—manage ROS levels by modulating antioxidant mechanisms. Transcription factors orchestrate the plant's stress response by activating stress-responsive genes, which play a critical role in mitigating cellular damage and enhancing the plant's overall resilience to adverse conditions. The miRNAs as noncoding RNAs, regulate gene expression by silencing at the transcription level either through cleavage or methylation and regulate the expression of genes involves in antioxidant mechanisms. Phytohormones also regulate seed germination, root and shoot growth, photosynthesis, and antioxidant systems under stress. The cross-talk of the trio, TFs-miRNA-phytohormones, becomes the key step to understand the involvement of key regulators under different stresses.

To understand the mechanism, omics study has emerged widely for in-depth visualization and interconnectivity at the cellular level to unravel defence strategies.

Through this functional genomics approaches many stress biological questions shall be answered with stress-specific mechanisms in species-specific manner that would help develop climate smart crops through conventional and new breeding technology platforms.

Acknowledgements This work was funded by the National Research, Development and Innovation Office (grants K131638 and TKP2021-NKTA-06). Prof. Sanjib Kumar Panda was supported by the MTA Distinguished Guest Scientist Fellowship Programme 2024 (grant VENDEGKUTATO2024-5) to visit the HUN-REN Centre for Agricultural Research. Kalpita Singh is supported by the the Stipendium Hungaricum program of the Tempus Public Foundation (SHE-079837-004/2022)

References

Ali B, Mwamba TM, Gill RA et al (2014) Improvement of element uptake and antioxidative defense in *Brassica napus* under lead stress by application of hydrogen sulfide. Plant Growth Regul 74:261–273. https://doi.org/10.1007/s10725-014-9917-9

Alvarez S, Hicks LM, Pandey S (2011) ABA-dependent and-independent G-protein signaling in Arabidopsis roots revealed through an iTRAQ proteomics approach. J Proteome Res 10:3107–3122

Alvarez S, Roy Choudhury S, Hicks LM, Pandey S (2013) Quantitative proteomics-based analysis supports a significant role of GTG proteins in regulation of ABA response in Arabidopsis roots. J Proteome Res 12:1487–1501

Argosubekti N (2020) A review of heat stress signaling in plants. In: IOP conference series: earth and environmental science. IOP Publishing, p 12041

Artlip T, McDermaid A, Ma Q, Wisniewski M (2019) Differential gene expression in non-transgenic and transgenic "M. 26" apple overexpressing a peach CBF gene during the transition from eco-dormancy to bud break. Hortic Res 6:86

Asati A, Pichhode M, Nikhil K (2016) Effect of heavy metals on plants. Int J Appl Innov Eng Manag 5:56–66. https://doi.org/10.13140/RG.2.2.27583.87204

Barrera-Figueroa BE, Gao L, Wu Z et al (2012) High throughput sequencing reveals novel and abiotic stress-regulated microRNAs in the inflorescences of rice. BMC Plant Biol 12:1–11

Batcho AA, Sarwar MB, Rashid B et al (2021) Heat shock protein gene identified from *Agave sisalana* (As HSP70) confers heat stress tolerance in transgenic cotton (*Gossypium hirsutum*). Theor Exp Plant Physiol 33:141–156

Berka M, Kopecká R, Berková V et al (2022) Regulation of heat shock proteins 70 and their role in plant immunity. J Exp Bot 73:1894–1909

Berni R, Luyckx M, Xu X et al (2019) Reactive oxygen species and heavy metal stress in plants: Impact on the cell wall and secondary metabolism. Environ Exp Bot 161:98–106. https://doi.org/10.1016/j.envexpbot.2018.10.017

Bittner A, van Buer J, Baier M (2020) Cold priming uncouples light-and cold-regulation of gene expression in *Arabidopsis thaliana*. BMC Plant Biol 20:1–19

Bourgine B, Guihur A (2021) Heat shock signaling in land plants: from plasma membrane sensing to the transcription of small heat shock proteins. Front Plant Sci 12:710801

Busch A, Deckena M, Almeida-Trapp M et al (2019) Mp TCP 1 controls cell proliferation and redox processes in *Marchantia polymorpha*. New Phytol 224:1627–1641

Cao J, Gulyás Z, Kalapos B et al (2019) Identification of a redox-dependent regulatory network of miRNAs and their targets in wheat. J Exp Bot 70:85–99. https://doi.org/10.1093/jxb/ery339

Cardon GH, Höhmann S, Nettesheim K et al (1997) Functional analysis of the *Arabidopsis thaliana* SBP-box gene SPL3: a novel gene involved in the floral transition. Plant J 12:367–377
Chae HB, Bae SB, Paeng SK et al (2023) The physiological role of thiol-based redox sensors in plant defense signaling. New Phytol 239:1203–1211. https://doi.org/10.1111/nph.19018
Chang J, Guo Y, Zhang Z et al (2020) CBF-responsive pathway and phytohormones are involved in melatonin-improved photosynthesis and redox homeostasis under aerial cold stress in watermelon. Acta Physiol Plant 42:1–12
Chen L, Luan Y, Zhai J (2015) Sp-miR396a-5p acts as a stress-responsive genes regulator by conferring tolerance to abiotic stresses and susceptibility to Phytophthora nicotianae infection in transgenic tobacco. Plant Cell Rep 34:2013–2025
Chen X, Jiang X, Sun X et al (2024) Gene editing and overexpression of soybean miR396a reveals its role in salinity tolerance and development. Crop J 12:1655–1665
Choudhary SP, Oral HV, Bhardwaj R et al (2012) Interaction of brassinosteroids and polyamines enhances copper stress tolerance in *Raphanus sativus*. J Exp Bot 63:5659–5675. https://doi.org/10.1093/jxb/ers219
Considine MJ, Foyer CH (2014) Redox regulation of plant development. Antioxid Redox Signal 21:1305–1326
de Larminat P (2016) Earth climate identification vs. anthropic global warming attribution. Annu Rev Control 42:114–125
Dong C-H, Pei H (2014) Over-expression of miR397 improves plant tolerance to cold stress in *Arabidopsis thaliana*. J Plant Biol 57:209–217
Dreyer A, Dietz K-J (2018) Reactive oxygen species and the redox-regulatory network in cold stress acclimation. Antioxidants 7:169
Dubey S, Shri M, Misra P et al (2014) Heavy metals induce oxidative stress and genome-wide modulation in transcriptome of rice root. Funct Integr Genomics 14:401–417. https://doi.org/10.1007/s10142-014-0361-8
Eremina M, Rozhon W, Poppenberger B (2016) Hormonal control of cold stress responses in plants. Cell Mol Life Sci 73:797–810
Fang Y, Liao K, Du H et al (2015) A stress-responsive NAC transcription factor SNAC3 confers heat and drought tolerance through modulation of reactive oxygen species in rice. J Exp Bot 66:6803–6817
Feng C, Zhang X, Du B et al (2023) MicroRNA156ab regulates apple plant growth and drought tolerance by targeting transcription factor MsSPL13. Plant Physiol 192:1836–1857
Fichman Y, Mittler R (2021) A systemic whole-plant change in redox levels accompanies the rapid systemic response to wounding. Plant Physiol 186:4–8. https://doi.org/10.1093/plphys/kiab022
Finkelstein R (2013) Abscisic acid synthesis and response. Arab Book/American Soc Plant Biol 11
Foyer CH, Kunert K (2024) The ascorbate-glutathione cycle coming of age. J Exp Bot 75:2682–2699. https://doi.org/10.1093/jxb/erae023
Ghassemian M, Lutes J, Chang H-S et al (2008) Abscisic acid-induced modulation of metabolic and redox control pathways in *Arabidopsis thaliana*. Phytochemistry 69:2899–2911
Giacomelli JI, Weigel D, Chan RL, Manavella PA (2012) Role of recently evolved miRNA regulation of sunflower HaWRKY6 in response to temperature damage. New Phytol 195:766–773
Gill SS, Tuteja N (2010) Reactive oxygen species and antioxidant machinery in abiotic stress tolerance in crop plants. Plant Physiol Biochem 48:909–930
Gilmour SJ, Sebolt AM, Salazar MP et al (2000) Overexpression of the Arabidopsis CBF3 transcriptional activator mimics multiple biochemical changes associated with cold acclimation. Plant Physiol 124:1854–1865
Gulyás Z, Székely A, Kulman K, Kocsy G (2023) Light-dependent regulatory interactions between the redox system and miRNAs and their biochemical and physiological effects in plants. Int J Mol Sci 24:8323

Hassan MU, Chattha MU, Khan I et al (2021) Heat stress in cultivated plants: Nature, impact, mechanisms, and mitigation strategies—a review. Plant Biosyst Int J Deal Asp Plant Biol 155: 211–234
He H, Van Breusegem F, Mhamdi A (2018) Redox-dependent control of nuclear transcription in plants. J Exp Bot 69:3359–3372
Hoang XLT, Nguyen NC, Nguyen Y-NH et al (2019) The soybean GmNAC019 transcription factor mediates drought tolerance in Arabidopsis in an abscisic acid-dependent manner. Int J Mol Sci 21:286
Hopkins BL, Neumann CA (2019) Redoxins as gatekeepers of the transcriptional oxidative stress response. Redox Biol 21:101104
Huang S, Zhou J, Gao L, Tang Y (2020) Plant miR397 and its functions. Funct Plant Biol 48:361–370
Jägermeyr J, Müller C, Ruane AC et al (2021) Climate impacts on global agriculture emerge earlier in new generation of climate and crop models. Nat Food 2:873–885
Jaiswal P, Sahi AN, Barthakur S (2022) Cytokinin seed priming mediated induction of terminal heat stress tolerance and expression profiling of SKP1 transcripts, a component of ubiquitin proteasome system in bread wheat (*Triticum aestivum* L.): a transgenerational analysis. Plant Growth Regul 98:259–280
Jiang D, Zhou L, Chen W et al (2019) Overexpression of a microRNA-targeted NAC transcription factor improves drought and salt tolerance in Rice via ABA-mediated pathways. Rice 12:1–11
Juneja S, Saini R, Adhikary A et al (2023) Drought priming evokes essential regulation of Hsp gene families, Hsfs and their related miRNAs and induces heat stress tolerance in chickpea. Plant Stress 10:100189
Khan MIR, Palakolanu SR, Chopra P et al (2021) Improving drought tolerance in rice: ensuring food security through multi-dimensional approaches. Physiol Plant 172:645–668
Klein P, Seidel T, Stöcker B, Dietz K-J (2012) The membrane-tethered transcription factor ANAC089 serves as redox-dependent suppressor of stromal ascorbate peroxidase gene expression. Front Plant Sci 3:247
Kocsy G, Tari I, Vanková R et al (2013) Redox control of plant growth and development. Plant Sci 211:77–91
Kohli SK, Handa N, Gautam V et al (2017) ROS signaling in plants under heavy metal stress BT. In: Khan MIR, Khan NA (eds) Reactive oxygen species and antioxidant systems in plants: role and regulation under abiotic stress. Springer, Singapore, pp 185–214
Kumar D, Chattopadhyay S (2018) Glutathione modulates the expression of heat shock proteins via the transcription factors BZIP10 and MYB21 in Arabidopsis. J Exp Bot 69:3729–3743
Kumar S, Dwivedi SK, Basu S et al (2020) Anatomical, agro-morphological and physiological changes in rice under cumulative and stage specific drought conditions prevailed in eastern region of India. Front Crop Res 245:107658
Laskowski MJ, Dreher KA, Gehring MA et al (2002) FQR1, a novel primary auxin-response gene, encodes a flavin mononucleotide-binding quinone reductase. Plant Physiol 128:578–590
Lee ES, Park JH, Wi SD et al (2021) Demyristoylation of the cytoplasmic redox protein trx-h2 is critical for inducing a rapid cold stress response in plants. Antioxidants 10:1287
Li T, Li H, Zhang Y-X, Liu J-Y (2011) Identification and analysis of seven H2O2-responsive miRNAs and 32 new miRNAs in the seedlings of rice (Oryza sativa L. ssp. indica). Nucleic Acids Res 39:2821–2833
Li M, Wang X, Cao Y et al (2013) Strength comparison between cold-inducible promoters of Arabidopsis cor15a and cor15b genes in potato and tobacco. Plant Physiol Biochem 71:77–86
Li W, Pang S, Lu Z, Jin B (2020) Function and mechanism of WRKY transcription factors in abiotic stress responses of plants. Plants 9:1515
Liang X, Luo G, Li W et al (2022) Overexpression of a *Malus baccata* CBF transcription factor gene, MbCBF1, increases cold and salinity tolerance in Arabidopsis thaliana. Plant Physiol Biochem 192:230–242

Lin J-S, Kuo C-C, Yang I-C et al (2018) MicroRNA160 modulates plant development and heat shock protein gene expression to mediate heat tolerance in Arabidopsis. Front Plant Sci 9:68
Liu Y, Ye N, Liu R et al (2010) H2O2 mediates the regulation of ABA catabolism and GA biosynthesis in Arabidopsis seed dormancy and germination. J Exp Bot 61:2979–2990
Liu Y, Zhang C, Chen J et al (2013) Arabidopsis heat shock factor HsfA1a directly senses heat stress, pH changes, and hydrogen peroxide via the engagement of redox state. Plant Physiol Biochem 64:92–98
Lu X, Dun H, Lian C et al (2017) The role of peu-miR164 and its target PeNAC genes in response to abiotic stress in *Populus euphratica*. Plant Physiol Biochem 115:418–438
Luo D, Raza A, Cheng Y et al (2023) Cloning and functional characterization of cold-inducible MYB-like 17 transcription factor in rapeseed (*Brassica napus* L.). Int J Mol Sci 24:9514
Ma H, Liu C, Li Z et al (2018) ZmbZIP4 contributes to stress resistance in maize by regulating ABA synthesis and root development. Plant Physiol 178:753–770
Ma Y, Xue H, Zhang F et al (2021) The miR156/SPL module regulates apple salt stress tolerance by activating MdWRKY100 expression. Plant Biotechnol J 19:311–323
Mansoor S, Ali A, Kour N et al (2023) Heavy metal induced oxidative stress mitigation and ROS scavenging in plants. Plants 12:1–17. https://doi.org/10.3390/plants12163003
Maryan KE, Lahiji HS, Farrokhi N et al (2021) Comparative leaf proteomics of *Brassica napus* genotypes with distinctive levels of early cold acclimation. Plant Mol Biol Report 39:317–334
Masson-Delmotte V, Zhai P, Pirani A et al (2021) Climate change 2021: the physical science basis. Contrib Work Gr I Sixth Assess Rep Intergov Panel Clim Chang 2:2391
Matsuo M, Johnson JM, Hieno A et al (2015) High REDOX RESPONSIVE TRANSCRIPTION FACTOR1 levels result in accumulation of reactive oxygen species in *Arabidopsis thaliana* shoots and roots. Mol Plant 8:1253–1273
Mittler R, Zandalinas SI, Fichman Y, Van Breusegem F (2022) Reactive oxygen species signalling in plant stress responses. Nat Rev Mol Cell Biol 23:663–679. https://doi.org/10.1038/s41580-022-00499-2
Miura K, Furumoto T (2013) Cold signaling and cold response in plants. Int J Mol Sci 14:5312–5337
Nagajyoti PC, Lee KD, Sreekanth TVM (2010) Heavy metals, occurrence and toxicity for plants: A review. Environ Chem Lett 8:199–216. https://doi.org/10.1007/s10311-010-0297-8
Nahar S, Kalita J, Sahoo L, Tanti B (2016) Morphophysiological and molecular effects of drought stress in rice. Ann Plant Sci 5:1409–1416
Nguyen KH, Mostofa MG, Li W et al (2018) The soybean transcription factor GmNAC085 enhances drought tolerance in Arabidopsis. Environ Exp Bot 151:12–20
Nguyen KH, Mostofa MG, Watanabe Y et al (2019a) Overexpression of GmNAC085 enhances drought tolerance in Arabidopsis by regulating glutathione biosynthesis, redox balance and glutathione-dependent detoxification of reactive oxygen species and methylglyoxal. Environ Exp Bot 161:242–254
Nguyen NC, Hoang XLT, Nguyen QT et al (2019b) Ectopic expression of *Glycine max* GmNAC109 enhances drought tolerance and ABA sensitivity in Arabidopsis. Biomolecules 9:714
Park S, Lee C, Doherty CJ et al (2015) Regulation of the Arabidopsis CBF regulon by a complex low-temperature regulatory network. Plant J 82:193–207
Pei L, Peng L, Wan X et al (2019) Expression pattern and function analysis of AtPPRT1, a novel negative regulator in ABA and drought stress responses in Arabidopsis. Int J Mol Sci 20:394
Ritonga FN, Ngatia JN, Wang Y et al (2021) AP2/ERF, an important cold stress-related transcription factor family in plants: a review. Physiol Mol Biol Plants:1–16
Riyazuddin R, Singh K, Iqbal N et al (2023) Unveiling the biosynthesis, mechanisms, and impacts of miRNAs in drought stress resilience in plants. Plant Physiol Biochem 202:107978
Rolly NK, Imran QM, Shahid M et al (2020) Drought-induced AtbZIP62 transcription factor regulates drought stress response in Arabidopsis. Plant Physiol Biochem 156:384–395

Schommer C, Bresso EG, Spinelli S V, Palatnik JF (2012) Role of microRNA miR319 in plant development. In: MicroRNAs plant development stress responses, pp 29–47
Shen J, Xing T, Yuan H et al (2013) Hydrogen sulfide improves drought tolerance in *Arabidopsis thaliana* by microRNA expressions. PLoS One 8:e77047
Shi Y, Ding Y, Yang S (2018) Molecular regulation of CBF signaling in cold acclimation. Trends Plant Sci 23:623–637
Shi X, Jiang F, Wen J, Wu Z (2019) Overexpression of Solanum habrochaites microRNA319d (sha-miR319d) confers chilling and heat stress tolerance in tomato (*S. lycopersicum*). BMC Plant Biol 19:1–17
Singh K, Gupta R, Shokat S et al (2024) Ascorbate, plant hormones and their interactions during plant responses to biotic stress. Physiol Plant 176:e14388
Soltész A, Smedley M, Vashegyi I et al (2013) Transgenic barley lines prove the involvement of TaCBF14 and TaCBF15 in the cold acclimation process and in frost tolerance. J Exp Bot 64: 1849–1862
Székely A, Gulyás Z, Balogh E et al (2023) Identification of ascorbate-and salicylate-responsive miRNAs and verification of the spectral control of miR395 in Arabidopsis. Physiol Plant 175: e14070
Tang Z, Zhang L, Xu C et al (2012) Uncovering small RNA-mediated responses to cold stress in a wheat thermosensitive genic male-sterile line by deep sequencing. Plant Physiol 159:721–738
Thomashow MF (1999) Plant cold acclimation: freezing tolerance genes and regulatory mechanisms. Annu Rev Plant Biol 50:571–599
Wang H, Wang H (2015) The miR156/SPL module, a regulatory hub and versatile toolbox, gears up crops for enhanced agronomic traits. Mol Plant 8:677–688
Wang S, Sun X, Hoshino Y et al (2014) MicroRNA319 positively regulates cold tolerance by targeting OsPCF6 and OsTCP21 in rice (*Oryza sativa* L.). PLoS One 9:e91357
Wang J, Mei J, Ren G (2019) Plant microRNAs: biogenesis, homeostasis, and degradation. Front Plant Sci 10:360
Wang Q, Shi H, Huang R et al (2021) AIR12 confers cold tolerance through regulation of the CBF cold response pathway and ascorbate homeostasis. Plant Cell Environ 44:1522–1533
Wang Z, Wang Q, Duan W et al (2023) Overexpression of Sly-miR164a in tomato decreases expression of NAC and delays pre-and postharvest ripening of fruit. Postharvest Biol Technol 195:112132
Wisniewski M, Norelli J, Bassett C et al (2011) Ectopic expression of a novel peach (*Prunus persica*) CBF transcription factor in apple (Malus× domestica) results in short-day induced dormancy and increased cold hardiness. Planta 233:971–983
Yang Z-T, Wang M-J, Sun L et al (2014) The membrane-associated transcription factor NAC089 controls ER-stress-induced programmed cell death in plants. PLoS Genet 10:e1004243
Yang Y, Liu J, Zhou X et al (2020) Identification of WRKY gene family and characterization of cold stress-responsive WRKY genes in eggplant. PeerJ 8:e8777
Yang Y, Huang J, Sun Q et al (2022) microRNAs: key players in plant response to metal toxicity. Int J Mol Sci 23:8642
Yin Z, Li Y, Yu J et al (2012) Difference in miRNA expression profiles between two cotton cultivars with distinct salt sensitivity. Mol Biol Rep 39:4961–4970
Zhang H, Gong W, Wu S, Perrett S (2022) Hsp70 in redox homeostasis. Cells 11:829
Zhou X, Wang G, Sutoh K et al (2008) Identification of cold-inducible microRNAs in plants by transcriptome analysis. Biochim Biophys Acta (BBA)-Gene Regul Mech 1779:780–788
Zhu Q, Zhang J, Gao X et al (2010) The Arabidopsis AP2/ERF transcription factor RAP2. 6 participates in ABA, salt and osmotic stress responses. Gene 457:1–12

Chapter 5
Genome Manipulation and Genetic Engineering of Sugarcane to Enhance Drought Tolerance

Malira Masoabi, Sandra J. Snyman, and Christell van der Vyver

Abstract Sugarcane is a vital crop plant with significant economic importance, cultivated globally for its role as a primary source of sucrose and increasingly recognized as a key feedstock for biofuel production. The growth of sugarcane hinges on precipitation and temperature, crucial factors for sustaining high yields. However, climate change has disrupted rainfall patterns, leading to frequent and prolonged drought periods that negatively affect sugarcane productivity. Consequently, the development of sugarcane with enhanced drought tolerance traits is crucial for ensuring sustainable production in the future. This review focuses on the biotechnological strategies employed to enhance drought tolerance in sugarcane plants. Through the application of diverse biotechnological tools such as mutagenic breeding and genetic modification techniques to alter the expression of key genes, followed by intense physiological and biochemical characterisation, our comprehension of the mechanisms governing the drought response in sugarcane has significantly advanced in recent years. This knowledge base is supplemented by transcriptomic and proteomic datasets, which primarily compare susceptible versus tolerant, or genetically altered versus wild-type sugarcane genotypes. Moreover, notable success has been achieved in introducing novel genetic diversity associated with enhanced drought tolerance, a feat often unattainable through conventional breeding methods.

Keywords Sugarcane · Saccharum · Drought · Biotechnology · Genetic modification · Mutation breeding

M. Masoabi · C. van der Vyver (✉)
Institute for Plant Biotechnology, Department of Genetics, University of Stellenbosch, Stellenbosch, South Africa
e-mail: cvdv@sun.ac.za

S. J. Snyman
South African Sugarcane Research Institute, Mount Edgecombe, South Africa

School of Life Sciences, University of KwaZulu-Natal, Durban, South Africa

S. K. Panda, M. A. Khan (eds.), *Plant Functional Genomics for Abiotic Stress Resilience*, Plant in Challenging Environments 7,
https://doi.org/10.1007/978-3-032-01704-8_5

5.1 Sugarcane, an Important Agricultural Crop

Sugarcane (*Saccharum* spp. hybrids) is a perennial grass characterised by tall fibrous stalks that accumulate high amounts of sucrose and lignocellulosic biomass, features that enable a highly productive source of sugar and bioethanol (Welbaum and Meinzer 1990; Waclawovsky et al. 2010). It is regarded as the world's largest crop by production quantity, annually yielding close to 1.9 billion tons of biomass. It is cultivated on approximately 26.9 million hectares of land across 106 countries and accounts for 76% of sugar produced worldwide (OECD-FAO Agricultural Outlook 2021–2030; Huang et al. 2020).

5.2 Environmental Requirements for Sugarcane Growth and Development

The perennial nature of sugarcane is determined by its growth cycle and the optimal outcome of each developmental phase is dependent on environmental conditions (Rajesh et al. 2016). Upon the planting of vegetative buds, sprouting is activated and completed during germination (Carr and Knox 2011). This is a 40-day process that requires temperatures of between 25 and 35 °C and ±300 mm/annum water in the form of rain or irrigation. After sprouting, the buds of primary shoots form secondary and tertiary shoots called tillers that subsequently form the stalk population. The tillering phase follows germination and lasts for approximately 120 days. This phase requires humidity of more than 40%, temperatures ranging between 25 and 30 °C, minimum rainfall (or irrigation) of ±550 mm/annum and adequate sunlight, particularly at the base of the plant. The availability of water is essential at this stage to ensure optimal crop yield (Delauney and Verma 1990; Zhao and Li 2015; Pandey et al. 2017; Ma et al. 2020). Once a stable tiller population is obtained, stalk elongation commences during the grand growth phase and yield build-up takes place. This takes up to 270 days and requires high humidity (80%) and temperatures of ±30 °C, as well as water of ±1000 mm/annum. Once the grand growth phase is completed, vegetative growth ceases and the ripening of the cane, where sucrose is accumulated, then proceeds. For optimal ripening, dry weather with temperatures between 30 and 35 °C and humidity of 40–65% are required. Unlike in the vegetative phases, limited supply of water is required for ripening because the assimilation of CO_2 for plant growth is deflected for sucrose production (Inman-Bamber 2004; Ferreira et al. 2017; Ma et al. 2020).

Since the development and growth of sugarcane is dependent on the availability of water, collective efforts to develop strategies to maintain crop productivity in water deficient conditions is an increasingly important research focus (Inman-Bamber and Smith 2005; Machado et al. 2009; Ferreira et al. 2017). This chapter discusses genetic manipulation and engineering strategies, along with molecular approaches used to enhance sugarcane's ability to withstand unfavourable environmental conditions, specifically drought, with the overall aim to sustain the production of this crop.

5.3 The Impact of Drought on Crop Functionality

Climate changes experienced across the globe have introduced irregular rainfall patterns, which have caused severe and frequent drought periods, consequently affecting growth and productivity of crops worldwide. Among the various environmental stresses brought about by climate change, water deficit is the most deleterious to sugarcane yield and productivity (Wang et al. 2003; Rampino et al. 2006; Ferreira et al. 2017). The extent to which drought impacts sugarcane is dependent on the developmental stage, the impact of high temperatures to the crop during these conditions as well as the degree and duration of the water deficit period (Zhao and Li 2015; Gomathi et al. 2020). Furthermore, when sugarcane is water stressed, it becomes more susceptible to additional pressures such as heat, salinity and pathogens and pests, which can occur either concurrently and/or successively with water restrictions (Smit and Singels 2006; Da Graça et al. 2010; Inman-Bamber et al. 2012; Pandey et al. 2017).

Drought stress in plants is characterised as a deficiency of water, both above- and below-ground, to a point where water sourced from the soil by roots cannot meet the atmospheric demand to regulate transpiration, resulting in the disruption of normal plant functionality and development (Levitt 1980). However, through evolution, plants such as sugarcane have developed C_4 metabolism, which adapts plants to the high photorespiration pressures that accompany water deficit stress by saturating CO_2 around the photosynthetic site within the bundle sheath cells to increase the likelihood of CO_2 fixation (Ehleringer et al. 1997; Sage 2004; Tipple and Pagani 2007). The C_4 metabolism is advantageous in water deficit conditions because it provides the crop with the ability to preserve soil and plant water by ensuring the sustenance of photosynthesis under low rates of incoming CO_2 resulting from limited stomatal conductance due to concurrent heat and water limiting stresses (Du et al. 1996; Endres et al. 2010; da Silva et al. 2012; Medeiros et al. 2013; Basnayake et al. 2015; Pandey et al. 2017; Ma et al. 2020; Verma et al. 2021).

5.4 Prominent Biochemical Adaptations in Sugarcane Upon Water Deficit Stress

Stimuli sensors influence modifications in biochemical and/or biological pathways which affect cellular, physiological and morphological responses of the plant (Levitt 1972). When experiencing drought, plants employ adaptation mechanisms to avoid and/or tolerate the effects of the stress (Blum 2005; Fang and Xiong 2015). Drought avoidance is the ability of a plant to preserve high water potential upon signals of water deficit (Levitt 1972) through: (i) rapid stomatal closure and leaf rolling (Blum 2005); (ii) enhancing soil water uptake by developing deep roots and enhancing water storage facilities aided by osmotic adjustment (OA) and cellular elasticity (Ogburn and Edwards 2010; Tardieu 2013); and (iii) accelerating/decelerating the conversion of energy between vegetative and reproductive growth to avoid complete termination of plant development, also referred to as drought escape (Mitra 2001; Luo 2010), but is somewhat restricted in sugarcane due to the crop's perennial development cycle. Drought avoidance facilitates the longevity of cell viability especially in moderate water deficit stress (Blum 2005; Cominelli et al. 2013; Kooyers 2015; Ma et al. 2020).

Drought tolerance, on the other hand, refers to the ability of a plant to sustain physiological activities crucial for the maintenance of the overall photosynthetic capacity under low tissue water potential resulting from severe water deficit stress (Blum 2005). Tolerance adaptations are comprised of the regulation of a wide range of genes and cascades of metabolic pathways that halt both vegetative and reproductive growth and instead further exacerbate OA through increased release of sugars (Fang and Xiong 2015). Unlike drought avoidance where water is still available to cells for viability, in tolerance adaptation, because of the scarcity of water in cells, they proportion solutes between intra- and extra-cellular spaces to maintain cellular hydration through osmosis (Fang and Xiong 2015). It is in this way that maintenance of relative stomatal conductance, transpiration and photosynthesis are sustained in plants under severe stress. As previously mentioned, transpiration is pivotal to a plant undergoing drought stress because it prevents heat stress whilst ensuring adequate photosynthesis capacity (Blum 2005; Cominelli et al. 2013; Ferreira et al. 2017; Nawae et al. 2020). In sugarcane particularly, moderate drought stress is favoured for the ripening phase because as vegetative growth is halted, sucrose accumulation is enhanced for OA, especially in the stalk tissues (Da Silva and Da Costa 2004; Soares et al. 2004; Silva et al. 2008; Ferreira et al. 2017; Sugiharto 2017).

Both avoidance and tolerance mechanisms serve to maintain some level of physiological functionality upon mild, moderate and severe water deficit in sugarcane (Ferreira et al. 2017). Prolonged water deficit however exhausts cell functionality and induces cell death, eventually leading to significant yield and production losses (Morison et al. 2008; Gosal et al. 2009; Ferreira et al. 2017). Therefore, understanding these mechanisms, together with the crop's water-related biological system(s), provides insights that can assist biotechnological approaches towards the protection of sugarcane yield against the effects of water deficit stress.

5.4.1 Activation of Phytohormones Responsive to Water Deficit Stimuli

Phytohormones including ABA, auxins, cytokinin (CK), gibberellins (GA), ethylene (ET), salicylic and jasmonic acids (SA and JA respectively) as well as brassinosteroids (BR) are known to regulate almost all plant and/or cell development aspects. They play an important role in regulating developmental aspects upon water deficit stress to oversee tolerance of crops during these conditions (Iqbal et al. 2022).

In general, ABA is involved in stomatal regulation, responsive gene upregulation, compatible osmolyte synthesis, photosynthetic upregulation, root growth and germination upon drought conditions (Seo and Koshiba 2011; Shi et al. 2015a, b). Auxins are involved in cell division, cell elongation, apical dominance and phyllotaxis to drive cellular adaptation responses as well as root branching to enhance water uptake in deep soils (Im Kim et al. 2013); Cytokinin is implicated in the growth of plant parts, the development of female gametes and embryos, as well as photomorphogenesis and leaf senescence (Zwack and Rashotte 2013; Terceros et al. 2020; Hönig et al. 2018; Wu et al. 2021). Gibberellins are key for cell elongation and enhancing the vegetative and reproductive stages in plants undergoing water deficit stress (Li et al. 2012a; Colebrook et al. 2014); Salicylic acid is involved in responses against progressive heat stress related to water deficit and stomatal closure (Munne-Bosch and Penuelas 2003; Dempsey et al. 2011); Ethylene has been shown to lower transpiration rates by limiting stomatal apertures and thinning the leaf cuticle (Zandalinas et al. 2016); the brassinosteroids are implicated in regulating growth to preserve yield in crops under drought stress (Li et al. 2012b); while JA is implicated in stomatal closure (De Ollas et al. 2013) and the increase of antioxidant activity during drought stress (Dong and Hwang 2014). All these plant growth regulators operate as a network, and it is their collaborative efforts that potentially drive the adaptation(s) of multiple traits that can preserve plant functionality under water deficit stress, making them prospective targets for modification to improve drought tolerance in plants (Iqbal et al. 2022). In sugarcane however, although phytohormones are associated with positive regulation of water deficit tolerance, research demonstrating the direct impact of hormones on water deficit tolerance is limited (Table 5.1).

5.4.2 The Importance of ABA in Response to Water Deficit Stimuli

When the atmospheric demands of transpiration outweigh the plant's capacity to absorb water from the soil, the plant's functional systems are signalled by a surge in abscisic acid (ABA), which in/effluxes through various tissues and cell types to trigger response adaptations (Himmelbach et al. 2003; Mittler and Blumwald 2015; Ferreira et al. 2017; Haworth et al. 2018a) (Fig. 5.1). The initial adaptation activated

Table 5.1 A summary of adaptations induced by changes in phytohormone levels in sugarcane exposed to water deficit

Genotype	Phytohormone and application	Drought experiment setup	Drought adaptations	Reference
ROC22 (tolerant)	ABA: Treated plants exposed to drought and foliar application of 15 μM ABA	Pot trial: 5-month-old plants exposed to 50% soil moisture content for 7 days	Treated plants: increased ABA, CAT, GPX and GR antioxidant enzyme activity; decreased MDA content and H_2O_2 accumulation	Li et al. (2014)
RB92579	Auxin: Before planting, plants were immersed in 1000 mg/L indolebutyric acid (IBA) solution	Pot trials: conducted 117 days after planting (DAP), 3 days without water (ww)	Treated plants: increased growth rates; biomass accumulation	Endres et al. (2016)
CoLk94184 CoPK05191	GA_3: Stressed plants treated with a solution of 35 ppm GA_3 applied on upper parts of the stalk using a controlled dispensing sprayer	Field trials: 120 DAP, withholding irrigation for 10 weeks	Both genotypes: increased RWC and chlorophyll content; expression of *pyroline-5-carboxylase synthetase (P5CS)* and *betaine-aldehyde dehydrogenase*; decreased proline content and antioxidant activity Only in CoPK05191: decreased proline oxidase/transporter	Tripathi et al. (2019)
XintaitangR22	Ethylene: Transformation with *tomato ethylene responsive factor 1* (TERF1) gene	Pot trials: 3-month-old plants, 21 days ww	Transgenic plants: increased proline, soluble sugars and glycine betaine; decreased MDA content and H_2O_2 accumulation	Rahman et al. (2021)
NCo310	Cytokinin: Exogenous application (foliar drench) with 100 μM N6-benzyladenine	Pot trials: 3-month-old treated plants, 20 days ww	Untreated plants: leaf senescence and wilting (10 days ww) Treated plants: maintained efficiency of PSII, high levels of chlorophyll and RWC (10 days ww), high CAT activity and proline content (20 days ww)	Masoabi et al. (2023a)

ww without water, *DAP* days after planting, *CAT* catalase, *GPX* glutathione peroxidase, *GR* glutathione reductase, *MDA* malondialdehyde, *PSII* photosystem II

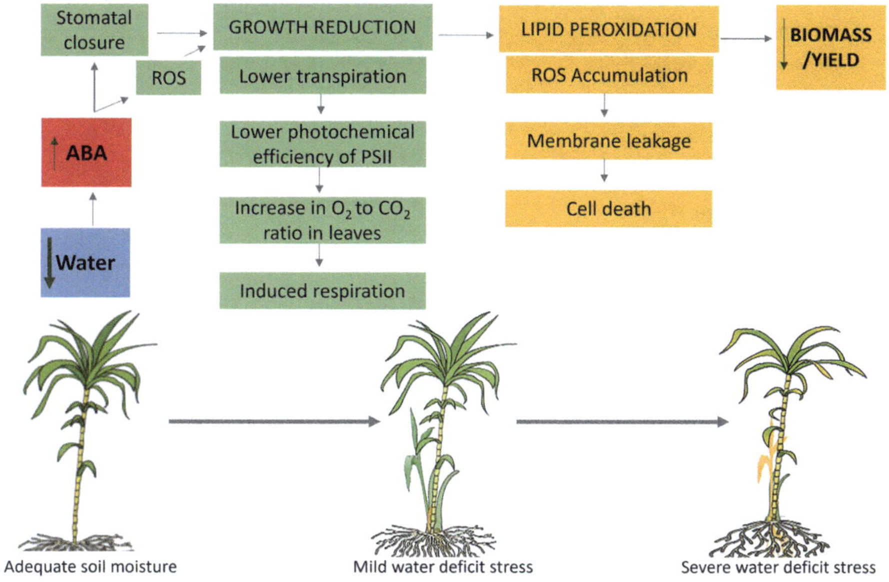

Fig. 5.1 Drought stress effects sugarcane. When water is limited, a surge of ABA occurs, leading to the production of ROS and restriction of stomatal conductivity. These initially trigger the restriction of growth to preserve high water potential in cells. Low water potential induced by severe drought further leads to peroxidation of lipids, and membrane damage due to overproduction of ROS, resulting in cell death, and decreased biomass and crop yield

by the ABA surge is the increase of reactive oxygen species (ROS) (Cruz de Carvalho 2008), which trigger the restriction of vegetative growth and reduction of stomatal conductance to limit the rate of plant water loss to the atmosphere (Davies and Zhang 1991; Mittler and Blumwald 2015; Marino et al. 2017; Ferreira et al. 2017; Haworth et al. 2018b) (Fig. 5.1). However, because C_4 mechanism causes accumulation of more CO_2 in leaves, the restriction of vegetative growth and stomatal conductance from the ABA surge allows sugarcane to use soil water efficiently, whilst the elevated CO_2 levels assist the crop to sustain photosynthesis, facilitating plant functionality under stress (Li et al. 2019).

5.4.3 *The Role of Reactive Oxygen Species During Water Deficit Stress*

The formation of ROS, as part of the electron transport chain (ETC) housed in the chloroplast that carry excited light energy to the bundle sheath cells for photosynthesis, is a normal and crucial occurrence in plants. ROS serve as signalling molecules in pathways that regulate growth and developmental processes, pathogen defence responses, stress hormone production and programmed cell death (Apel and

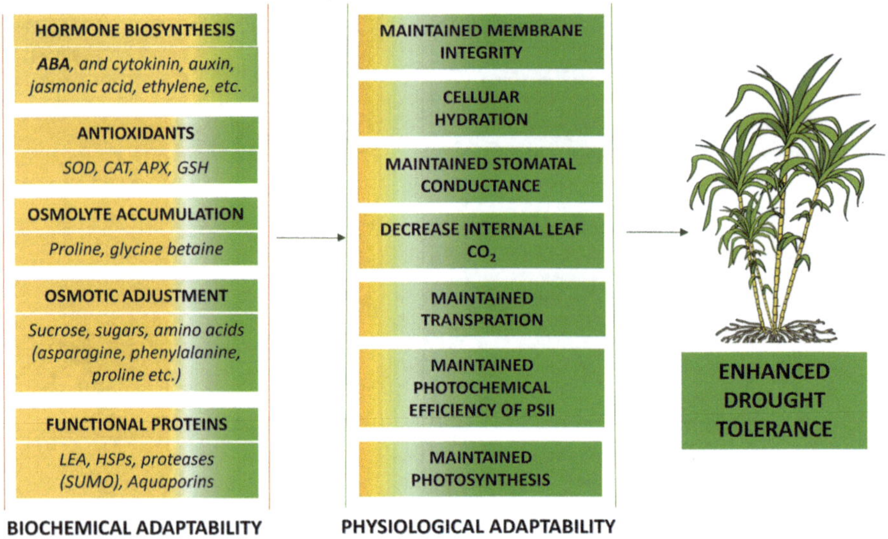

Fig. 5.2 Drought stress response in sugarcane. Upon water deficit stress, the ABA upsurge triggers the regulation of other hormones, the production of antioxidants, osmolytes, sugars, amino acids and responsive proteins to protect cell functionality. These then enhance physiological adaptations including maintained photosynthesis, cellular hydration, transpiration and stomatal conductance to enhance tolerance to water deficit

Hirt 2004; Tripathy and Oelmüller 2012; Luis 2015). Their occurrence is normally accompanied by a complex antioxidant system that functions to dissipate excess amount of the ROS upon completion of their functions (Tripathy and Oelmüller 2012; dos Santos and de Almeida Silva 2015; Kumar and Trivedi 2018) (Fig. 5.2). The antioxidant system encompasses two pathways, the non-enzymatic pathway, which is dependent on the activity of low molecular weight molecules of hydrophilic (ascorbic acid and reduced glutathione (GSH)) and lipophilic (α-tocophenol and carotenoids) nature, as well as the enzymatic pathway. This one involves the activity of enzymes such as superoxide dismutase (SOD), catalase (CAT), ascorbate peroxidase (APX), non-specific peroxidase (POX), glutathione reductase (GR), glutathione *S*-transferase (GST) and monodehydroascorbate reductase (MDHA) (Tripathy and Oelmüller 2012; dos Santos and de Almeida Silva 2015; Kumar and Trivedi 2018) (Fig. 5.2).

Should water deficit stress be prolonged to a point where ROS overproduction outweighs the capabilities of its dissipation machinery, the plants undergo oxidative stress, which results in the peroxidation of cell membrane lipids, DNA and proteins, ultimately inducing necrosis (Cruz de Carvalho 2008; Gill and Tuteja 2010; Tripathy and Oelmüller 2012; Ferreira et al. 2017; Hussain et al. 2018) (Fig. 5.1). The excitation of $^3O_2$ forms the reactive singlet oxygen ($^1O_2$) as well as other derivatives such as superoxide ($O^-{}_2$), hydrogen peroxide (H_2O_2) and the hydroxyl radical ($^{\bullet}OH$). The damage caused by photoinhibition is determined by the balance between the rate of its damage to the photosystem II (PSII) and the rate at which the PSII is repaired. Environmental stresses do not directly damage the PSII but rather supress its repair capabilities (Takahashi and Murata 2008; Pimentel 2014). The inability to neutralise a ROS surge and repair the damage it can cause, is where stresses such as drought and heat exacerbate cell death, especially in sugarcane cultivating areas where drought and heat stresses are prone to co-occur (Nawae et al. 2020).

Manimekalai et al. (2018) validated the impact of a ROS surge in sugarcane by applying exogenous H_2O_2 to sugarcane plants and demonstrated induced oxidative stress and its effect on stress-responsive genes. The study demonstrated an up-regulation of glutathione *S*-transferase, APX and CAT upon increased H_2O_2. In relation to drought, dos Santos and de Almeida Silva (2015) illustrated the significant role played by the increased activity of antioxidants in protecting cells from oxidative stress as well as maintaining the photochemical efficiency of the PSII along with the chlorophyll pigment upon water deficit stress. Similarly, several studies exposing a drought tolerant and a more susceptible sugarcane cultivar to water deficit stress, demonstrated the importance of antioxidants such as CAT, APX, GR and SOD, in enhancing drought tolerance in sugarcane (Sales et al. 2013; Boaretto et al. 2014; Vilela et al. 2017). Furthermore, upon overexpressing the sugarcane drought responsive 1 (*Scdr1*) gene in tobacco plants, Begcy et al. (2012) characterised the gene's ability to confer enhanced tolerance to drought, salinity and oxidative stresses by displays of reduced accumulation of ROS with increased photochemical efficiency in PSII and chlorophyll content. On a physiological level, the transpiration rate, net photosynthesis, stomatal conductance and the internal leaf CO_2 concentration, were less affected by the stresses in the *Scdr1* transgenic lines. Additionally, a recent study conducted by de Araujo et al. (2023), showed that sugarcane plants treated with selenium could potentially mitigate abiotic stress and enhance growth due to enhanced ROS scavenging by antioxidants like SOD and APX, as well as enhanced N assimilation leading to increased chlorophyll, CO_2 assimilation, stomatal conductance and internal CO_2 levels.

5.4.4 *The Role of Osmotic Adjustment During Water Deficit Stress*

In addition to the antioxidant system triggered by ROS overproduction upon prolonged or severe depletion of tissue water potential, the ABA and ROS surges encourage osmotic adjustment (OA) in cells (Lopes et al. 2011) (Fig. 5.2). The primary aim of this adaptation is to maintain cellular hydration using influx of small molecular weighed solutes including proline, glycine betaine (GB) and soluble carbohydrates, including sucrose, into cells to encourage the limited extracellular water to be taken up by cells through osmosis and retain cell turgidity and protection of relative cellular functionalities (Farooq et al. 2009; Zivcak et al. 2016; Yang et al. 2021) (Fig. 5.2).

The importance of maintaining the integrity of cell membranes in the survival of sugarcane during drought stress has been illustrated by Abbas et al. (2014) when they detected an increase in the concentration of biochemical markers of membrane protectants (proline and GB), lipid peroxidation (malondialdehyde) and antioxidants in sugarcane varieties tolerant to polyethylene glycol-induced osmotic stress. Furthermore, de Oliveira Maia Júnior et al. (2020) demonstrated the importance of the same membrane protectants in minimising the effects of drought stress in sugarcane. In that study, under drought, lines with higher proline and GB concentrations displayed lower electrolyte leakage and marginal declines in both the regulation of electron transport and efficiency of the PSII. All these minimised damages of the net photosynthetic apparatus and contributed to the maintenance of cell turgidity and protected the membrane structures of chloroplasts in the sugarcane plants with enhanced drought tolerance.

5.5 Prominent Sugarcane Physiological Adaptations to Water Deficit

5.5.1 Root Architecture Changes Upon Water Deficit

Upon the ABA surge, another primary response of plants, including sugarcane, is the development of deep and large root systems to enhance extraction of moisture from soils (Blum 2005; Zingaretti et al. 2012; Sah et al. 2016; Ferreira et al. 2017; Hussain et al. 2018). This characteristic is mediated by auxins, where on their own, or through interacting with other hormones such as ET, CK, SA, JA and BR, in addition to ABA, regulate root development (Zaman et al. 2016). In sugarcane, auxins have been demonstrated as a positive regulator for adventitious root formation (Li et al. 2020), which increased in flood-tolerant cultivar exposed to water logging (Bamrungrai et al. 2021). A water stress tolerant sugarcane genotype displaying higher root length and density with better field performance under water deficit stress has been reported by Endres et al. (2010). Furthermore, Kumar et al. (2014) and Raza et al. (2016) through transgenesis, were able to demonstrate that sugarcane plants surviving drought stress displayed, in addition to maintained photosynthesis, longer and profuse root systems than non-GM controls.

5.5.2 Regulation of Stomatal Conductance in Sugarcane During Water Deficit Stress

Sugarcane genotypes capable of maintaining higher stomatal conductance under severe and/or prolonged water scarcity can result in drought tolerance to ensure the maintenance of transpiration and photosynthetic capabilities and are directly proportional to biomass production (Flexas et al. 2012; Hadi et al. 2016; Ferreira et al. 2017; Hussain et al. 2020) (Table 5.2). In this context, Endres et al. (2010) conducted a drought observational study of sugarcane varieties grown in fields subject to different ambient conditions and showed that drought tolerant lines maintained high stomatal conductance under low leaf water potential, which resulted in high transpiration and photosynthetic rates. de Oliveira Maia Júnior et al. (2019) further validated these results when varieties that survived severe drought stress adapted physiological traits such as maintenance of relatively high stomatal conductance, internal CO_2 concentration as well as transpiration and photosynthesis rates, all of which resulted in marginal reduction of biomass (Table 5.2). Furthermore, Nawae et al. (2020) investigated the differential gene expression between drought-tolerant and -sensitive sugarcane varieties under drought stress and demonstrated that tolerance was accounted for by the up-regulation of genes involved in stomatal regulation, water retention, secondary metabolite biosynthesis, oxidative and osmotic protection as well as photosynthesis. Interestingly, Verma et al. (2020) applied silicon (Si) to sugarcane plants exposed to drought stress and examined the effect of the element in facilitating tolerance to drought stress. They showed that Si increased the stomatal conductivity as well as the photosynthesis and transpiration rates, which resulted in significant protection from drought stress and increased plant growth development and biomass/yield (Table 5.2).

Since the regulation of stomatal apertures is reliant on osmotic potential, and it controls leaf temperature, the maintenance of sustainable stomatal conductance can therefore lower the potential onset of concurrent stresses such as salinity (Hussain et al. 2020) and heat (Sharma et al. 2015; Hussain et al. 2019) upon drought stress (Table 5.2). This adaptation inhibits the overproduction of stress signalling molecules such as ABA and ROS, preserving cell viability and plant functionality in water deficit conditions. Although there is a lack of investigation in the simultaneous occurrence of numerous stresses, particularly in monocots, de Almeida Silva et al. (2011) showed that tolerant lines of four commercial sugarcane genotypes (two susceptible and two tolerant to drought stress) displayed higher photochemical efficiency of the PSII, chlorophyll and relative water content, with lower leaf temperature than the susceptible ones under water deficit. However, leaf temperature was significantly affected in the susceptible genotypes and this they attributed to stomatal closure (Table 5.2).

5.6 Sugarcane Genome Complexity Hampers Breeding Efforts to Enhance Drought Tolerance

Factors such as arable land and water availability are a growing concern for sustainable sugarcane production (Dlamini 2021). Although conventional breeding is a viable option to develop sugarcane with desirable traits, it requires approximately 11–15 years to select an improved sugarcane variety (Butterfield et al. 2001;

Table 5.2 A summary of drought induced biochemical, physiological and morphological adaptations reported in sugarcane exposed to water deficit stress

Sugarcane genotype	Method of stress induction	Main adaptation(s) in tolerant plants	Reference
RB72545, SP79-1011 and RB92579	Field trials: measurements during dry seasons at different fields along the Brazil coastal region	Higher root density Maintained high stomatal conductance, photosynthetic rate	Endres et al. (2010)
HOCP01-523 and TCP89-3505 (drought tolerant) and TCP87-3388 and HOCP93-776 (susceptible)	Field trials with water restriction: 0 for control, 45 days for moderate stress, and 90 days for severe stress	High stomatal conductance and efficiency of PSII Increased chlorophyll content Low leaf temperatures	de Almeida Silva et al. (2011)
RB92579, SP83-2847, SP81-3250, and IAC91-5155	Pot trials: irrigation suspended for 15 days	Increase in SOD, CAT and APX activity Maintained efficiency of PSII and high chlorophyll content	dos Santos and de Almeida Silva (2015)
Yuetang 93-159 (drought susceptible) and Yunzhe 05-51 (tolerant)	Pot trials with varying levels of soil moisture: 70 ± 5% field capacity (FC) controls; 50±5% FC; 30 ± 5% FC	Enhanced antioxidant enzyme activities Higher photosynthesis rate Enhanced osmolyte accumulation	Zhang et al. (2020)
Thirteen Pakistan genotypes	In vitro PEG-induced osmotic stress	Increased membrane protectants (proline and GB) Higher antioxidant activity	Abbas et al. (2014)
RB72910, RB99382, RB72454, RB92579, RB855536 and RB931011	Pot trials: 8 days without water (ww) Field trials: soil available water capacity (SAWC) at: 80–100% (controls); 40–60% and 0–20%	Increased proline and GB concentrations Lower electrolyte leakage Maintained high stomatal conductance and leaf internal CO_2 concentration	de Oliveira Maia Júnior et al. (2019, 2020)
PS01-12 (drought tolerant) and UT12 (susceptible)	Field trials: irrigation suspended for 14 days	Maintained high stomatal regulation Increased photosynthesis rate Increased antioxidant biosynthesis	Nawae et al. (2020)
GT 42	Pot trials: FC at 100–95% (control), 55–50% and 35–30%, with added Si	Increased stomatal conductance Increased	Verma et al. (2020)

(continued)

Table 5.2 (continued)

Sugarcane genotype	Method of stress induction	Main adaptation(s) in tolerant plants	Reference
		photosynthesis and transpiration rate	
Fourteen varieties	Pot trials: From 14 to 150 DAP treatments at 100% and 30% of FC	Increased stomatal conductance Increased leaf vapour pressure deficits Increased stem height	Chandiposha et al. (2023)

Mahlanza et al. 2013; Sugiharto 2017; Dlamini 2021). However, this timeframe is subject to several factors including the complexity of the trait(s) of interest, whether the trait of interest is endogenous to the crop (or its wild relatives) and/or the complexity of the crop's genome.

Modern commercial sugarcane varieties (*Saccharum* spp.) are complex hybrids derived from interspecific hybridisations between *S. officinarum* and *S. spontaneum* species (Screenivasan et al. 1987; D'Hont et al. 1998; Garsmeur et al. 2018). *Saccharum officinarum* (2n = 80; n = 10) has a high sucrose content and is domesticated from the wild species *S. robustum* (2n = 60–200) (Grivet et al. 2006; Garsmeur et al. 2018), while *S. spontaneum* is a wild species with various cytotypes and aneuploid forms (5n = 40 to 16n = 128; n = 8) (Panje and Babu 1960; D'Hont et al. 1998) that serves as the parental source for disease resistance, vigour and adaptability (Screenivasan et al. 1987; Garsmeur et al. 2018). In modern sugarcane varieties, high sucrose production was recovered by backcrossing the first hybrid to the sucrose-contributing parent and then enhanced through 2n chromosome transmission by the female *S. officinarum* (Bremer 1961). The resulting varieties are thus highly heterozygous and have 100–130 chromosomes, most of which are from *S. officinarum*, 10–20% from *S. spontaneum* and approximately 10% from interspecific recombinations (D'Hont et al. 1996; Piperidis et al. 2010).

The production of sugarcane genotypes with enhanced drought tolerance traits has proven to be more difficult than anticipated due to the limited knowledge of the crop's genetic makeup (Grivet and Arruda 2002; Souza et al. 2011). The sequencing of sugarcane hybrid cultivars was introduced to improve knowledge of the crop's genome for the development of complex traits. Riaño-Pachón and Mattiello (2017) was first to present a draft sequence of a model sugarcane cultivar used for genomic analyses in Brazil using bacterial artificial chromosomes (BAC) libraries and allowed the identification of 373,869 genes including homologous copies. Furthermore, Garsmeur et al. (2018) exploited the co-linearity of sugarcane and sorghum to assemble a 382 Mb BAC-based sugarcane monoploid genome sequence. The reference was strategically extracted from gene-rich, repeat-poor and recombinationally active genome regions of the sorghum genome. This assembly serves as a base reference for the ongoing sequencing and assembling of the 10 Gb polyploid version (Garsmeur et al. 2018; Thirugnanasambandam et al. 2018; D'Hont et al. 2023) and as an essential template for sequencing-related studies including genotyping by sequencing, whole genome sequencing and RNA-Seq to explore homo(eo)logous allelic variations. Furthermore, the annotation of the sequence will also facilitate the identification of candidate genes/loci in regions found to be associated with targeted phenotypic traits (Garsmeur et al. 2018). These will allow the unravelling of complications hampering the efficiency of genetic engineering to improve traits related to drought tolerance in sugarcane.

Adding to the ground-breaking sequencing of the modern sugarcane monoploid genome, Zhang et al. (2018) sequenced the haploid *S. spontaneum* genome and facilitated the assembly of 32 pseudo-chromosomes comprising 8 homologous groups of 4 members each, bearing 35,525 genes with defined alleles. This study illuminates the hereditary blueprint and evolution of modern sugarcane. Furthermore, because *S. spontaneum* is the parental bearer of disease resistance, vigour and adaptability in modern cultivars, this research forms a pivotal reference for all research relating to climate change and its effects on the crop. In these terms, Trujillo-Montenegro et al. (2021) performed RNA-Seq in root tissue of drought tolerant and sensitive sugarcane genotypes subjected to drought stress and revealed 11,264 differentially expressed genes, from which 2832 were exclusive to the tolerant genotype (Trujillo-Montenegro et al. 2021). Additionally, Trujillo-Montenegro et al. (2021) also assembled a 903.2 Mbp monoploid genome of a Columbian sugarcane hybrid and generated 10 pseudo-chromosomes, 44 scaffolds and 35,035 contigs. They also annotated 63,724 protein coding genes and performed a reconstruction and comparative analyses of the sucrose metabolism pathway to investigate the genetic drivers of sucrose synthesis and metabolism in sugarcane.

In addition to the complexity of its genome, another factor that hinders conventional breeding of sugarcane is the unstable nature of its pollen's viability (Brett 1950). The chromosomal structural differences between the parental species can result in meiotic instabilities that produce aneuploid gametes, increasing the probability of sterile pollen production (Rutherford et al. 2014; Dlamini 2021). However, in cases where pollen is harvested, sugarcane pollen rapidly desiccates following dehiscence, has a half-life of 12 min, and is viable for 35 min under natural conditions. This decreases the pollen's fertility attributes and poses a challenge to conventional breeding, particularly when breeding for complex traits (Sugiharto 2017).

It is evident that the complexity of the trait(s) required for drought tolerance and of the sugarcane genome, coupled with long selection periods, lack of genetic diversity within the selection pool and variable viable pollen production, greatly delays the efforts of conventional breeding in developing commercial varieties with enhanced drought tolerance. However, the growing demand for this crop and the products derived from its cultivation drives the need to utilise and/or develop alternative approaches to ensure its sustainable production under challenging environmental conditions.

5.7 Biotechnology Advances for Drought Tolerance in Sugarcane

The water deficit stress response in sugarcane is considered a complex process due to its multigenic impact; however, biotechnological approaches such as induced mutagenesis and genetic engineering show promise to achieve desirable traits in this crop.

5.7.1 *Mutagenesis*

Mutagenesis is a biotechnological approach that emerged in parallel to conventional breeding and employs either physical or chemical treatments to induce novel heritable genetic variations in a plant's genome that result in desirable trait (s) (Oladosu et al. 2016). Physical mutagenesis (e.g. irradiation) can be administered either with non-ionising agents (ultra-violet (UV) light), ionising agents (X and gamma rays, alpha and beta rays) or with high linear energy transfer. The induction of physical mutations normally results in large-scale deletions, high chromosomal aberrations, and high levels of non-repairable DNA damage in plant genomes (Parry et al. 2009). On the other hand, chemically induced mutations in plants are administered using mutagens such as ethylmethane sulfonate (EMS), methylmethane sulfonate (MMS), hydrogen fluoride (HF), sodium azide (NaN_3), N-methyl-N-nitrosourea (MNU) and hydroxylamine (Oladosu et al. 2016). These agents randomly induce mostly single nucleotide DNA mutations at high frequencies and result in low mortalities. Additionally, they are easy to administer and do not require specialised expensive equipment (Koch et al. 2012).

Recently, studies focusing on epigenetic mutations induced by chemical mutagens have also been reported (Kakoulidou et al. 2021). Epigenetic gene regulation is a procedure by which genome expression is regulated without altering the genome sequence mostly through methylation or demethylation of DNA (Tonosaki et al. 2022). Particularly in terms of abiotic stress tolerance, manipulating DNA methylation patterns can be useful. This can create novel binding of transcription factors (TF) or alter TF binding in promoter regions, in addition to enhancing gene splicing capacity of the plant to handle stress via transcriptome plasticity (Brenet et al. 2011; Wang et al. 2016). Mutagenesis however needs to be coupled with selection methods to select mutants displaying the desired characteristics (Koetle et al. 2022).

In sugarcane, mutations are generally induced in callus cells, which are cultured on media containing agent(s) that lower water potential in extracellular regions and induce cellular osmotic stress in vitro (Masoabi et al. 2017; Khalil et al. 2018; Dlamini et al. 2020; Hartati et al. 2020; Weksanthia et al. 2021). Plants that survive this selection are then exposed to controlled drought trials in conditions such as glasshouse and in pots (Masoabi et al. 2017; Khalil et al. 2018; Koetle et al. 2022). These trials further select plants for water deficit tolerance because water deficit impacts various traits additional to osmotic stress such as photosynthetic capacity, transpiration rate, stomatal conductivity, production of ROS and their antioxidant systems as well as membrane integrity. These controlled trials indicate promising lines to further induce drought stress in-field and validate tolerance traits in uncontrolled environments (Masoabi et al. 2023a, b, c). In addition, other sugarcane mutation studies, particularly those that employ EMS, further use high-throughput technologies such as Targeting of Induced Local Lesions IN Genomes (TILLING) to detect chemically induced point mutations in genomes using reverse genetics (Khalil et al. 2018)

A review of the FAO/International Atomic Energy Agency's (IAEA) Mutant Varieties Database (http://mvgs.iaea.org/Search.%20aspx; accessed on 10 May 2023), reported only 13 induced sugarcane mutant accessions. These approved sugarcane mutant varieties comprise eight lines resistant to biotic stresses and five lines contributing to the improvement of yield, while none have been reported for tolerance to abiotic stresses. Nevertheless, the FAO/IAEA mutation database reports grass species where mutation breeding has resulted in drought tolerant genotypes and these include maize (three genotypes), rice (15 genotypes), barley (9 genotypes), sorghum (two genotypes) and the polyploid crop, wheat, with 28 genotypes. However, a number of studies have reported the successful use of mutagenesis to develop sugarcane genotypes with enhanced drought tolerance. In terms of drought stress, Masoabi et al. (2017) used EMS to induce mutations in sugarcane callus cells, which were selected in vitro on growth media with polyethylene glycol (PEG) 6000 to identify cells that can regenerate under osmotic stress. The regenerated mutant plants were subjected to preliminary ex vitro drought pot trials to select plants with enhanced tolerance to water deficit. The study identified seven lines that outlived the control, while one was able to recover from the stress. Khalil et al. (2018) replicated the concept of this procedure and supplemented it with physiological, molecular and enzymatic analyses of the mutant lines that outlived the control in the ex-vitro drought trials. They were able to recover mutant lines promising for drought resistance breeding due to their better adaptation to drought stress. Furthermore, characterisation of an EMS-derived mutant line, M9.2 with beneficial drought tolerance traits for potential exploitation in sugarcane breeding programs was conducted (Masoabi et al. 2023a, b, c). Furthermore, Koetle et al. (2022) produced ten sugarcane epimutants using 5-Azacytidine demethylation, followed by re-methylation in the presence of imidacloprid (IMI). Heat stress and PEG were used as selection pressures to identify drought tolerant epimutants. The epimutant lines displayed increased photosynthetic performance resulting in increased stalk height, better maintained RWC and a stay-green phenotype when exposed to drought. These epimutant sugarcane lines were then further analysed using in vitro chimera dissolution and near-infrared spectroscopy (NIRS). From this analysis, they identify two epilines with persistent tolerant of osmotic, heat and drought stresses in pot-based trials (Koetle et al. 2023).

Traits including number of stalks or millable cane; stalk diameter, height and weight; length of mature internodes; bud prominence; leaf colour, length and width as well as sucrose yield, are routinely used to assess yield in sugarcane (Akhtar et al. 2006; Nikam et al. 2015; Brasileiro et al. 2016; Barreto et al. 2021a, b). Khan et al.

(2000) applied radiation to sugarcane and demonstrated improvement of plant/stalk height, number of tillers/stalks, cane thickness/diameter and number of green leaves. Yasmeen et al. (2020) investigated the influence of gamma rays on stomata and assessed the effects of this on sugarcane yield traits. This study showed that stomatal length, width, and density increased at low radiation dosages, while sugar-quality attributes such as Brix%, fibre, sucrose, sugar and biomass yield as well as sugar purity, gradually increased with radiation doses. Furthermore, biomass contributing traits including plant height, number of tillers, tiller weight, stalk thickness, number of internodes, internode length, leaf length and width, directly correlated with the stomatal length, width and density. In the context of abiotic stresses, Nikam et al. (2015) exposed sugarcane callus to gamma radiation and exposed it to salt stress to demonstrate the importance of environment on the crop's yield-traits. Upon maturation, mutant clones exhibited improved sugar yield and number of millable cane stalks.

5.7.2 New Mutation Technology

Genome editing is the latest genetic modification technique and harbours the capacity to simultaneously modify the expression of multiple targeted genes (Li et al. 2013; Mao et al. 2013; Mohan 2016; Holme et al. 2019). In genome editing, mutations are induced by insertions, deletions and/or replacements of DNA in the genome of an organism using engineered nucleases. These nucleases are designed to create double strand breaks at specific sites of the genome. The induced breaks are then repaired though non-homologous end joining (NHEJ) or homologous recombination (HR), resulting in desired mutations. There are currently four widely used families of engineered nucleases available, the meganucleases, zinc finger nucleases (ZFN), transcription activator-like effector nucleases (TALEN) and the type II clustered regularly interspaced short palindromic repeat/CRISPR-associated 9 (CRISPR/Cas9) system (Mohan 2016; Shi et al. 2017). Although all three techniques are used, the latter is the latest development and far more advantageous owing to three important attributes.

The ZFN and TALEN have repeated tandems in their DNA-binding domains that can be engineered to recognise specific DNA sequences and the resulting chimeric nucleases are then guided to their genomic target sequences to generate the double-strand breaks. For each target site, a new ZFN/TALEN chimeric protein also needs to be engineered to recognise its respective target (Mao et al. 2013). This has been recorded as a major hurdle in the use of the two systems. Nevertheless, Mohan et al. (2021) elegantly elaborated on the potential impact of genome editing in sugarcane in the context of the crop's biofuel production. As it currently stands, the recalcitrance of lignin digestion has been the greatest challenge in harnessing lignocellulosic biomass for fuel production. However, the review highlights important studies that provide the importance and feasibility of genome editing to counteract this challenge. In fact, they highlight the first sugarcane genome editing report of Jung and Altpeter (2016), where they used TALEN to knockout the *COMT* gene in sugarcane which resulted in 99% mutation frequency in transgenic plants representing 29–32% reduction of lignin content when compared with the control plants. Under field trials when Kannan et al. (2018) observed the agronomic characteristics of these lines, the mutant lines exhibited a reduction of 19.7% lignin without compromising biomass production, therefore demonstrating the relevance of these techniques, regardless of the incomplete genomic knowledge of the crop.

The CRISPR system uses a simple and efficient principle, where a 20-nucleotide sequence in an engineered single guide (sg) RNA molecule determines specificity of the target DNA strand, and the resulting double-stranded hybrid will then be cleaved by the Cas9 enzyme (the Cas9 nuclease degrades DNA that matches the sgRNA), where the NHEJ/HR will then intervene to create mutations (Mohan 2016). The efficiency and simplicity of this principle makes it easier for the mutation of multiple genes simultaneously, which has been demonstrated in *Arabidopsis* (Li et al. 2013). This advantage is important because of its potential to reduce the time it can take to alter traits in crops with complex genomes. Lastly, ZFN and TALEN function as dimers and their vector construction and plant transformation processes are highly intricate compared with the simple RNA/DNA hybridisation (Mao et al. 2013; Mohan 2016).

The CRISPR technique has been demonstrated to be applicable in many plants, as summarised in recent reviews (Cardi et al. 2023; Chen et al. 2024). There is however one study that has been reported for the CRISPR system in monocots targeting a drought related gene. In 2015, Shi et al. overexpressed the Auxin Regulated Gene involved in Organ Size (ARGOS)8—a negative regulator of ethylene response gene in *Arabidopsis* and maize. Both cases demonstrated sensitivity to ethylene, which subsequently conferred their improved tolerance to drought stress. Following this study, aiming to create novel ARGOS8 variants with other beneficial traits in addition to their improved drought tolerance nature, Shi et al. (2017) designed the system to replace the promoter of the ARGOS8 gene with the G0/G1 Switch gene (GOS) 2 promoter, both native to maize, to produce ubiquitous and elevated expression of the gene across multiple tissues at different developmental stages. The modified plants displayed increased grain yield under drought stress conditions, and there were no yield penalties when tested under well-watered conditions.

The knowledge of genomes and/or their functionality are paramount requirements for CRISPR as designing the sgRNA to specify the target DNA to be mutated is necessary. Despite having minimal information on the sugarcane genome, the CRISPR technique has been explored for herbicide tolerance, where Oz et al. (2021) used the technology to mediate multi-allelic mutations in the crop that were able to confer tolerance to selected herbicides. Furthermore, Eid et al. (2021) was also able to edit 49 allele copies of the magnesium chetalase subunit (MgCh) in sugarcane using this technique. In this case, the aim of the study was to validify the precision of the technique in a complex genome that is only partially characterised. Because MgCh is involved in the biosynthesis of chlorophyll, the mutated plants demonstrated decreased chlorophyll and yellowing of leaves, yet still maintained normal growth. Furthermore, very recently Laksana et al. (2024) reported the successful use of the CRISPR/Cas9 system for the site-directed mutation of the *SoLIM* transcription factor (LIM domain TF gene) to reduce lignin content in sugarcane. Additionally, Brant et al. (2024) edited the LIGULELESS1 gene in sugarcane, which influence the leaf inclination angle to manipulate biomass yield. All the abovementioned cases demonstrate the feasibility of the technique in sugarcane, despite its genomic complexity.

Although the application of mutagenesis is limited in sugarcane, particularly in the context of drought tolerance, records of drought tolerance conferred in other plant species through mutagenesis on the FAO/IAEA Mutant Varieties Database and other related research as elaborated above, demonstrate the potential impact of mutagenesis for the development of this trait in sugarcane.

5.7.3 Genetic Engineering

An increasing number of studies on the successful enhancement of water deficit stress tolerance in sugarcane through the modification of target genes have been reported. However, although single gene modifications have demonstrated drought tolerant traits in transgenic sugarcane, approaches that can simultaneously target and modify the expression of multiple genes in genomes may augment the impact of the approach by collectively improving numerous traits in crops. In these terms, recent studies have focussed on the transcriptome and proteome to unravel the connection pathway(s) from the receipt of the stress signal to the differentially regulated transcriptome and its impact on the functionality of the encoded proteome, all the way to the morphological adaption(s) of the plant to survive the changing environment (Table 5.3). These holistic perspectives, rather than the impact of single genes, have not only improved understanding of the genetic network(s) involved in the drought stress response of this crop, but have also been reliable sources to identify target genes for GM approaches in the development of improved genotypes.

The fundamental reference for all sugarcane transcriptomic research was established by the Brazilian Sugarcane Expressed Sequence Tags (ESTs) Sequencing Project (SUCEST) in 2003. The project sequenced over 238,000 ESTs sourced from different tissues and cultivars. It generated 43,141 assembled sequences which were estimated to represent 33,600 unique genes (Vettore et al. 2003). This enabled the evaluation of differential transcript and protein expression profiles of genotypes exposed to drought stress and has accentuated the importance of viewing both the transcriptome and proteome as networks. Currently, the concept of identifying differentially expressed gene transcripts enhanced in sugarcane during drought stress includes the induction of water deficit stress in sugarcane, then the expressed genes are detected and/or isolated for identification (using strategies such as enrichment). Identification of genes encompasses RNA sequencing (RNA-Seq) and their expression is validated by quantitative real-time PCR (qRT-PCR). Through this channel, transcriptome analyses on sugarcane genotypes contrasting in their tolerance to water deficit show differences in expression of genes mainly involved in cellular

Table 5.3 Differentially expressed genes among sugarcane genotypes exposed to water deficit stress identified by transcriptome analysis. A subset of genes identified, as highlighted by the independent studies and which display a high level of differential expression, are listed in the table

Genotypes	Experimental setup	Differentially expressed genes (DEG)	Gene ontology/KEGG pathways—DEG	References
SP90-1638 (drought susceptible)	Pot trial: 90-day-old plants, 24, 72 and 120 h without water (ww) cDNA SUCEST clones amplified and spotted; microarray analysis	Upregulated—Stressed: MYB and WRKY TFs, low temperature induced (LT1) proteins, AP2/DREB2 TF, delta-12 oleate desaturase, S-adenosylmethionine decarboxylase, PP2C-like protein phosphatase, fatty acid desaturase 2, nitrilases, auxin-binding protein, auxin response factor, AUX-IAA TF, cane Osmotic stress-activated protein kinases-2/8/13/14 Down regulated—stressed: TF from the AUX/IAA category	Selected 1545 genes for array—linked to signal transduction components	Rocha et al. (2007)
SP83-5073 (tolerant) SP90-1638 (susceptible)	Pot trials: 60-day-old plants, 2, 8 and 10 days ww Macro-array analysis	Upregulated—Tolerant/Severe stress: Cinnamoyl coA reductase protein, PPR repeat containing protein-like, HSP17, S-adenosylmethionine decarboxylase, SHR5-receptor-like kinase, receptor-mediated endocytosis 1 isoform, IS1 transposase, wound induced protein Upregulated—Susceptible/Severe stress: Vacuolar proton-ATPase, serine/theonine protein kinase, PP2C, NLI interacting factor family protein-like, bifunction nuclease, ERD4, MYB TF-like, glycine-rich	3575 ESTs from sugarcane leaf libraries. Metabolic functions in amino acid and DNA metabolism; bioenergetics and photosynthesis; signal transduction; mobile elements; stress response; transport	Rodrigues et al. (2009, 2011)

		RNA binding protein, ribosomal protein L7, proteinase inhibitor; Phosphoribosylanthranilate transferase		
Q117	Pot trial: 5-month-old plants, 15 days ww; stem and leaf tissue RT-qPCR analysis	Upregulated—Stress/Older internodes: LEA, dehydrin, proline biosynthesis, OAT, proline oxidase, trehalase, SPDS, asparagine synthase; TFs of bZIP, MYB and ERF families; sugar transporters PST5/7/2a/2b Downregulated—Stressed/Older internodes: Sucrose transporter ShSUT1	Gene subset (literature based): osmoprotection, water and ion movement, chaperones, sugar transporters, TF, stress-related genes	Iskandar et al. (2011)
CTC15, CTC6, SP83-2847, SP83-5073 (tolerant) CTC9, CTC13, SP90-3414, SP90-1638 (susceptible)	Pot trial: 24 h ww; pooled 4× tolerant, 4× susceptible genotypes; Root tissue SuperSAGE libraries, Solexa sequencing (Illumina)	Upregulated—Tolerant: Glycine-rich RNA binding protein (GRP) superfamily, CoA-thioester hydrolase, GST, SHMT, PPIase, Viviparous14, BCATs, AOS, Na+/J+ antiporter, glutathione peroxidase, SAPK1, P5CS, ZIM motif, chromatin-remodelling factor CHD3, APETALA 2/ethylene response element binding protein, AP2/ERF domain containing TF	Only annotated ESTs (63.5%): Water deprivation, heat, cold, osmotic stress, oxidative stress, ABA stimulus, jasmonic acid stimulus	Kido et al. (2012)
Co-740 (tolerant) Co-775 (susceptible)	Pot trials: withheld water 60–150 days after planting RT-PCR analysis	Upregulated—Tolerant: IGS (indole 3-glycerol phosphate synthase), PIN1, SOD (superoxide dismutase), cAPB (cytosolic ascorbate peroxidase), GST, DREB, Prokin, LEA3	17 selected genes involved in drought tolerance: auxin and redox mediation, dehydration responsive elements, drought responsive regulatory factors, genes coding for protective factors	Hemaprabha and Simon (2012)

(continued)

Table 5.3 (continued)

Genotypes	Experimental setup	Differentially expressed genes (DEG)	Gene ontology/KEGG pathways—DEG	References
SP90-1638 (susceptible)	Pot trials: 90-day-old plants, 24, 72 and 120 h ww Microarray analysis	Upregulated: Aconitate hydratase, photosystem II polypeptide, dironuclease S-like RNase, NAC TF ATAF1, dehydrin DHN1, alkaline alpha galactosidase 1, Downregulated: Pyrophosphate-fructose 6-phosphate 1-phosphotransferase α-subunit, phosphoglycerate kinase, ADP-glucose pyrophosphorylase, β-amylase, PSI reaction centre, H/ACA ribonucleoprotein complex subunit 1-like protein	21 901 Microarray probes (SUCEST). Function categories: signal transduction, carbohydrate metabolism, cell wall and amino acid and nitrogen metabolism, transporters; redox and protein metabolism	Lembke et al. (2012)
RB86-7515 (tolerant) RB8-55536 (susceptible)	Field trials: 7-month-old plants, with or without water Small RNA deep sequencing, Solexa	13 differentially expressed mature miRNAs	Predicted miRNA targets: TFs, transporters, proteins associated with senescence and flower development	Gentile et al. (2013)
RB867515 (tolerant) SP86-155 (susceptible)	Pot trial: 175-days-old plants; 1, 3, 5 and 10 days ww cDNA-AFLP	Upregulated—Tolerant: disulfide isomerase protein; HSP70, trehalose	Large number of unidentified proteins. Annotated genes: stress-related, communication pathways, signal transduction, bioenergetics, secondary metabolism, growth and development, dynamics and regulations, carbohydrate metabolism, protein metabolism	Vantini et al. (2015)
GT21	Water-deficit at mild, moderate, severe levels during the elongation stage Microarray technology	Upregulated—Stressed: Purine metabolism, phenylpropanoid biosynthesis, glycolysis; starch, sucrose, pyruvate, cysteine and methionine metabolism;	GO matches (20%, 22.6% and 22.5% of DEG at the different levels of stress): cellular component organisation, cellular process and response to stimulus; pathways	Li et al. (2016)

		spliceosome, carbon fixation and photosynthesis, phenylalanine	include biosynthesis of secondary metabolites, carbon metabolism, biosynthesis of amino acids, hormone signal transduction	
SP81-3250 (tolerant) RB855453 (susceptible)	Pot trial: 60-day-old plants, water-deficit at moderate and severe levels for prolonged stress (30–90 days) Illumina RNA -seq	Upregulated—Tolerant: AKR2 (ankyrin) and APX3; MYB; E3 SUMO-protein ligase SIZ2; 4CL (4-cumarate: coenzyme A ligase); cullin-1-like protein; aquaporin Upregulated—Susceptible: Several kinases-related transcripts; HLH (basic helic-loop-helix TF); ACO/ACC (aminocyclopropane-1-carboxylate oxidase)	Maximum annotation of 40.38% of genes, mostly involve in binding and catalytic molecular functions, cellular and metabolic processes, and cell and cell part	Belesini et al. (2017)
S. officinarum var. Mex 69-290	*In vitro* plantlets exposed to 40% PEG8000 (−4.5 MPa); leaf and root tissue Illumina sequencing	Upregulated—Leaves: bZIP, ZF, HSF, NAC, PHR1 AP2/ERF and MYB TFs; isoflavone 2-hydroxylase, chalcone synthase 1, auxin-responsive SAUR36, laccase-25-like, LEA14-A, PP2C-59 Upregulated—Roots: NAC, HSF, MYB, bHLH and AP2/ERF TFs; dehydrin DHN4 and COR410; cationic peroxidase SPC4, desiccation-related PCC13-62, LEA3, BAX-inhibitor, anthocyanidin reductase Downregulated—Leaves: ERF014 and ERF-27 (ethylene-responsive TF), dehydration-responsive element-binding 1G Downregulated—Roots: sugar	GO terms: Organic substance metabolic and cellular processes, primary metabolic process; cellular and nitrogen compound metabolic processes; intracellular and membrane-bounded organelle categories; organic cycling and heterocyclic compound binding Main metabolic functional categories: photosynthesis and carbon fixation, nitrogen metabolism, glycolysis and gluconeogenesis, purine and glutathione metabolism, starch and sucrose metabolism; heat response, water deprivation, ABA, flavonoids biosynthesis, transcriptional regulation, and glutathione metabolism	Pereira-Santana et al. (2017)

(continued)

Table 5.3 (continued)

Genotypes	Experimental setup	Differentially expressed genes (DEG)	Gene ontology/KEGG pathways—DEG	References
		transporters, casparian strip membrane 2, wall-associated receptor kinase 2, Root hair defective 3, cellulose synthase E6, cytosine-5-methltransferase 1		
Saccharum narenga	Pot trial: 22 days ww Illumina HiSeq 2000	Upregulated—Stressed: Transposon TF2-9, lipid-transfer protein 2, early light-inducible protein HV60, dehydrin DHN1, LEA, low temp-induced Lt101.2, ABS transporter, auxin response factor, chlorophyll a-b binding, galactinol-sucrose galactosyltransferase	Response to blue light; chloroplast part, negative regulation of defence response, circadian rhythm; apoptosis, plant hormone signal transduction, flavoid biosynthesis, sesquiterpenoid and triterpenoid biosynthesis	Liu et al. (2018)
Co7336, Co0315, Co99004, Co06015, Co06022 (tolerant) Co8021 (susceptible)	Pot trials: 60-day-old plants, 10 days ww and recovery Semi-quantitative reverse transcription and real time quantitative PCR analysis	Upregulated—Early response: LEA; dehydrin, invertase, SPS, ERD4 (early responsive to dehydration), calmodulin, SNF1-related protein kinase Upregulated—Late response (tolerant varieties): Dehydrin, TPS (trehalose 6-phoshate synthase), DREB2, SNAC1, ERF3 Upregulated—Late response (all varieties): Calmodulin, ABRE, SNF1 Upregulated—Recovery: TPS (tolerant variety): invertase Down regulated—Late response: LEA	Selected candidate genes potentially leading to drought tolerance	Devi et al. (2019)

KPS01-12 (tolerant) UT12 (susceptible)	Pot trials: 2-month-old plants, 7 and 14 days ww Ion S5XL sequencing	Upregulated—Tolerant: SINAT3, MAX2 (stomatal closure); KCS-6, TPS2, CHS3 and FLS1 (biosynthesis of terpenoids and flavonoids); ACA4, RFS2, beta-amylase (BAM1); AAE3 (oxalyl-CoA synthetase); CYB561 Upregulated—Susceptible: GDSL-lipase-like (water retention); α-GAL2; CHUp1; NAC100 Down regulated—Tolerant: 5-glucosyltransferase, GST, alpha-amylase (AMY3), photosynthesis, carbon fixation; a bZIP and MYBAS1-like TFs Downregulated—Susceptible: phosphoenolpyruvate carboxykinase (pckA), NADP-MPH and FBPase-like	Water retention, secondary metabolite biosynthesis, stress response, photosynthesis, growth, molecular transport, signalling, TFs	Nawae et al. (2020)
Co06022 (tolerant) Co8021 (susceptible)	Pot trial: 60-day-old plants, 2, 6 and 10 days ww and recovery Illumina Nextseq500	Upregulated—Tolerant: Cytochrome p450, peroxidase, HSP20 (heat shock protein), HS TFc-1b, HSP70, RPM1, soluble inorganic pyrophosphatase-like isoform X2, GST (glutathione S-transferase) Upregulated—Susceptible: HSP20; Upregulated—Tolerant/Recovery: HS TFc-1b, LEA, HS70, ZIM, WRKY TF 40, protein app 1 precursor Upregulated—Susceptible/Recovery: MAPF, jasmonate-induce	Plant hormone signal transduction, spliceosome, MAPK signalling, ribosome	Selvi et al. (2020)

(continued)

Table 5.3 (continued)

Genotypes	Experimental setup	Differentially expressed genes (DEG)	Gene ontology/KEGG pathways—DEG	References
		protein, probable protein phosphatase 2c-32, katE Downregulated—Tolerant: Microtubule associated protein futsch (MAPF) Downregulated—Susceptible: Cytochrome p450, peroxidase Downregulated—Susceptible/Recovery: Heat stress TF c-1b, LEA, HS70, peroxidase, carboxypeptidase		
S. spontaneum Clone 13-13 (tolerant) Clone 84-261 (susceptible)	Pot trials: elongation growth stage, 7 days ww Illumina HiSeq 2500 sequencing	DEG up/down regulated in both clones, stress and non-stress: GST, thioredoxin, aldehyde dehydrogenase, SOD, catalase, epoxide hydrolase, phosphoesterases, protein kinase, cytochrome P450, MYBs, HLTF, Hex, MEIS1s, E2F3s, bZIP, NAC	GO terms: biological process of single-organism and metabolism; association with cytoplasm, chloroplast and plastid; ion and metal ion binding Metabolic pathways: hormone signal transduction, glycolysis and gluconeogenesis, starch and sucrose metabolism, photosynthesis and carbon fixation, and oxidative phosphorylation. Osmotic regulation, ROS removal, toxin-degrading enzymes, secondary metabolism, signalling, cell wall, and TF	Wang et al. (2020)
S. spontaneum GXS87-16	Pot trial: 60-day-old plants, 9 days ww and recovery; top visible dewlap leaf tissue Illumina HiSeq4000 platform	Upregulated—Stressed: Cycloartenol synthase genes; MAPK (mitogen-activated protein kinase); trehalose phosphatase; PP2C, SnRK2, bHLH-type TF, IAA	GO terms: Metabolic and cellular processes and stimulus response; cell part and organelle and membrane categories; catalytic activity, binding, transport, electron carrier,	Li et al. (2021)

		synthase Upregulated—Recovery: LRR-type receptor kinase, wall-associated kinases, cysteine proteinase, LCAT (lecithin-cholesterol acyltransferase), proline-rich protein gene Downregulated—Stressed: Flavone and flavonol biosynthesis gene, 4-coumarate-CoA ligase; proline-rich protein gene	nucleic acid binding TF, antioxidant, molecular transducer, receptor, enzyme regulator, structural molecule and nutrient reservoir activities	
ROC22	Pot trial: seedlings at 8-leaf stage, 10 days ww RNA-seq, UPLC-ESI-MS/MS	Upregulated—Stressed: D-lactic acid, L-malic acid, pyruvic acid, 2-propylmalic acid and 2-isopropylmalic acid, of which D-lactic acid	12 275 DEGs in both stressed and control groups: 2-oxocarboxylic acid metabolism, ABC transporters, biosynthesis of amino acids, glucosinolate biosynthesis and valine, leucine and isoleucine biosynthesis, metabolome, pyruvate and phenylalanine metabolism	Yang et al. (2022)
ROC22 (tolerant) Badila (susceptible)	Hydroponic setup: seedlings at 3-5 leaf stage, exposed to PEG6000 for 0, 4, 8, 16 and 32 h RNA-seq, DEG annotation and KEGG enrichment	Upregulated—Stressed/Tolerant: SsPIP1;2, ACS and ACO, RBOHF, DWARF4, CAT1, FBA3, ADH1, ALDH, HXK8, sucrose phosphate synthase, β-amylase, β-glucosidase, SsNAC87 and SsJAMYB Downregulated—Stressed/Tolerant: BRI1, 9S-LOX Downregulated—Stressed/Susceptible*: 9S-LOX*	16 662 and 4 423 DEG identified in two stressed genotypes: ABA-activated signalling pathway, response to water deprivation, response to salt stress and photosynthesis-related processes	Li et al. (2022)
IACSP94- 2094 (tolerant)	Field trials: at mild (soil potential -60 kPa) and severe (soil potential	Upregulated—Stressed/Tolerant: EL3, IAA, Stellacyanin, GSTU, DHN1, CesA, dirigent-jacalin	622 DEG between the two genotypes Susceptible: citrate cycle, fructose and mannose metabolism, pyruvate	Contiliani et al. (2022)

(continued)

Table 5.3 (continued)

Genotypes	Experimental setup	Differentially expressed genes (DEG)	Gene ontology/KEGG pathways—DEG	References
IACSP97-7065 (susceptible)	-75 kPa) water deficit stress Microarray analysis	(*ShDJ*), 40S and 60S ribosomal protein genes Upregulated—Stressed/Susceptible: EL3 and LOX2 and SOX-1, PIP2-6, EREBP and 60S ribosomal proteins Downregulated—Stressed/Susceptible: ZFP, Stellacyanin, PYRB, GSTU, and Metacaspase-1 gene expression Downregulated—Stressed/Tolerant: PIP2-6, MYB1 and PSII10	metabolism, glyoxylate and dicarboxylate metabolism, amino sugar and nucleotide sugar metabolism, glutathione metabolism, cyanoamino acid metabolism, selenocompound metabolism, carbon fixation in photosynthetic organisms, fatty acid elongation, and arachidonic acid metabolism Tolerant genotype: photosynthesis (antenna proteins), plant-pathogen interaction, spliceosome and metabolism of amino acids, glycerophospholipid, and nitrogen. Where the Exclusive -Tolerant: oxidoreductase activity, transcriptional regulation, metabolism of amino acids, and translation	
IACSP97-7065 (susceptible) IACSP94- 2094 (tolerant)	Pot trial: 7-months-old plants, 21 days ww RNA-seq, transcriptome mapping, gene count and expression	Downregulated—Exclusive/Tolerant/Stressed: TFs including NF-Y, DELLA, WRKY1, ORR23, DOF2, SCL6, TGAL6-like, ILR3-like, CDF1-like, and CBP60G-1. Photosynthesis-related: PPDK1 and RSH1. Solute transport: including GAT1, PHT1, NHX2, SWEET2b-like, ZIF/TOM, MOT1, MCU2, G3P1, TauE/SafE4 and NIPA4 Upregulated—Exclusive/Tolerant/	1 585 DEGs between the two genotypes Tolerant genotype: protein binding, DNA-binding TF activity, carbohydrate metabolic, protein metabolic, response to water deprivation, unfolded protein binding, water transport, water channel activity, ATP binding, DNA binding, nucleotide binding, protein binding, biosynthetic process, protein	Contiliani et al. (2023)

		Stressed: TFs including HOX29-like, bZIP23, DREB1C, ZFP1-like, bHLH035, RAP2-10- like, ERF060 and bZIP/ RISBZ5. Photosynthesis-related: CAB4, CP29.1, CPN10, CPN20, CPN60, CURT, PPIL1, psa, psaO, psb27, psbX and rbcL. Redox homeostasis: TRX-H, RBOHB and GSTT. Solute transport: NRT1/ PTR6.4, OEP23, PIP2-1, KT10, MCU, ERD4, and PIP2-5	modification process, kinase activity, hydrolase activity, transport, response to stress, transporter activity, lipid metabolic process transmembrane transport and catabolic process	

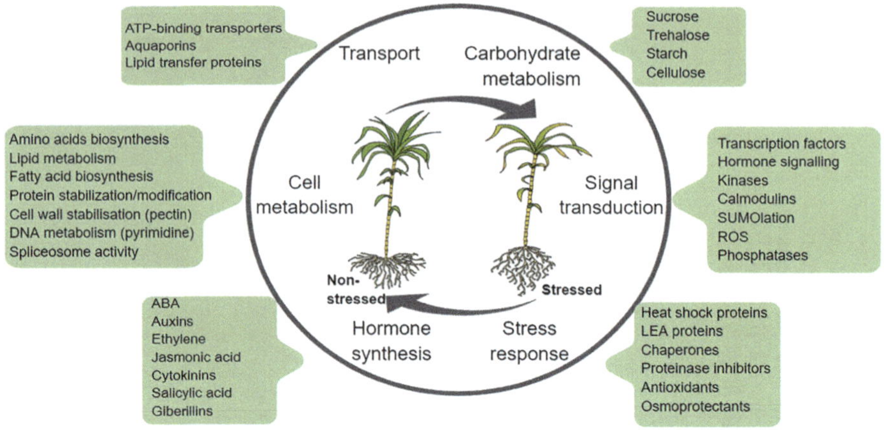

Fig. 5.3 Gene classes differentially expressed in sugarcane under water deficit stress conditions. The complexity of gene regulation under stress depends on the experimental setup and genotypes used, and the length and severity of the induced stress. Data was compiled from studies listed in Table 5.3

metabolism, transport, signal transduction, hormone biosynthesis and stress responses (Ferreira et al. 2017; Fig. 5.3). These studies are summarised in Table 5.3.

Although the transcriptomic make-up is vital, it is the modification of synthesised proteins that accounts for the final function of the protein, which is a direct reflection of a trait. This characteristic is beyond the scope of transcriptomics and thus identifies proteomics as the most direct approach to characterise a gene's final product/function (Yahoueian et al. 2021). Therefore, targeting the proteomic make-up is important in developing drought tolerant genotypes because drought affects multiple traits. Moreover, post translational modifications (PTM) are impacted by the severity of the stress as they decide the final function of protein products, thus affecting tolerance by providing physiological and morphological adaptations to stress. Most studies reporting the protein make-up of sugarcane with enhanced tolerance to water deficit stress analyse sugarcane with contrasting genotypes to water deficit or osmotic stress and determine differentially abundant proteins accumulated during stress. They then isolate/enrich proteins (either through 2-dimensional (2-D) electrophoresis or protein digestion with sequencing), identify them (through mass spectrometry) and functionally annotate the identified proteins, which groups them into functional classes based on their gene ontology. Examples of comparative proteome analyses of sugarcane plants with contrasting tolerance to water deficit stress are given in Table 5.4.

Table 5.4 A summary of comparative proteome analyses in sugarcane genotypes exposed to water deficit or osmotic stress. Independent studies identify differentially abundant and/or exclusive proteins up/downregulated during stress

Genotype	Experimental setup	Differentially abundant proteins (DAP)	Functional annotations using DAP	References
K86-161 (tolerant) Khon Kaen 1 (susceptible)	Pot trial: 3-month-old plants, 3 weeks without water (ww) Proteomics: MALDI-TOF-MS and 2-D electrophoresis	Exclusive—Tolerant/Stressed: Serine protease inhibitor and replication protein A1 Upregulated—Tolerant/Stressed: chalcone isomerase, RuBisCO activase small isoform, putative chelatase subunit, anthocyanidin synthase, S-adenosylmethionine decarboxylase proenzyme, ferredoxin, ATP synthase CF1 beta chain, proteinase inhibitor type-2 K precursor, ubiquitin containing H+-transporting two-sector ATPase beta chain, ATP synthase alpha chain, actin/8	128 DAP—both genotypes: cellular process and metabolism	Jangpromma et al. (2010)
ROC22	Pot trial: sugarcane setts treated with 10% PEG 6000 for 14 h Proteomics: 2-D electrophoresis and MALDI-TOF-MS	Exclusive—Stressed: drought inducible 22-kDa protein, Rubisco small subunit, ATP synthase delta chain and IRLs	DAP—Stressed: protein degradation, photosynthesis, energy metabolism and antioxidant activity	Zhou et al. (2012)
K86-161 (tolerant) B34-164 (susceptible)	Pot trial: 3-week-old plants, 5 days ww Proteomics: 2-D electrophoresis and LC-MS/MS	Upregulated—Tolerant/Stressed: phosphoribulokinase, oxygen-evolving enhancer protein, 2-Cys peroxiredoxin BAS1, fructose-bisphosphate aldolase, 14-3-3-like protein and SOD Downregulated—Susceptible/Stressed: phosphoribulokinase, oxygen-evolving enhancer protein,	DAP—both stressed genotypes: energy and metabolism, photosynthesis, antioxidant, and defence protein	Khueychai et al. (2015)

(continued)

Table 5.4 (continued)

Genotype	Experimental setup	Differentially abundant proteins (DAP)	Functional annotations using DAP	References
		2-Cys peroxiredoxin BAS1, fructose-bisphosphate aldolase, 14-3-3-like protein and ferredoxin NADP reductase		
Transgenic sugar-cane—*TERF1*	Pot trial: 7-month-old transgenic and WT plants treated with 20% PEG 8000 for 21 days Proteomics: 2-D electrophoresis and MALDI—MS/MS	Upregulated—Transgenics/Stressed: NAD (P) H-quinone oxidoreductase subunit M, cellulose synthase-like protein D2, 14-3-3-like protein 1, phosphoenol pyruvate carboxyl-ase, RuBisCO large chain, ferredoxin-NADP reductase, proba-ble fructose-biphosphate aldolase I, malate dehydrogenase, glyceralde-hyde-3-phosphate- dehydrogenase A and B, ATPase 3, plasma membrane-type and phosphoglycerate kinase	DAP—both transgenic and WT plants: energy, disease/defence, metabolism, protein synthesis, tran-scription, intracellular traffic and signal transduction	Rahman et al. (2015)
F172 (tolerant) YL6 (susceptible)	Pot trial: 2-month-old plants, 7 days ww Proteomics: 2-D electrophoresis and MALDI-TOF-TOF/MS	Exclusive—Susceptible/Stressed: Os03g0279950, Os03g0592500, ribulose bisphosphate carboxylase, OS06G0603000, ferredoxin and ribulose-1, 5-bisphosphate carboxyl-ase/oxygenase large subunit Upregulated—Tolerant/Stressed: oxygen-enhanced and PSII proteins, fructose-15, 1-bisphosphoaldolase precursors, triose phosphate isomer-ase, adenylate kinase, glutamine synthetase, cell growth and division-related proteins, inorganic	DAP—both stressed genotypes: free radical scavenging, photosynthesis, cell growth and division, metabo-lism, defence protection, protein processing and signal transduction	Do et al. (2017)

		pyrophosphatase, ghrelin-binding protein ABP4 precursor, protein involved in protein- processing-related 20S proteasome α-subunit and a drought inducible 16 kD protein Upregulated—Susceptible/Stressed: ATP synthase CF18 α subunit, glutamine synthetase and the cell growth and division-related protein auxin-binding protein ABP17 precursor		
RB867515 (tolerant) RB855536 (susceptible)	Pot trial: 4-month-old plants exposed to moderate (−0.5 MPa) and severe water deficit stress (−1 MPa) Proteomics: 2-D electrophoresis, MALDI-TOF-TOF and MS/MS	Upregulated—Tolerant/Stressed: PSII polypeptide subunits, glyceraldehyde-3-phosphate dehydrogenase, fructokinase, aconitate hydratase, S-glutathione transferase, cell wall biosynthesis proteins and β-glucanase Downregulated—Tolerant/Stressed: Light-harvesting chlorophyll binding complex and S-glutathione transferase	130 DAP—Tolerant/Stressed: protein metabolism, amino acid metabolism, and TCA/org. transformation, photosynthesis and redox 84 DAP—Susceptible/Stressed: redox, amino acid metabolism and TCA/org. transformation Exclusive—Tolerant/Stressed: stress, secondary metabolism, signalling, OPP and biodegradation of xenobiotics Proteins exclusively in the stressed susceptible genotype: Protein, RNA and glycolysis	Vital et al. (2017)
CTC15 (tolerant) SP90-3414 (susceptible)	Pot trial: 5-month-old plants, 4 and 12 days ww. Soil moisture maintained at 100% (controls), 40% (−286 KPa), and 25% (−850 KPa) of FC Proteomics: reverse-phase ultra-	Downregulated—Tolerant/Stressed: zeaxantin epoxidase, polygalacturonanse precursor, ndph-dependent reductase, two ribosomal proteins, glycine cleavage h-protein, plasma membrane atpase, ribosomal protein, elongation factor and	49 DAP—Tolerant: metabolic processes, translation, catabolic processes and biosynthetic processes 59 DAP—Susceptible: metabolic process, cellular process, generation of precursor metabolites and energy,	Budzinski et al. (2019)

(continued)

Table 5.4 (continued)

Genotype	Experimental setup	Differentially abundant proteins (DAP)	Functional annotations using DAP	References
	performance liquid chromatography/MS and GO annotation	fumarylacetoacetate hydrolase domain Exclusive—Tolerant/Stressed: glutathione-s-transferase, ferredoxin-sulfite reductase, lipoxygenase, NADH-ubiquinone oxidoreductase, vacuolar atpase b subunit, endoplasmin and 60s ribosomal protein Differentially abundant—Susceptible/Stressed: ABA stress ripening protein, glutathione s-transferase, glutaredoxin, 60s ribosomal protein, light-induced protein, and mRNA binding precursor, thylakoid membrane phosphoprotein, chlorophyll a-b binding protein and PSI reaction center subunit	carbohydrate metabolic process and response to stress	
IACSP 95–5000	Pot trial: 6-month-old plants, 7 days ww (30% FC) Proteomics: LC-MS/MS and GO annotation	Exclusive—Stressed: PLT2, DNA-binding storekeeper protein, zinc finger C-x8-C-x5-C-x3-H type family protein, MYBS2, homeobox protein 22 and 7 Control: cellulose synthase interactive protein, COBRA-like protein, chitinase-like protein, IRX10, glycosyl hydrolases, a glycosyltransferase, a pectin methyl esterification protein and a LRR family protein	41 DAP—Stressed: biosynthesis of amino acids, oxidation–reduction process and pyridoxal phosphate clusters 74 DAP—Control plants: biosynthesis, organization and deposition of the secondary cell wall	Salvato et al. (2019)

MEX69290	Pot trial: plants 10 days ww Proteomics: Nano LC-Mass Spectrometry and GO annotation	Exclusive—Stressed: LEA proteins, dehydrins, RAB28, an osmotin, cold response kinases, heat shock proteins, superoxide dismutases, glutathione-S-transferases, thioredoxin, plastocyanin proteins, proteins with domain PF00538 from the family of junction histones H1 and H5, methyl-CpG binding proteins, fasciclins, cupins, proteases with papain-like domain and chaperonins	GO enriched proteins—Cellular Component: the membrane and the extracellular space, cell wall and apoplast Molecular Function: oxidoreductase activity, lipid binding, metal binding and chaperone activity Biological Process: response to stress, detoxification, or ROS metabolism	Zamora-Briseño et al. (2023)
EMS-derived tolerant N19 mutant: M9.2	Pot trial: 3.5-months-old mutant plants, 21 days ww Proteomics: 2D-DIGE and LC-MS	Upregulated—Stressed: Pyruvate orthophosphate dikinase, un-fragmented ribulose-1,5-bisphosphate carboxylase/oxygenase large subunit, chlorophyll a/b binding protein 3, sucrose synthase, S-adenosyl-homocysteine hydrolase (SAHH), ATP synthase F1 alpha, ribosomal protein S3 and nucleoside diphosphate kinase	Upregulated—Mutant: Photosynthesis, carbohydrate metabolism, methylation homeostasis, energy metabolism and DNA repair	Masoabi et al. (2023b)
Transgenic sugarcane NCo310 transformed with *OTS1*	Pot trial: 3-months-old transgenic plants, 21 days ww Proteomics: LC-MS/MS and GO annotation	Exclusive—Stressed: chloroplastic ATP-dependent zinc metalloprotease FTSH2, chlorophyll a/b binding protein 1B-21, NADP-dependent malic enzyme, Rubisco large subunit, ATP-dependent metalloprotease FTSH2, the alpha-glucan water dikinase chloroplastic isoform X1, Hsp70-Hsp90 organising protein, elongation factor G-2, proteasome subunit alpha type-5, vacuolar	Exclusive—Stressed/Transgenics: regulation of ABA, mediation of transcription of ABA-responsive genes, ROS neutralisation, stomatal closure, formation of thylakoids, regulation of photosynthesis, root growth, protein assembly, pre-protein processing, protein synthesis, faulty protein degradation, post-translational modification, ubiquitination, protein	Masoabi et al. (2023c)

(continued)

Table 5.4 (continued)

Genotype	Experimental setup	Differentially abundant proteins (DAP)	Functional annotations using DAP	References
		protein sorting-associated protein 2 homologue 1, ATP-dependent caseinolytic protease (Clp) proteolytic subunit, ubiquitin-40S ribosomal protein S27a, vacuolar protein sorting-associated protein 2 and 1, HSP70, HSP90, HSP 81–1, coatomer subunit alpha-3, coatomer subunit beta'-3, phosphoribosylaminoimidazole-succinocarboxamide synthase, UDP-glucose pyrophosphorylase, monodehydroascorbate reductase, RAN GTPase-activating protein 2, GDP-mannose 3,5-epimerase 1 and 2, adenylate kinase 4, probable aldo-keto reductase 1 and 60S ribosomal protein L23	transportation, protein homeostasis), ATP production, source-sink energy production, cell division, cell membrane synthesis and turgor, DNA restructuring and transcription, ROS antioxidation	

5.7.4 Genetic Modification of Single Target Genes

As previously mentioned, increasing number of studies on the successful enhancement of water deficit stress tolerance in sugarcane through genetic modification, particularly the overexpression of target genes from other species, have been reported and are listed in Table 5.5. To briefly describe, transgenic sugarcane producing high levels of trehalose due to the overexpression of a *Grifola frondosa* trehalose synthase gene displayed improved drought tolerance through protected physiological indexes such as membrane permeability, and integrity, total chlorophyll content, and the antioxidant system (Wang et al. 2005; Zhang et al. 2006). Sugarcane transformed with a heterologous *P5CS* (*Δ1-pyrroline-5-carboxylate synthetase*) gene under the control of a stress-inducible promoter enhanced drought tolerance by increasing proline levels during dehydration that acted as an antioxidant than an osmoprotectant (Molinari et al. 2007). Overexpressing a *P5CS* gene from *Vigna aconitifolia* in sugarcane resulted in enhanced salt tolerance in the transgenic plants, with high levels of proline protecting the plant's photosynthetic apparatus and preventing oxidative damage (Guerzoni et al. 2014).

Augustine et al. (2015a) isolated the *HSP70* gene from *Erianthus arundinaceus*, a wild relative of the *Saccharum* genus, and overexpressed it in sugarcane, compared with the wildtype, the transgenics maintained photosynthetic machinery and chlorophyll content for longer upon drought stress. Furthermore, Augustine et al. (2015b) overexpressed the *Ea*DREB2 transcription factor in sugarcane, then further pyramided it with an overexpressed *DNA helicase* gene from pea (*PDH45*). In both cases, transgenic sugarcane demonstrated enhanced tolerance to drought and salinity stresses by protecting the photosynthetic efficiency and membrane stability under water deficit stress (Augustine et al. 2015a, b). Furthermore, in two separate studies Kumar et al. (2014) and Raza et al. (2016) constitutively overexpressed the *Arabidopsis vacuolar* H^+ *pyrophosphatase* gene (*AVP1*) in sugarcane. In comparison with the wildtypes, the transgenic plants displayed longer and more profuse root systems, maintained photosynthesis and were able to withstand water deficit for longer.

Additional target genes that resulted in drought tolerant phenotypes in transgenic sugarcane plants included the overexpression of a late embryogenesis abundant protein (LEA) gene from *Xerophyta humulis*, *XhLEA* (Barker 2021). When compared with the control plants, the photosynthetic efficiency and membrane integrity were maintained in the transgenic plants during water deficit stress. However, Barker (2021) also demonstrated that the constitutive expression of the *XhLEA* transgene negatively impacted sugarcane growth and sucrose accumulation. Furthermore, overexpression of an *Arabidopsis* BAX inhibitor (*BI-I*) enhanced drought tolerance in transgenic cane by suppressing endoplasmic reticulum (ER) stress in the plant roots (Ramiro et al. 2016). Lastly, Saravanan et al. (2018) isolated the *Brassica carinata Zinc Finger* of *Arabidopsis thaliana 12* (*BcZAT12*) gene, transformed it into sugarcane and analysed its effect on drought and salt tolerance. Their analyses demonstrated that growth and vigor, relative water content, photosynthesis rate,

Table 5.5 A summary of genes transformed into sugarcane to enhance tolerance to drought stress

Gene source	Transgene	Experimental setup	Drought tolerance characteristics	References
Grifola frondosa	*Trehalose synthase*	In vitro PEG-induced osmotic stress	Transgenics: stray-green till day 7. Controls: Turning yellow at day 3, and wilted and dry by day 7	Wang et al. (2005)
Grifola frondosa	*Trehalose synthase*	Field trials: Irrigation restricted for 15 days	Transgenics: accumulated high levels of trehalose, improved stress indexes for membrane permeability, malondialdehyde and chlorophyll content, and activity of antioxidant enzymes	Zhang et al. (2006)
Vigna aconitifolia	*Δ1-pyrroline-5-carboxylate synthetase (P5CS)*	Pot trials: 9 days of restricted water	Transgenics: Proline content increased 2.5-fold; higher photosynthetic rate; display protection from oxidative stress; higher biomass yield	Molinari et al. (2007)
Arabidopsis thaliana	*Vacuolar pyrophosphatase (VP1)*	Pot trials: 14 days of restricted water	Transgenics: longer and profuse root systems; survived water deficit longer than the controls	Kumar et al. (2014)
Arabidopsis thaliana	*Vacuolar H+-pyro-phosphatase (H+-PPase) (AVP1-H+)*	Field trials: 6 months with 50% reduced irrigation	Transgenics: increased RWC, improved osmotic and turgor potentials; photosynthetic rate, stomatal conductance, and transpiration rate were less affected by water-deficit stress	Raza et al. (2016)
Erianthus arundinaceus	*Heat Shock Protein 70 (HSP70)*	Pot trials: 10 days of restricted water	Transgenics: increased cell membrane thermostability, RWC, gas exchange parameters, chlorophyll content and	Augustine et al. (2015a)

(continued)

Table 5.5 (continued)

Gene source	Transgene	Experimental setup	Drought tolerance characteristics	References
			photosynthetic efficiency. Up regulation of stress-related genes *DREB2, DNA helicase 45, LEA, RD29, ERD, ERF, Cor15* and *BRICK* when compared to controls	
Erianthus arundinaceus	*Dehydration Responsive Element Binding 2 (DREB2)*	Pot trials: 10 days of restricted water	Transgenics: Higher transcript level of *LEA, RD29, ERD, ERF, COR15, HSP70 and BRICK*; increased membrane stability, photosynthetic rate, RWC and chlorophyll content	Augustine et al. (2015b)
Pisum sativum	*DNA Helicase 45 (DH45)*	Pot trials: 10 days of restricted water	Transgenics: *DREB2, HSP70, RD29, LEA, ERD, ERF, COR15, BRICK* and *PDH45* were up regulated; higher RWC, chlorophyll content and increased photosynthetic rate	Augustine et al. (2015b)
Saccharum spp. hybrid cv. GT 21	*Δ^1-pyrroline-5-carboxylate synthase (P5CS)*	Pot trials: 3 days without water (ww)	Transgenics: Higher proline, ABA SOD and RWC, lower MDA and chlorophyll (SPAD value) decreased	Li et al. (2018)
Xerophyta humulis	*Late embryogenesis abundant protein (LEA)*	Pot trials: 7 (low), 14 (moderate) and 21 (severe) days ww	Transgenics—Drought-induced: low electrolyte leakage, high RWC, maintained higher stomatal conductance and PSII efficiency; up regulation of *PPDK, RuBisCo* and *PEPC* Transgenics—Constitutively: growth	Barker (2021)

(continued)

Table 5.5 (continued)

Gene source	Transgene	Experimental setup	Drought tolerance characteristics	References
			impairment and reduced sucrose accumulation	
Arabidopsis thaliana	*Cell death suppressor Bax Inhibitor-1 (BI-1)*	Pot trials: 21 days ww	Transgenics: display tolerance to long-term (>20 days) water deficit stress, robust root system; decreased stress to the endoplasmic reticulum (ER), up regulation of the protein glycosylation inhibitor tunicamycin	Ramiro et al. (2016)
Brassica carinata	*Zinc-finger protein ZAT12*	Field trials: Irrigation restricted with analysis on days 15, 30, 45, 60 and 75	Transgenics: enhanced tolerance, high level of *Bc*ZAT12 expression; increased growth and vigor, relative RWC, photosynthetic rate, transpiration rate, stomatal conductance, chlorophyll content, proline and glycine betaine content	Saravanan et al. (2018)
Arabidopsis thaliana	*DREB2A-CA*	Pot trials: 21 days ww	Transgenics: up regulation of *LEA, Dehydrins, Scdr2* and *4, P5CS* and *galactinol synthase (GOLS)*; higher RWC, leaf water potential and photosynthetic rates. Displayed increased sucrose levels and improved bud sprouting	Reis et al. (2014)
Arabidopsis thaliana	*DREB2A-CA*	Field trials: no irrigation/dependent on rainfall; during dry season—irrigated once a week (60 mm water)	Transgenics: increased soluble solids (°Brix), sugar content in culm juice (Pol%), tons of cane per hectare	de Souza et al. (2019)

(continued)

Table 5.5 (continued)

Gene source	Transgene	Experimental setup	Drought tolerance characteristics	References
			(TCH), and tons of Pol% per hectare (TPH); under drought conditions comparable to the high-yield and drought-tolerant Brazilian elite variety CTC9001	
Arabidopsis thaliana	*DREB2A-CA*	Pot trials: 8-month-old transgenic plants 6 days ww	Transgenics: increased sucrose content and bud sprouting rate; higher stomatal conductance and photosynthetic and transpiration rates	de Souza et al. (2020)
Eriathus arundinaceus	*Expansin (EaEXPA1)*	Pot trials: 90-day-old transgenic plants 10 days ww	Transgenics: higher RWC, cell membrane thermostability, chlorophyll content, photosynthetic efficiency and gas-exchange	Narayan et al. (2021)
Arabidopsis thaliana	*B-box 29* (*BBX29*)	Pot trials: 7, 14 and 21 days ww	Transgenics: higher RWC and better protected photosynthetic machinery; accumulated more proline and enzymatic antioxidants; reduced ROS levels and decreased oxidative damage	Mbambalala et al. (2021)
Tripidium arundinaceum	*Dehydration-responsive element binding (TaDREB2B)*	Field and pot trials: 4-leaf-stage transgenic plants 15 days ww	Transgenics: improved yield, maintained sucrose content, enhanced water retention ad reduced membrane damage	Xiao et al. (2022)
Saccharum spp. hybrid	*Cytokinin oxygenase/dehydrogenase (CKX)* silencing	Pot trials: 3-month-old transgenic plants 21 days ww	Transgenics: less H_2O_2 accumulation, high SOD and CAT activities, proline, chlorophyll and photosynthetic rate. Low salicylic acid	Masoabi et al. (2023a)

(continued)

Table 5.5 (continued)

Gene source	Transgene	Experimental setup	Drought tolerance characteristics	References
			levels. Transgenic line with highest level of CKX silencing: also displayed delayed leaf senescence and maintained high RWC	
Arabidopsis thaliana	*Overly Tolerant to Salt 1 (AtOTS1)*	Pot trials: 3-month-old transgenic plants exposed to 21 days ww	Transgenics: Maintained high RWC and photo-synthesis rate, maintained high stomatal conduc-tance and chloro-phyll content, and less MDA and elec-trolyte leakage	Masoabi et al. (2023c)
Saccharum spontaneum	*BRK1*—a homo-log of human HSPC300	Pot trials: 90-day-old trans-genic plants 10 days ww	Transgenics: improved cell membrane thermo-stability due to for-mation of interlocking mar-ginal lobes in the epidermal leaf cells. Maintained high RWC, enhanced transpiration rate, stomatal conduc-tance, and photo-synthesis rate	Narayan et al. (2023)

transpiration rate, stomatal conductance, chlorophyll content, proline and glycine betaine were higher in transgenic than in control plants, resulting in enhanced tolerance to drought.

5.8 Concluding Remarks

Years of research have increased our knowledge of the morphological, physiological, biochemical, and genetic systems that drive drought tolerance in sugarcane. This knowledge is essential to ensure the sustainable production of this crop under current and predicted future changing climatic conditions. Several research programs to development water stress tolerant sugarcane are currently underway through traditional and modern breeding technologies. However, the genomic complexity of commercial sugarcane hybrids still makes this process challenging even though significant progress has been made in recent years in obtaining sugarcane genome and transcriptome datasets. Research outlined in this review demonstrates that the evolution of techniques developed thus far can be applied in the quest to introduce drought stress tolerant traits into the complex genome of this crop.

Acknowledgements We are grateful for the funding provided by the South African Sugar Association (Grant Agreement No S007246) and National Research Foundation (SJS Grant No 119472) that supported the review of the data and literature.

References

Abbas SR, Ahmad SD, Sabir SM, Shah AH (2014) Detection of drought tolerant sugarcane genotypes (Saccharum officinarum) using lipid peroxidation, antioxidant activity, glycine-betaine and proline contents. J Soil Sci Plant Nutr 14(1):233–243

Akhtar M, Jamil M, Ahmad S (2006) Agronomic traits and morphological characteristics of some exotic varieties of sugarcane. Pak J Agric Res 19:70–77

Apel K, Hirt H (2004) Reactive oxygen species: metabolism, oxidative stress, and signal transduction. Annu Rev Plant Biol 55:373–399

Augustine S, Narayan A, Syamaladevi DP, Chinnaswamy A, Mohan C, Ravichandran V, Subramonian N (2015a) *Erianthus arundinaceus* HSP70 (EaHSP70) overexpression increases drought and salinity tolerance in sugarcane (*Saccharum* spp hybrid). Plant Sci 232:23–34

Augustine SM, Narayan JA, Syamaladevi DP, Appunu C, Chakravarthi M et al (2015b) Overexpression of EaDREB2 and pyramiding of EaDREB2 with the pea DNA helicase gene (PDH45) enhance drought and salinity tolerance in sugarcane (*Saccharum* spp hybrid). Plant Cell Rep 34:247–263

Bamrungrai J, Tubana B, Tre-loges V et al (2021) Effects of water stress and auxin application on growth and yield of two sugarcane cultivars under greenhouse conditions. Agriculture 11(7):613

Barker AM (2021) The overexpression of a late embryogenesis abundant protein in *Saccharum* species hybrid to improve osmotic stress tolerance under water-deficit. MSc Thesis, Stellenbosch University

Barreto FZ, Balsalobre TWA, Chapola RG, Garcia AAF, Souza AP et al (2021a) Genetic variability, correlation among agronomic traits, and genetic progress in a sugarcane diversity panel. Agriculture 11:533

Barreto FZ, Balsalobre TWA, Chapola RG et al (2021b) Genetic variability, correlation among agronomic traits, and genetic progress in a sugarcane diversity panel. Agriculture 11:533

Basnayake J, Jackson PAN, Inman-Bamber G, Lakshmanan P (2015) Sugarcane for water-limited environments Variation in stomatal conductance and its genetic correlation with crop productivity. J Exp Bot 66:3945–3958

Begcy K, Mariano ED, Gentile A, Lembke CG, Zingaretti SM, Souza GM et al (2012) A novel stress-induced sugarcane gene confers tolerance to drought, salt and oxidative stress in transgenic tobacco plants. PLoS ONE 7(9):e44697

Belesini AA, Carvalho FM, Telles BR, de Castro GM, Giachetto PF et al (2017) De novo transcriptome assembly of sugarcane leaves submitted to prolonged water-deficit stress. Genet Mol Res 16(2):25

Blum A (2005) Drought resistance, water-use efficiency, and yield potential—are they compatible, dissonant, or mutually exclusive? Crop Pasture Sci 56:1159–1168

Boaretto LF, Carvalho G, Borgo L, Creste S, Landell MG, Mazzafera P, Azevedo RA (2014) Water stress reveals differential antioxidant responses of tolerant and non-tolerant sugarcane genotypes. Plant Physiol Biochem 74:165–175

Brant EJ, Eid A, Kannan B, Baloglu MC, Altpeter F (2024) The extent of multiallelic, co-editing of LIGULELESS1 in highly polyploid sugarcane tunes leaf inclination angle and enables selection of the ideotype for biomass yield. Plant Biotechnol J 22(10):2660–2671

Brasileiro BP, Peternelli LA, Silveira LCI, Barbosa MHP (2016) Importance of agronomic traits in the individual selection process of sugarcane as determined using logistic regression. Acta Scientiarum Agron 38:289–297

Bremer G (1961) Problems in breeding and cytology of sugarcane. Euphytica 10:59–78

Brenet F, Moh M, Funk P, Feierstein E, Viale AJ, Socci ND, Scandura JM (2011) DNA methylation of the first exon is tightly linked to transcriptional silencing. PLoS One 6:e14524

Brett PGC (1950) Investigations on sugarcane breeding in Natal during 1949. Proc S Afr Sugarcane Technol Assoc 24:99–103

Budzinski IGF, de Moraes FE, Cataldi TR et al (2019) Network analyses and data integration of proteomics and metabolomics from leaves of two contrasting varieties of sugarcane in response to drought. Front Plant Sci 10:1524

Butterfield MK, D'Hont A, Berding N (2001) The sugarcane genome: a synthesis of current understanding and lessons for breeding and biotechnology. Proc S Afr Sugarcane Technol Assoc 75:1–5

Cardi T, Murovec J, Bakhsh A et al (2023) CRISPR/Cas-mediated plant genome editing: outstanding challenges a decade after implementation. Trends Plant Sci 28(10):1144–1165

Carr MKV, Knox JW (2011) The water relations and irrigation requirements of sugar cane (*Saccharum officinarum*): a review. Exp Agric 47(1):1–25

Chandiposha M, Zharare GE et al (2023) Screening of sugarcane varieties for tolerance to water deficiency using containers. Int J Agron. https://doi.org/10.1155/203/5705785

Chen F, Chen L, Yan Z, Xu J, Feng L, He N, Guo M, Zhao J, Chen Z, Chen H, Yao G, Liu C (2024) Recent advances of CRISPR-based genome editing for enhancing staple crops. Front Plant Sci 15:1478398

Colebrook EH, Thomas SG, Phillips AL et al (2014) The role of gibberellin signalling in plant responses to abiotic stress. J Exp Biol 217:67–75

Cominelli E, Conti L, Tonelli C, Galbiati M (2013) Challenges and perspectives to improve crop drought and salinity tolerance. New Biotechnol 30:355–361

Contiliani DF, de Oliveira Nebó JFC, Ribeiro RV et al (2022) Leaf transcriptome profiling of contrasting sugarcane genotypes for drought tolerance under field conditions. Sci Rep 12:9153

Contiliani DF, de Oliveira Nebó JFC, Ribeiro RV et al (2023) Drought triggered leaf transcriptional responses disclose key molecular pathways underlying leaf water use efficiency in sugarcane (*Saccharum* spp). Front Plant Sci 14:1182461

Cruz de Carvalho MH (2008) Drought stress and reactive oxygen species: production, scavenging and signalling. Plant Signal Behav 3:156–165

D'Hont A, Grivet L, Feldmann P, Rao S, Berding N, Glaszmann JC (1996) Characterisation of the double genome structure of modern sugarcane cultivars (*Saccharum* spp) by molecular cytogenetics. Mol Gen Genomics 250:405–413

D'Hont A, Ison D, Alix K, Roux C, Glaszmann JC (1998) Determination of basic chromosome numbers in the genus Saccharum by physical mapping of ribosomal RNA genes. Genome 41: 221–225

D'Hont A, Garsmeur O, Piperidis N (2023) Sugarcane genome architecture revealed by chromosome-specific probes, implication for chromosome nomenclature. Proc Int Soc Sugar Cane Technol 31:266–273

Da Graça J, Rodrigues F, Farias J, Oliveira M, Hoffmann-Campo C, Zingaretti S (2010) Physiological parameters in sugarcane cultivars submitted to water deficit. Braz J Plant Physiol 22: 189–197

Da Silva ALC, Da Costa WAJM (2004) Varietal variation in growth, physiology and yield of sugarcane under two contrasting water regimes. Trop Agric Res 16:1–12
da Silva PP, Soares L, da Costa JG, Viana LDS, Farias de Andrade JC et al (2012) Path analysis for selection of drought tolerant sugarcane genotypes through physiological components. Ind Crop Prod 37:11–19
Davies WJ, Zhang JH (1991) Root signals and the regulation of growth and development of plants in drying soil. Annu Rev Plant Physiol Plant Mol Biol 42:55–76
de Almeida Silva M, Jifon JL, Sharma V et al (2011) Use of physiological parameters in screening drought tolerance in sugarcane genotypes. Sugar Tech 13:191–197
De Araujo MA, Melo AAR, Silva VM, Reis ARD (2023) Selenium enhances ROS scavenging systems and sugar metabolism increasing growth of sugarcane plants. Plant Physiol Biochem 201:107798
de Oliveira Maia Júnior S, Andrade J, dos Santos C, Silva JAC, Santos KPO, Silva J, Endres L (2019) Leaf thickness and gas exchange are indicators of drought stress tolerance of sugarcane. Emirates J Food Agric 31:29–38
de Oliveira Maia Júnior S, de Andrade JR, dos Santos CM et al (2020) Osmoregulators' accumulation minimizes the effects of drought stress in sugarcane and contributes to the recovery of photochemical efficiency in photosystem II after rewatering. Acta Physiol Plant 42:62
De Ollas C, Hernando B, Arbona V et al (2013) Jasmonic acid transient accumulation is needed for abscisic acid increase in citrus roots under drought stress conditions. Phys Plantarum 147:296–306
de Souza WR, de Oliveira NG, Vinecky F et al (2019) Field evaluation of AtDREB2A CA overexpressing sugarcane for drought tolerance. J Agro Crop Sci 205:545–553
de Souza WR, de Oliveira NG, Vinecky F, Ribeiro AP, Basso MF et al (2020) Development of drought-tolerant sugarcane overexpressing the AtDREB2A CA gene In: Nakashima K, Urao T (eds) Development of biotechnologies and biotech crops for stable food production under adverse environments and changing climate conditions. JIRCAS Working Report 91 (3–3), pp 89–98
Delauney AJ, Verma DPS (1990) A soybean gene encoding Δ^1-pyrroline-5-carboxylate reductase was isolated by functional complementation in *Escherichia coli* and is found to be osmoregulated. Mol Gen Genomics 221:299–305
Dempsey DA, Vlot AC, Wildermuth MC et al (2011) Salicylic acid biosynthesis and metabolism. Arabidopsis Book 9:e0156
Devi K, Prathima PT, Gomathi R, Manimekalai R, Lakshmi K, Selvi A (2019) Gene expression profiling in sugarcane genotypes during drought stress and rehydration. Sugar Tech 21:717–733
Dlamini PJ (2021) Drought stress tolerance mechanisms and breeding effort in sugarcane: a review of progress and constraints in South Africa. Plant Stress 2:100027
Dlamini PJ, Schultz-Viljoen T, Ntuli NR (2020) Morphological and physiological screening of ethyl methanesulfonate treated sugarcane calli and plantlets for *in vitro* mannitol-induced osmotic stress tolerance. Plant Cell Biotechnol Mol Biol 21(37, 38):107–120
Do T-T, Li J, Zhang F-J et al (2017) Analysis of differential proteome in relation to drought resistance in sugarcane. Acta Agron Sin 43(9):1337–1346
Dong T, Hwang I (2014) Contribution of ABA UDP-glucosyltransferases in coordination of ABA biosynthesis and catabolism for ABA homeostasis. Plant Signal Behav 9:e28888
dos Santos CM, de Almeida Silva M (2015) Physiological and biochemical responses of sugarcane to oxidative stress induced by water deficit and paraquat. Acta Physiol Plant 37:172
Du Y, Kawamitsu Y, Nose A, Hiyane S, Murayama S et al (1996) Effects of water stress on carbon exchange rate and activities of photosynthetic enzymes in leaves of sugarcane (*Saccharum* sp). Aust J Plant Physiol 23:719–726
Ehleringer J, Cerling T, Helliker B (1997) C4 photosynthesis, atmospheric CO_2, and climate. Oecologia 112:285–299

Eid A, Mohan C, Sanchez S, Wang D, Altpeter F (2021) Multiallelic, targeted mutagenesis of magnesium chelatase with CRISPR/Cas9 provides a rapidly scorable phenotype in highly polyploid sugarcane. Front Genome Edit 3:654996

Endres L, Silva JV, Ferreira VM, Barbosa GVS (2010) Photosynthesis and water relations in Brazilian sugarcane. Open Agric J 11:31–37

Endres L, Wanderley Filho HC, Clemente PR, Bezerra BKL, Ferreira VM (2016) Root-promoting substances enhance sugarcane's drought tolerance. Afr J Agric Res 11(52):5162–5170

Fang Y, Xiong L (2015) General mechanisms of drought response and their application in drought resistance improvement in plants. Cell Mol Life Sci 72:673–689

Farooq M, Wahid A, Kobayashi N et al (2009) Plant drought stress: effects, mechanisms and management. Agron Sustain Dev 29:185–212

Ferreira THS, Tsunada MS, Bassi D, Araújo P, Mattiello L et al (2017) Sugarcane water stress tolerance mechanisms and its implications on developing biotechnology solutions. Front Plant Sci 8:1077

Flexas J, Gallé A, Galmés J, Ribas-Carbo M, Medrano H (2012) The response of photosynthesis to soil water stress. In: Aroca R (ed) Plant responses to drought stress: from morphological to molecular features. Springer, Berlin/Heidelberg, pp 129–144

Garsmeur O, Droc G, Antonise R, Grimwood J, Potier B et al (2018) A mosaic monoploid reference sequence for the highly complex genome of sugarcane. Nat Commun 9:2638

Gentile A, Ferreira TH, Mattos RS, Dias LI, Hoshino AA et al (2013) Effects of drought on the microtranscriptome of field-grown sugarcane plants. Planta 237:783–798

Gill SS, Tuteja N (2010) Reactive oxygen species and antioxidant machinery in abiotic stress tolerance in crop plants. Plant Physiol Biochem 48:909–930

Gomathi R, Krishnapriya V, Arunkumar R et al (2020) Physiological traits imparting drought stress tolerance to promising sugarcane (*Saccharum* spp) clones. Plant Physiol Rep 25:509–515

Gosal SS, Wani SH, Kang MS (2009) Biotechnology and drought tolerance. J Crop Improv 23:19–54

Grivet L, Arruda P (2002) Sugarcane genomics: depicting the complex genome of an important tropical crop. Curr Opin Plant Biol 5:122–127

Grivet L, Glaszmann JC, D'Hont A (2006) Darwin's harvest: new approaches to the origins, evolution and conservation of crops (Motley TJ, ed). Columbia University Press, New York, pp 49–66

Guerzoni JTS, Belintani NG, Moreira RMP et al (2014) Stress-induced Δ1-pyrroline-5-carboxylate synthetase (*P5CS*) gene confers tolerance to salt stress in transgenic sugarcane. Acta Physiol Plant 36:2309–2319

Hadi P-A, Armin S-M, Hassan P, Mohammad P (2016) Stomatal responses to drought stress In: Ahmad P (ed) Water stress and crop plants: a sustainable approach. Wiley Online Library. https://doi.org/10.1002/9781119054450ch3

Hartati R, Suhesti S, Ardana I et al (2020) In-vitro selection of sugarcane (Saccharum officinarum L.) putative mutant for drought stress. IOP Conference Series: Earth and Environmental Science 653:012135. https://doi.org/10.1088/1755-1315/653/1/012135

Haworth M, Marino G, Brunetti C, Killi D, De Carlo A, Centritto M (2018a) The impact of heat stress and water deficit on the photosynthetic and stomatal physiology of olive (*Olea europaea* L)—a case study of the 2017 heat wave. Plants 7:76

Haworth M, Cosentino SL, Marino G, Brunetti C, Riggi E et al (2018b) Increased free abscisic acid during drought enhances stomatal sensitivity and modifies stomatal behaviour in fast growing giant reed (*Arundo donax* L). Environ Exp Bot 147:116–124

Hemaprabha G, Simon S (2012) Genetic diversity and selection among drought tolerant genotypes of sugarcane using microsatellite markers. Sugar Tech 14:327–333

Himmelbach A, Yang Y, Grill E (2003) Relay and control of abscisic acid signalling. Curr Opin Plant Biol 6:47–479

Holme IB, Gregersen PL, Brinch-Pedersen H (2019) Induced genetic variation in crop plants by random or targeted mutagenesis: convergence and differences. Front Plant Sci 10:1468

Hönig M, Plíhalová L, Husičková A, Nisler J, Doležal K (2018) Role of cytokinins in senescence, antioxidant defence and photosynthesis. Int J Mol Sci 19(12):4045

Huang J, Khan MT, Perecin D, Coelho ST, Zhang M (2020) Sugarcane for bioethanol production: potential of bagasse in Chinese perspective. Renew Sust Energ Rev 133:110296

Hussain HA, Hussain S, Khaliq A, Ashraf U, Anjum SA, Men S, Wang L (2018) Chilling and drought stresses in crop plants: implications, cross talk, and potential management opportunities. Front Plant Sci 9:393

Hussain HA, Men S, Hussain S, Chen Y, Ali S et al (2019) Interactive effects of drought and heat stresses on morpho-physiological attributes, yield, nutrient uptake and oxidative status in maize hybrids. Sci Rep 9:3890

Hussain T, Koyro H-W, Zhang W, Liu X, Gul B, Liu X (2020) Low salinity improves photosynthetic performance in *Panicum antidotale* under drought stress. Front Plant Sci 11:481

Im Kim J, Baek D, Park HC, Chun HJ et al (2013) Overexpression of ArabidopsisYUCCA6 in potato results in high-auxin developmental pheno-types and enhanced resistance to water deficit. Mol Plant 6:337–349

Inman-Bamber N (2004) Sugarcane water stress criteria for irrigation and drying off. Field Crop Res 89:107–122

Inman-Bamber N, Smith D (2005) Water relations in sugarcane and response to water deficits. Field Crop Res 92:185–202

Inman-Bamber N, Lakshmanan P, Park S (2012) Sugarcane for water limited environments: theoretical assessment of suitable traits. Field Crop Res 134:95–104

Iqbal S, Wang X, Mubeen I et al (2022) Phytohormones trigger drought tolerance in crop plants: outlook and future perspectives. Front Plant Sci 12:799318

Iskandar HM, Casu RE, Fletcher AT et al (2011) Identification of drought-response genes and a study of their expression during sucrose accumulation and water deficit in sugarcane culms. BMC Plant Biol 11:12

Jangpromma N, Kitthaisong S, Lomthaisong K et al (2010) A proteomics analysis of drought stress responsive proteins as biomarker for drought-tolerant sugarcane cultivars. Am J Biochem Biotech 6:89–102

Jung JH, Altpeter F (2016) TALEN mediated targeted mutagenesis of the caffeic acid O-methyltransferase in highly polyploid sugarcane improves cell wall composition for production of bioethanol. Plant Mol Biol 92:131–142

Kakoulidou I, Avramidou EV, Baranek M et al (2021) Epigenetics for crop improvement in times of global change. Biology 10:766

Kannan B, Jung JH, Moxley GW et al (2018) TALEN-mediated targeted mutagenesis of more than 100 COMT copies/alleles in highly polyploid sugarcane improves saccharification efficiency without compromising biomass yield. Plant Biotechnol J 16:856–866

Khalil F, Naiyan X, Tayyab M, Pinghua C (2018) Screening of EMS-induced drought-tolerant sugarcane mutants employing physiological, molecular and enzymatic approaches. Agronomy 8:226

Khan SJ, Khan HU, Khan RD, Iqbal MM, Zafar Y (2000) Development of sugarcane mutants through in vitro mutagenesis. Pak J Biol Sci 3:1123–1125

Khueychai S, Jangpromma N, Daduang S et al (2015) Comparative proteomic analysis of leaves, leaf sheaths, and roots of drought-contrasting sugarcane cultivars in response to drought stress. Acta Physiol Plant 37:88

Kido ÉA, Ferreira Neto JR, Silva RL, Pandolfi V, Guimarães AC et al (2012) New insights in the sugarcane transcriptome responding to drought stress as revealed by super SAGE. Sci World J 2012:1–15

Koch AC, Ramgareeb S, Rutherford RS, Snyman SJ, Watt MP (2012) An in vitro mutagenesis protocol for the production of sugarcane tolerant to the herbicide imazapyr. In Vitro Cell Dev Biol Plant 4:417–427

Koetle MJ, Snyman SJ, Rutherford RS (2022) Ex vitro morpho-physiological screening of drought tolerant sugarcane epimutants generated via 5-azacytidine and imidacloprid treatments. Trop Plant Biol 15:288–300

Koetle MJ, Snyman SJ, Rutherford RS (2023) Rapid identification of drought tolerant sugarcane epimutants via in vitro chimera dissolution and near infrared screening *ex vitro*. Plant Stress 9: 100200

Kooyers NJ (2015) The evolution of drought escape and avoidance in natural herbaceous populations. Plant Sci 234:155–162

Kumar S, Trivedi PK (2018) Glutathione S-transferases: role in combating abiotic stresses including arsenic detoxification in plants. Front Plant Sci 9

Kumar T, Uzma KMR, Abbas Z, Ali GM (2014) Genetic improvement of sugarcane for drought and salinity stress tolerance using Arabidopsis vacuolar pyrophosphatase (AVP1) gene. Europe PMC 56:199–209

Laksana C, Sophiphun O, Chanprame S (2024) Lignin reduction in sugarcane by performing CRISPR/Cas9 site-directed mutation of SoLIM transcription factor. Plant Sci 340:111987

Lembke CG, Nishiyama MY, Sato PM et al (2012) Identification of sense and antisense transcripts regulated by drought in sugarcane. Plant Mol Biol 79:461–477

Levitt J (1972) Responses of plants to environmental stresses. Academic, New York

Levitt J (1980) Responses of plants to environmental stresses. In: Water, radiation, salt, and other stresses, vol 2. Academic, New York

Li J, Sima W, Ouyang B et al (2012a) Tomato SlDREB gene restricts leaf expansion and internode elongation by downregulating key genes for gibberellin biosynthesis. J Exp Bot 63:6407–6420

Li Q-F, Wang C, Jiang L et al (2012b) An interaction between BZR1 and DELLAs mediates direct signaling crosstalk between brassinosteroids and gibberellins in Arabidopsis. Sci Signal 5:72–72

Li C-N, Yang L-T, Srivastava MK et al (2014) Foliar applicaton of abscisic acid improves drought tolerance of sugarcane plant under severe water stress. Int J of Agriculture Innovations and Research. 3(1): 101–107

Li JF, Norville JE, Aach J, McCormack M, Zhang D et al (2013) Multiplex and homologous recombination-mediated genome editing in *Arabidopsis* and *Nicotiana benthamiana* using guide RNA and Cas9. Nat Biotechnol 31:688–691

Li C, Nong Q, Solanki MK, Liang Q, Xie J et al (2016) Differential expression profiles and pathways of genes in sugarcane leaf at elongation stage in response to drought stress. Sci Rep 6:25698

Li J, Phan TT, Li YR et al (2018) Isolation, transformation and overexpression of sugarcane SoP5CS gene for drought tolerance improvement. Sugar Tech 20:464–473

Li P, Li B, Saman S, Zong Y, Han Y, Hao X (2019) Photosynthesis and yield response to elevated CO_2, C4 plant foxtail millet behaves similarly to C3 species. Plant Sci 285:239–247

Li A, Lakshmanan P, He W et al (2020) Transcriptome profiling provides molecular insights into auxin-induced adventitious root formation in sugarcane (*Saccharum* spp. Interspecific Hybrids) microshoots. Plants 9(8):931

Li C, Wang Z, Nong Q et al (2021) Physiological changes and transcriptome profiling in *Saccharum spontaneum* L leaf under water stress and re-watering conditions. Sci Rep 11:5525

Li P, Lin P, Zhao Z et al (2022) Gene co-expression analysis reveals transcriptome divergence between wild and cultivated sugarcane under drought stress. Int J Mol Sci 23(1):569

Liu X, Zhang R, Ou H, Gui Y, Wei J et al (2018) Comprehensive transcriptome analysis reveals genes in response to water deficit in the leaves of *Saccharum narenga* (Nees ex Steud) hack. BMC Plant Biol 18:250

Lopes MS, Araus JL, van Heerden PDR et al (2011) Enhancing drought tolerance in C4 crops. J Exp Bot 62(9):3135–3153

Luis A (2015) ROS and RNS in plant physiology: an overview. J Exp Bot 66:2827–2837

Luo LJ (2010) Breeding for water-saving and drought-resistance rice (WDR) in China. J Exp Bot 61(3):3509–3517

Ma Y, Dias MC, Freitas H (2020) Drought and salinity stress responses and microbe-induced tolerance in plants. Front Plant Sci 11:591911

Machado R, Ribeiro R, Marchiori P, Machado D, Machado E, Landell M (2009) Biometric and physiological responses to water deficit in sugarcane at different phenological stages. Pesq Agrop Brasileira 44:1575–1582

Mahlanza T, Rutherford RS, Snyman SJ, Watt MP (2013) In vitro generation of somaclonal variant plants of sugarcane for tolerance to *Fusarium sacchari*. Plant Cell Rep 32:249–262

Manimekalai R, Narayanan J, Ranjini R et al (2018) hydrogen peroxide-induced oxidative stress in sugarcane and response expression pattern of stress-responsive genes through quantitative RT-PCR. Sugar Tech 20:681–691

Mao Y, Zhang H, Xu N, Zhang B, Gou F, Zhu JK (2013) Application of the CRISPR-Cas system for efficient genome engineering in plants. Mol Plant 6:2008–2011

Marino G, Brunetti C, Tattini M, Romano A, Biasioli F et al (2017) Dissecting the role of isoprene and stress-related hormones (ABA and ethylene) in *Populus nigra* exposed to unequal root zone water stress. Tree Physiol 37:1637–1647

Masoabi M, Lloyd J, Kossmann J, van der Vyver C (2017) Ethyl methanesulfonate mutagenesis and in vitro polyethylene glycol selection for drought tolerance in sugarcane (*Saccharum* spp). Sugar Tech 20:50–59

Masoabi M, Pols S, Hills P, Snyman S, van der Vyver C (2023a) Response of sugarcane plants with modified cytokinin homeostasis under water deficit conditions. Plant Stress 10:100240

Masoabi M, Snyman S, van der Vyver C (2023b) Characterization of an ethyl methanesulfonate-derived drought tolerant sugarcane mutant line. Ann Appl Biol:1–18

Masoabi M, Burger F, Botha A-M, Vlok M, Snyman S, van der Vyver C (2023c) Overexpression of Small Ubiquitin-Like Modifier proteases enhance drought tolerance in sugarcane (*Saccharum* spp. hybrid). Plant Biol 25(7):1121–1141

Mbambalala N, Panda SK, van der Vyver C (2021) Overexpression of AtBBX29 improves drought tolerance by maintaining photosynthesis and enhancing the antioxidant and osmolyte capacity of sugarcane plants. Plant Mol Biol Report 39:419–433

Medeiros DB, da Silva EC, Mansur Custodio Nogueira RJ, Teixeira MM, Buckeridge MS (2013) Physiological limitations in two sugarcane varieties under water suppression and after recovering. Theor Exp Plant Physiol 25:213–222

Mitra J (2001) Genetics and genetic improvement of drought resistance in crop plants. Curr Sci 80(6):758–763

Mittler R, Blumwald E (2015) The roles of ROS and ABA in systemic acquired acclimation. Plant Cell 27:64–70

Mohan C (2016) Genome editing in sugarcane: challenges ahead. Front Plant Sci 7:1542

Mohan C, Easterling M, Yau YY (2021) Gene editing technologies for sugarcane improvement: opportunities and limitations. Sugar Tech 15(1-17):8

Molinari HBC, Marur CJ, Daros E, De Campos MKF, De Carvalho JFR et al (2007) Evaluation of the stress-inducible production of proline in transgenic sugarcane (*Saccharum* spp): osmotic adjustment, chlorophyll fluorescence and oxidative stress. Physiol Plant 130:218–229

Morison JIL, Baker NR, Mullineaux PM, Davies WJ (2008) Improving water use in crop production. Philos Trans R Soc B Biol Sci 363:639–658

Munne-Bosch S, Penuelas J (2003) Photo-and antioxidative protection, and a role for salicylic acid during drought and recovery in field-grown *Phillyrea angustifolia* plants. Planta 217:758–766

Narayan JA, Chakravarthi M et al (2021) Overexpression of expansin EaEXPA1, a cell wall loosening protein enhances drought tolerance in sugarcane. Ind Crop Prod 159:113035

Narayan JA, Manoj VM, Nerkar G, Chakravarthi M, Dharshini S et al (2023) Transgenic sugarcane with higher levels of BRK1 showed improved drought tolerance. Plant Cell Rep 42(10): 1611–1628

Nawae W, Shearman JR, Tangphatsornruang S, Punpee P, Yoocha T et al (2020) Differential expression between drought-tolerant and drought-sensitive sugarcane under mild and moderate water stress as revealed by a comparative analysis of leaf transcriptome. Peer J 8:e9608

Nikam AA, Devarumath RM, Ahuja A, Babu H, Shitole MG, Suprasanna P (2015) Radiation-induced in vitro mutagenesis system for salt tolerance and other agronomic characters in sugarcane (Saccharum officinarum L). Crop J 3:46–56

Ogburn RM, Edwards EJ (2010) The ecological water-use strategies of succulent plants. In: Kader J-C, Delseny M (eds) Advances in botanical research, vol 55. Academic, London, pp 179–225

Oladosu Y, Rafii MY, Abdullah N, Hussin G, Ramli A, Rahim HA, Miah G, Usman M (2016) Principle and application of plant mutagenesis in crop improvement: a review. Biotechnol Biotechnol Equip 30(1):1–16

Oz MT, Altpeter A, Karan R, Merotto A, Altpeter F (2021) CRISPR/Cas9-Mediated multi-allelic gene targeting in sugarcane confers herbicide tolerance. Front Genome Edit 3:673566

Pandey P, Irulappan V, Bagavathiannan MV, Senthil-Kumar M (2017) Impact of combined abiotic and biotic stresses on plant growth and avenues for crop improvement by exploiting physio-morphological traits. Front Plant Sci 8:537

Panje R, Babu C (1960) Studies in *Saccharum spontaneum* Distribution and geographical association of chromosome numbers. Cytologia 25:152–172

Parry MAJ, Madgwick PJ, Bayon C, Tearall K, Hernandez-Lopez A et al (2009) Mutation discovery for crop improvement. J Exp Bot 60:2817–2825

Pereira-Santana A, Alvarado-Robledo EJ, Zamora-Briseño JA et al (2017) Transcriptional profiling of sugarcane leaves and roots under progressive osmotic stress reveals a regulated coordination of gene expression in a spatiotemporal manner. PLoS ONE 12:e0189271

Pimentel C (2014) Photoinhibition in a C4 plant, Zea mays L.: a minireview. Theor Exp Plant Physiol 26:157–165

Piperidis G, Piperidis N, D'Hont A (2010) Molecular cytogenetic investigation of chromosome composition and transmission in sugarcane. Mol Gen Genomics 284:65–73

Rahman MA, Ren L, Wu W, Yan Y (2015) Proteomic analysis of PEG-induced drought stress responsive protein in TERF1 overexpressed sugarcane (*Saccharum officinarum*) leaves. Plant Mol Biol Report 33:716–730

Rahman MA, Wu W, Yan Y et al (2021) Overexpression of TERF1 in sugarcane improves tolerance to drought stress. Crop Pasture Sci 72:268–279

Rajesh M, Geetika S, Diva B, Narendra S, Baxla AK, Singh KK (2016) Managing impact of extreme weather events in sugarcane in different agro-climatic zones of Uttar Pradesh. Mausam 6761:233–250

Ramiro DA, Melotto-Passarin DM, Barbosa MA, Santos F, Gomez SGP et al (2016) Expression of Arabidopsis Bax Inhibitor-1 in transgenic sugarcane confers drought tolerance. Plant Biotechnol J 14:1826–1837

Rampino P, Pataleo S, Gerardi C, Mita G, Perrotta C (2006) Drought stress response in wheat: physiological and molecular analysis of resistant and sensitive genotypes. Plant Cell Environ 29: 2143–2152

Raza G, Alı K, Ashraf M, Mansoor S, Javıd M, Asad S (2016) Overexpression of an H+-PPase gene from Arabidopsis in sugarcane improves drought tolerance, plant growth, and photosynthetic responses. Turk J Biol 40:109–119

Reis R, da Cunha B, Martins P, Martins M, Alekcevetch J et al (2014) Induced over-expression of *At*DREB2A CA improves drought tolerance in sugarcane. Plant Sci 221:59–68

Riaño-Pachón DM, Mattiello L (2017) Draft genome sequencing of the sugarcane hybrid SP80-3280. F1000Research 6:861

Rocha FR, Papini-Terzi FS, Nishiyama MY et al (2007) Signal transduction-related responses to phytohormones and environmental challenges in sugarcane. BMC Genomics 8:71

Rodrigues FA, de Laia ML, Zingaretti SM (2009) Analysis of gene expression profiles under water stress in tolerant and sensitive sugarcane plants. Plant Sci 176:286–302

Rodrigues FA, Graça JP, Laia ML, Nhani-Jr A, Galbiati JA et al (2011) Sugarcane genes differentially expressed during water deficit. Biol Plant 55:43–53

Rutherford RS, Snyman SJ, Watt MP (2014) *In vitro* studies on somaclonal variation and induced mutagenesis: progress and prospects in sugarcane (*Saccharum* spp)—a review. J Hortic Sci Biotechnol 89:1–16
Sage RF (2004) The evolution of C4 photosynthesis. New Phytol 161(2):341–370
Sah SK, Reddy KR, Li J (2016) Abscisic acid and abiotic stress tolerance in crop plants. Front Plant Sci 7:571
Sales CR, Ribeiro RV, Silveira JA, Machado EC, Martins MO, Lagôa AM (2013) Superoxide dismutase and ascorbate peroxidase improve the recovery of photosynthesis in sugarcane plants subjected to water deficit and low substrate temperature. Plant Physiol Biochem 73:326–336
Salvato F, Loziuk P, Kiyota E et al (2019) Label-free quantitative proteomics of enriched nuclei from sugarcane (Saccharum ssp) stems in response to drought stress. Proteomics 19:1900004
Saravanan S, Kumar KK, Raveendran M, Sudhakar D, Arul L et al (2018) Genetic engineering of sugarcane for drought and salt tolerant transgenic plants expressing the BcZAT12 gene. Int J Curr Microbiol App Sci 7:1594–1613
Screenivasan TV, Ahloowalia BS, Heinz DJ (1987) Sugarcane improvement through breeding. Cytogenetics 5:211–253
Selvi A, Devi K, Manimekalai R, Prathima PT (2020) Comparative analysis of drought-responsive transcriptomes of sugarcane genotypes with differential tolerance to drought. 3 Biotech 10:236
Seo M, Koshiba T (2011) Transport of ABA from the site of biosynthesis to the site of action. J Plant Res 124:501–507
Sharma DK, Andersen SB, Ottosen C-O, Rosenqvist E (2015) Wheat cultivars selected for high Fv/Fm under heat stress maintain high photosynthesis, total chlorophyll, stomatal conductance, transpiration and dry matter. Physiol Plant 153:284–298
Shi L, Guo M, Ye N, Liu Y, Liu R, Xia Y, Cui S, Zhang J (2015a) Reduced ABA accumulation in the root system is caused by aba exudation in upland rice (*Oryza sativa* L. var. Gaoshan1) and this enhanced drought adaptation. Plant Cell Physiol 56(5):951–964
Shi J, Habben JE, Archibald RL, Drummond BJ, Chamberlin MA et al (2015b) Overexpression of ARGOS genes modifies plant sensitivity to ethylene, leading to improved drought tolerance in both Arabidopsis and maize. Plant Physiol 169:266–282
Shi J, Gao H, Wang H, et al (2017) ARGOS8 variants generated by CRISPR-Cas9 improve maize grain yield under field drought stress conditions. Plant Biotechnology Journal. 15: 207–216. https://doi.org/10.1111/pbi.12603
Silva MDA, Goncalves da Silva JA, Enciso J, Sharma V, Jifon J (2008) Yield components as indicators of drought tolerance of sugarcane. Sci Food Agric 65:620–627
Smit M, Singels A (2006) The response of sugarcane canopy development to water stress. Field Crop Res 98:91–97
Soares RAB, Oliveira PFM, Cardoso HR, Vasconcelos ACM, Landell MGA, Rosenfeld U (2004) Effect of irrigation on the development and productivity of two varieties of sugarcane harvested at the beginning of the harvest. STAB Sugar Alcohol By-Products 22:38–41
Souza GM, Bergès H, Bocs S, Casu R, D'Hont A et al (2011) The sugarcane genome challenge: strategies for sequencing a highly complex genome. Trop Plant Biol 4:145–156
Sugiharto B (2017) Biotechnology of drought-tolerant sugarcane. In: Sugarcane—technology and research. IntechOpen. https://doi.org/10.5772/intechopen72436
Takahashi S, Murata N (2008) How do environmental stresses accelerate photoinhibi□on?. Trends in Plant Science. 13(4):178–182. https://doi.org/10.1016/j.tplants.2008.01.005
Tardieu F (2013) Plant response to environmental conditions: assessing potential production, water demand, and negative effects of water deficit. Front Physiol 18(4):17
Terceros GC, Resentini F, Cucinotta M, Manrique S, Colombo L, Mendes MA (2020) The Importance of cytokinins during reproductive development in Arabidopsis and beyond. Int J Mol Sci 21(21):8161
Thirugnanasambandam PP, Hoang NV, Henry RJ (2018) The challenge of analysing the sugarcane genome. Front Plant Sci 9:616

Tipple BJ, Pagani M (2007) The early origins of terrestrial C4 photosynthesis. Annu Rev Earth Planet Sci 35:435–461
Tonosaki K, Fujimoto R, Dennis ES, Raboy V, Osabe K (2022) Will epigenetics be a key player in crop breeding? Front Plant Sci 13:958350
Tripathy BC, Oelmüller R (2012) Reactive oxygen species generation and signalling in plants. Plant Signal Behav 7:1621–1633
Tripathi P., Chandra A. and Prakash J. (2019). Physio-biochemical assessment and expression analysis of genes associated with drought tolerance in sugarcane (Saccharum spp. hybrids) exposed to GA3 at grand growth stage. Plant Biology Journal. 21: 45–53. https://doi.org/10.1111/plb.12919
Trujillo-Montenegro JH, Rodríguez Cubillos MJ, Loaiza CD et al (2021) Unravelling the genome of a high yielding Colombian sugarcane hybrid. Front Plant Sci 12:694859
Vantini JS, Dedemo GC, Jovino Gimenez DF, Fonseca LFS, Tezza RID et al (2015) Differential gene expression in drought-tolerant sugarcane roots. Genet Mol Res 14:7196–7207
Verma KK, Song XP, Zeng Y, Li DM, Guo DJ et al (2020) Characteristics of leaf stomata and their relationship with photosynthesis in *Saccharum officinarum* under drought and silicon application. ACS Omega 5(37):24145–24153
Verma KK, Song XP, Verma CL et al (2021) Functional relationship between photosynthetic leaf gas exchange in response to silicon application and water stress mitigation in sugarcane. Biol Res 54:15
Vettore AL, da Silva FR, Kemper EL, Souza GM, da Silva AM et al (2003) Analysis and functional annotation of an expressed sequence tag collection for tropical crop sugarcane. Genome Res 13: 2725–2735
Vilela RD, Bezerra BKL, Froehlich A, Endres L (2017) Antioxidant system is essential to increase drought tolerance of sugarcane. Ann Appl Biol 171:451–463
Vital CE, Giordano A, de Almeida Soares E et al (2017) An integrative overview of the molecular and physiological responses of sugarcane under drought conditions. Plant Mol Biol:94577–94594
Waclawovsky AJ, Sato PM, Lembke CG, Moore PH, Souza GM (2010) Sugarcane for bioenergy production: an assessment of yield and regulation of sucrose content. Plant Biotech J 8:263–276
Wang W, Vinocur B, Altman A (2003) Plant responses to drought, salinity and extreme temperatures: towards genetic engineering for stress tolerance. Planta 218:1–14
Wang Z-Z, Zhang S-Z, Yang B-P, Li Y (2005) Trehalose synthase gene transfer mediated by *Agrobacterium tumefaciens* enhances resistance to osmotic stress in sugarcane. Sugar Tech 7: 49–54
Wang Y, Cheng X, Shan Q, Zhang Y, Liu J et al (2014) Simultaneous editing of three homoeoalleles in hexaploid bread wheat confers heritable resistance to powdery mildew. Nat Biotechnol 32:947–951
Wang X, Hu L, Wang X, Li N, Xu C, Gong L, Liu B (2016) DNA methylation affects gene alternative splicing in plants: an example from rice. Mol Plant 9:305–307
Wang TJ, Wang XH, Yang QH (2020) Comparative analysis of drought-responsive transcriptome in different genotype *Saccharum spontaneum* L. Sugar Tech 22:411–427
Weksanthia N, Chaisan T, Wannarat W, et al. (2021) Mutagenesis and identification of sugarcane mutants using survival on polyethylene glycol and leaf damage under managed water stress. International Journal of Agronomy. 2021: 4387696. https://doi.org/10.1155/2021/4387696
Welbaum GE, Meinzer FC (1990) Compartmentation of solutes and water in developing sugarcane stalk tissue. Plant Physiol 93:1147–1153
Wu W, Du K, Kang X et al (2021) The diverse roles of cytokinins in regulating leaf development. Hortic Res 8:118
Xiao S, Wu Y, Xu S, Jiang H, Hu Q, Yao W, Zhang M (2022) Field evaluation of TaDREB2B-ectopic expression sugarcane (*Saccharum* spp hybrid) for drought tolerance. Front Plant Sci 13: 963377

Yahoueian SH, Bihamta MR, Babaei HR et al (2021) Proteomic analysis of drought stress response mechanism in soybean (Glycine max L.) leaves. Food Sci Nutr 9(4):2010–2020

Yang X, Lu M, Wang Y et al (2021) Response mechanism of plants to drought stress. Horticulturae 7(3):50

Yang S, Chu N, Zhou H et al (2022) Integrated analysis of transcriptome and metabolome reveals the regulation of chitooligosaccharide on drought tolerance in Sugarcane (Saccharum spp. hybrid) under drought stress. Int J Mol Sci 23(17):9737

Yasmeen S, Khan MT, Khan IA (2020) Revisiting the physical mutagenesis for sugarcane improvement: a stomatal prospective. Sci Rep 10:16003

Zaman M, Kurepin LV, Catto W et al (2016) Evaluating the use of plant hormones and biostimulators in forage pastures to enhance shoot dry biomass production by perennial ryegrass (*Lolium perenne* L.). J Sci Food Agric 96:715–726

Zamora-Briseño JA, Ruíz-May E, Elizalde-Contreras JM et al (2023) Aggregation-resistant proteins are adjusted during drought in sugarcane plants. Acta Physiol Plant 45:43

Zandalinas SI, Rivero RM, Martínez V et al (2016) Tolerance of citrus plants to the combination of high temperatures and drought is associated to the increase in transpiration modulated by a reduction in abscisic acid levels. BMC Plant Biol 16:1–16

Zhang S-Z, Yang B-P, Feng C-L, Chen R-K, Luo J-P, Cai W-W, Liu F-H (2006) Expression of the Grifola frondosa trehalose synthase gene and improvement of drought tolerance in sugarcane (*Saccharum officinarum L*). J Integr Plant Biol 48:453–459

Zhang J, Zhang X, Tang H et al (2018) Allele-defined genome of the autopolyploid sugarcane *Saccharum spontaneum* L. Nat Genet 50:1565–1573

Zhang Y-B, Yang S-L, Dao J-M et al (2020) Droughtinduced alteratons in photosynthetc, ultrastructural and biochemical traits of contrastng sugarcane genotypes. PLoS ONE. 15(7): e0235845. https://doi.org/10.1371/journal.pone.0235845

Zhao D, Li Y-R (2015) Climate change and sugarcane production: potential impact and mitigation strategies. Int J Agron 2015:547386

Zhou G, Yang L-T, Li Y et al (2012) Proteomic analysis of osmotic stress-responsive proteins in sugarcane leaves. Plant Mol Biol Report 30:349–359

Zingaretti SM, Rodrigues FA, da Graça JP, de Matos Pereira L, Lourenço MV (2012) Sugarcane responses at water deficit conditions. Water Stress 13:255–276

Zivcak M, Brestic M, Sytar O (2016) Osmotic adjustment and plant adaptation to drought Stress. In: Hossain M, Wani S, Bhattacharjee S, Burritt D, Tran LS (eds) Drought stress tolerance in plants. Springer, p 1. https://doi.org/10.1007/978-3-319-28899-4_5

Zwack PJ, Rashotte AM (2013) Cytokinin inhibition of leaf senescence. Plant Signal Behav 8(7): e24737

Chapter 6
Crop Transcriptome Response for Abiotic Stress Tolerance

Behnam Derakhshani, Ki-Hong Jung, and Youko Oono

Abstract Abiotic stresses extensively reduce agricultural crop production globally. For functional genomics studies, it is important to understand the dynamic expression profiles of transcribed genes in different tissues, stages of development and in response to environmental stimuli. Transcriptomics studies have contributed significantly to the understanding of the intricate dynamics of gene expression during stress conditions, unearthing novel stress-responsive candidate genes and signaling pathways. RNA sequencing (RNA-Seq) provides a far more precise measurement of levels of transcripts and their isoforms than other methods. Furthermore, comparative transcriptomics have provided more information about plants' response to diverse stresses. Herein, we review studies that have used RNA-Seq profiling mainly in rice, barley and wheat to focus on their responses to the abiotic stresses including drought, salinity, heat, cold and heavy metals, which are important for developing tolerance mechanisms and designing climate-smart crops. Meanwhile, the utilization of pan-genomes in RNA-Seq profiling of rice and the available online RNA-Seq databases for the above-mentioned crops and their effective utilization are outlined. The acquired knowledge from RNA-Seq studies can support breeding for the identification of good agronomic traits in crop genome, to improve abiotic stress tolerance and crop yield.

B. Derakhshani
Graduate School of Green Green-Bio Science and Crop Biotech Institute, Kyung Hee University, Yongin, Republic of Korea
e-mail: behnam.derakhshani@khu.ac.kr

K.-H. Jung
Graduate School of Green Green-Bio Science and Crop Biotech Institute, Kyung Hee University, Yongin, Republic of Korea

Research Center for Plant Plasticity, Kyung Hee University, Yongin, Republic of Korea

Y. Oono (✉)
Institute of Crop Sciences, National Agriculture and Food Research Organization, Tsukuba, Ibaraki, Japan
e-mail: oono.youko090@naro.go.jp

S. K. Panda, M. A. Khan (eds.), *Plant Functional Genomics for Abiotic Stress Resilience*, Plant in Challenging Environments 7,
https://doi.org/10.1007/978-3-032-01704-8_6

Keywords RNA-Seq · Comparative transcriptomics · Crop · Abiotic stress · Databases · Pan-genome

6.1 Introduction

Plants are sessile, so they experience various inescapable abiotic stresses in their ecological habitat. Abiotic stressors, such as drought, soil salinity, heat, cold, and heavy metals are major limiting factors affecting crop production both qualitatively and quantitatively (Shahid et al. 2015). In order to enhance stress tolerance and increase the plant productivity, the focus of research has already shifted toward understanding the key molecular targets, regulators, and their signaling involved in plant interactions with the environment (Mosa et al. 2017).

Transcriptomic analyses have provided an in-depth knowledge of the cellular and molecular responses underlying plant adaptation to environmental stresses (Matsui et al. 2008; Rasheed et al. 2016). RNA-Seq, the latest Next Generation Sequencing (NGS) technique for investigating genome-wide transcriptomes, helps to examine the expressional variation in genes in contrasting sets of samples or panels subjected to different stress treatments and to pinpoint the potential candidate genes to provide a better crop productivity (Zhang et al. 2014). Moreover, it has advantages such as the ability to identify novel genes/transcripts, detect low abundance transcripts, identify genetic variants, and detect more differentially expressed genes (DEGs) with higher fold-change in comparison with microarray. Thus, RNA-Seq analysis has become an efficient tool for analyzing the transcriptomic responses of many plant species to various abiotic stresses and studying the associated molecular mechanisms (Onyemaobi et al. 2022; Rangan et al. 2020; Shan et al. 2013; Zhang et al. 2012). Third-generation long-read sequencing technologies, including Pacific Biosciences (PacBio) and Oxford Nanopore Technology (ONT), provide huge advantages in sequence length, transcript identification, and genetic information richness compared with second-generation Illumina RNA-Seq technology (Heikema et al. 2020; Zhang et al. 2020). An analysis of plant transcriptomes with these new powerful methods is expected to uncover novel gene regulatory mechanisms that control biological outcomes during plant development and in response to various stresses.

The number of RNA-Seq data from the plant community has been increasing exponentially in recent years. However, most RNA-Seq studies in plants remain underutilized and inaccessible due to the use of disparate transcriptome references and the lack of skills and resources to analyze and visualize these data. The availability of online databases and archives enables users to perform transcriptome-wide analysis of plant's stress response combined with an intuitive and interactive interface (Mehta et al. 2019). Here we present an RNA-Seq overview based on scientific articles covering food crops (rice, barley and wheat) abiotic stress responses, their available RNA-Seq based databases, and rice pan-genome usage in RNA-Seq studies. As shown in Fig. 6.1, an overview for the various applications of RNA-Seq analysis and the role of RNA-Seq analysis for molecular breeding

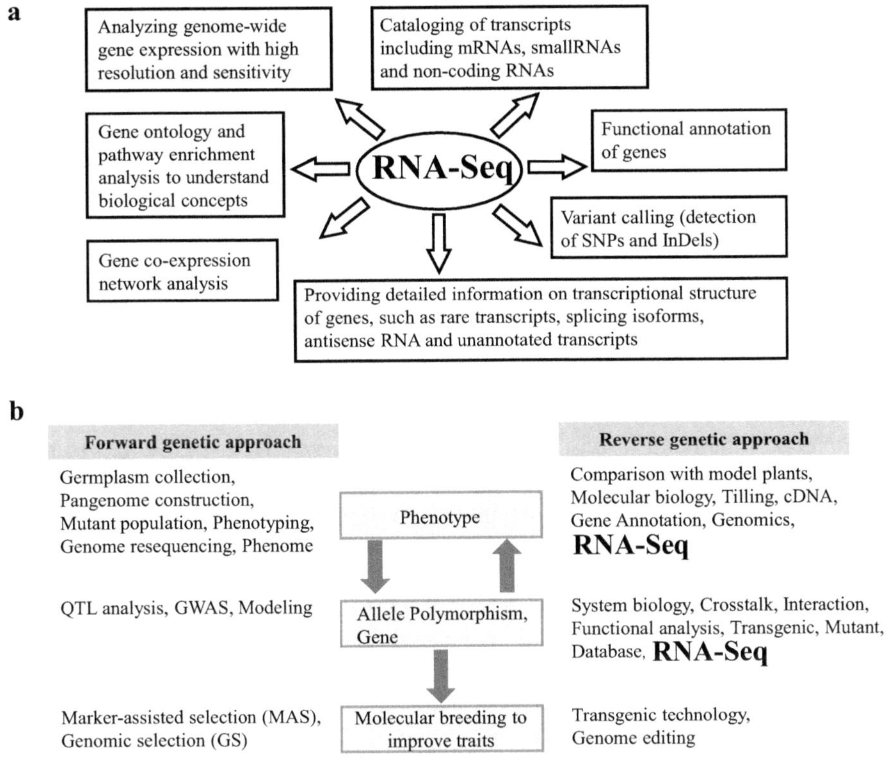

Fig. 6.1 (**a**) Application of crop transcriptome analysis under abiotic stress using RNA-seq, (**b**) Conceptual scheme and keywords for molecular breeding approaches to improve agronomical traits in abiotic stress

approaches to improve agronomical traits in abiotic stress is provided. We hope this chapter provides information helping breeders make good decisions favoring the improvement of crops worldwide.

6.2 RNA-Seq Profiling Under Abiotic Stress in Rice

Rice (*Oryza sativa* L.) is consumed as a main staple food by >50% of the global population. With a genome size of ~390 megabases (Mb), rice has the smallest genome among the domesticated cereals and the primary reason why it was the first crop genome to be sequenced 19 years ago (Sasaki and International Rice Genome Sequencing 2005). Although many global transcriptomic studies have been performed during the reproductive stage in rice, very limited information is available for the grain-filling stage. Recently, Liang et al. conducted comparative RNA-Seq analysis of a moderately tolerant and susceptible varieties during the grain-filling stage and demonstrated that genes and gene families related to transcription factors (TFs), drought tolerance genes, and reactive oxygen species (ROS) scavengers were significantly altered (Liang et al. 2021). The summary of their study is shown in Table 6.1. Tiwari et al. carried out time-based comparative transcriptomic profiling of two varieties that showed contrasting drought tolerance. Comparatively, a significant proportion of the identified DEGs related to phytohormone signaling, stress-responsive genes (*Late Embryogenesis Abundant Protein* (*LEA*), *Dehydration-Responsive Element-Binding Protein* (*DREB*)), TFs (*APETALA2/Ethylene-Responsive Factor* (*AP2/ERF*), *Myeloblastosis* (*MYB*), *WRKY*, *Basic Helix-Loop-Helix* (*bHLH*)), and genes involved in photosynthesis and antioxidative mechanisms were higher in the drought tolerant Heena variety (Tiwari et al. 2021). Integration of rice genome-scale metabolic network and transcriptomic data of a drought-tolerant rice line, DK151, helps in identifying critical TFs such as *AP2/ERF*, *basic leucine zipper* (*bZIP*), *MYB*, and (*NAM* (*no apical meristem*), *ATAF1−2* (*Arabidopsis thaliana Transcription Activator Factor 1/2*), *CUC2* (*cup-shaped cotyledon 2*)) *NAC* that control the important nodes in the gene regulatory pathway (Mohanty et al. 2016). Co-localization analysis of DEGs against QTL hotspots identified several putative candidate genes including a zinc finger domain containing protein, a *bZIP* TF, a *MYB* factor and an *Oryza Sativa MINICHROMOSOME MAINTENANCE FACTOR 1, AGAMOUS, DEFICIENS, Serum Response Factor* (*OsMADS57*), which were found to be highly up-regulated in root of drought tolerant cv. Nootripathu (Muthurajan et al. 2018).

Salinity tolerance in rice is a very important trait, especially in areas that are affected by soil salinity, such as tsunami-devastated areas and coastal regions in rice-producing countries. Soil salinity is first detected by the roots being in direct contact with the soil; thus, it is important to understand how roots contribute to salinity tolerance in agricultural crops. Cartagena et al. conducted a comparative RNA-Seq analysis on salt tolerant (cv. Mulai) and susceptible (cv.IR29) roots. More transporters, such as ion- and sugar-related transporters were identified in Mulai roots, and they play a role in the regulation of salt tolerance (Cartagena et al. 2021). Rice salt tolerance at the bud burst stage directly affects the seedling survival rate and the final yield in the direct seeding cultivation model. RNA-Seq for cv. Weiguo (salt-tolerant) and cv. IR36 (salt-susceptible) after 36 h salt stress at the bud burst stage, identified five differentially upregulated genes in the major QTL *qRSL7* region. Further sequence analysis revealed a 1-bp insertions/deletion (InDel) difference in *LOC_Os07g38240* (*stress-associated protein 16; OsSAP16*) between cv. IR36 and cv. Weiguo, which is useful for gene cloning and for improving the salt tolerance of rice varieties (Lei et al. 2020). Melatonin has been implicated in the salt tolerance of rice. Xie et al. reported that the AP2/EREBP-HB-WRKY cascade is activated by stress, imparting melatonin-mediated salt tolerance in rice (Xie et al. 2021).

High temperature anthesis is one of the most serious stress factors for rice production, causing irreversible yield losses and reducing grain quality. It is estimated that rice yields declined by 8.1% for each 1 °C increase in mean temperature during the growing season (Song et al. 2022). A transcriptional profile analysis under different time courses of heat treatment on the anther of thermotolerant

Table 6.1 Transcriptomic studies on analyzing abiotic stress responses in rice

Stress	Keywords	Tissue	Stage	Cultivars	References
Drought	TFs, Drought tolerance genes, and Reactive oxygen species (ROS) scavengers	Grain	Grain-filling stage	1. cv. 4610, *Indica* long-grain (PI615037) introduced from China (Dilday et al. 2001), moderate (T)[a] 2. cv. Rondo, developed by mutation breeding from the Chinese *Indica* germplasm '4484' (PI615022), (S)[b]	Liang et al. (2021)
Drought	Phytohormone signaling, Stress-responsive genes (*LEA, DREB*), TFs (*AP2/ERF, MYB, WRKY, bHLH*), genes involved in photosynthesis and antioxidative mechanisms	Whole root	Ten-day-old seedlings	1. cv. Heena, (both cultivars obtained from different parts of Northern India), (T) 2. cv. Kiran, (S)	Tiwari et al. (2021)
Salt	Stress-associated protein 16 (*OsSAP16; LOC_Os07g38240*)	Buds	Bud burst stage	1. cv. Weiguo, *Japonica*, (T) 2. cv. IR36, *Indica*, (S)	Lei et al. (2020)
Heat	Agmatine coumarin-acyltransferase gene (*OsACT; LOC_Os03g47860*)	Mature anthers	Anthesis stage	1. cv. SDWG005, an African landrace, *Indica*, (T) 2. cv. 9311, *Indica*, widely used restorer line in China, (S)	Liu et al. (2020b)
Heat	*Auxin-responsive Aux/IAA gene family member* (*OsIAA2*; *LOC_Os01g09450*), *Squalene synthetase* (*LOC_Os03g59040*) and *Cysteine synthase* (*LOC_Os12g42980*)	Leaves	15-day-old seedlings	–	Wei et al. (2021)
Heat	VQ domain containing protein (*OsVQ30; LOC_Os07g48710*)	Shoot	13-day-old seedlings	1. cv. FACAGRO 64, both cultivars are from the 3K Rice Genome Project, *Indica*, (T) 2. cv. PUILLIPINA KATARI, *Indica*, (S)	Li et al. (2023a)
Cold	*CATALASE C* (*OsCATC*) and *LOC_Os03g49440* encoding a protein similar to a sugar-starvation induced protein (fragment)	Leaves	30-day-old seedlings	1. cv. Koshihikari, *Japonica*, (T) 2. cv. Nipponbare, *Japonica*, (T) 3. cv. Fujisaka5, *Japonica*, (T) 4. cv. Lemont, *Japonica*, (T) 5. cv. Nanjing11, *Indica*, (S)	Zhong et al. (2024)

(continued)

Table 6.1 (continued)

Stress	Keywords	Tissue	Stage	Cultivars	References
				6. cv. 9311, *Indica*, (S) 7. cv. MADINIKA 1329::GERVEX 8366-C1, *indica*, (S) 8. cv. NONA_BOKRA, *Indica*, (S) 9. cv. TEQING, *Indica*, (S) 10. cv. KOGONI 91-1::C1, *Indica*, (S) 11. cv. Css12, *Indica*, (S)	
Cold	*Heat shock factors* (*HSF*; *LOC_Os09g28200* and *LOC_Os09g28354*), *bHLH* (*LOC_Os09g28210*), Lysine motif-containing protein (*LYP4; LOC_Os09g27890*), emp24/gp25L/p24 family protein (*LOC_Os09g27980*) and Retrotransposon protein (*LOC_Os09g28370*)	Seeds	Germination stage	1. cv. DF104, *Japonica*, (T) 2. cv. DN430, *Japonica*, (S)	Yang et al. (2023)
Cd	*OsNRT1.5A* (*LOC_Os02g46460*), *OsNramp5* (*LOC_Os07g15370*), *OsIRT1* (*LOC_Os03g46470*), *OsVIT2* (*LOC_Os09g23300*) and *OsABCC1* (*LOC_Os04g52900*)	Panicle node and the first node under panicle (node I)	Grain-filling	1. cv. Xiangwanxian No. 12, low Cd accumulation 2. cv. Yuzhenxiang, high Cd accumulation Both cultivars are Chinese, obtained from the germplasm recourses bank of Hunan rice research institute, *Indica*.	Liu et al. (2020a)

[a]*T* tolerant, [b]*S* susceptible

cv. SDWG005 at anthesis were conducted and identified *agmatine coumarin-acyltransferase* gene (*OsACT*; *LOC_Os03g47860*) that may involve in cv. SDWG005 thermotolerance (Liu et al. 2020b). The combination of novel methods is a useful tool for discovering candidate genes linked with complex characters. Li et al. identified the candidate gene encoding *Valine-glutamine (VQ) domain containing protein* (*OsVQ30*; *LOC_Os07g48710*) in the *qHT7* region by employing genome-wide association study (GWAS) and transcriptomics in rice (Li et al. 2023a). Yang et al. detected 11 promising candidate genes for heat stress tolerance in rice by combining GWAS and transcriptome data. *LOC_Os05g07050* encodes a putative pre-mRNA processing splicing factor and harbors a mutation in tolerant types (Leu replaced Ile) relative to cv. Nipponbare. *LOC_Os03g16460* encodes a putative uncharacterized protein and harbors seven mutation sites. Plants carrying a single mutation displayed relatively low heat tolerance, whereas plants with all seven mutation sites in *LOC_Os03g16460* showed the highest heat tolerance (Yang et al. 2022). Similarly, combined QTL mapping, GWAS, and transcriptome analysis facilitated the identification of candidate genes responsive to high temperature stress in rice, including *Auxin-responsive Aux/IAA gene family member, expressed* (*OsIAA2*; *LOC_Os01g09450*), *squalene synthetase, putative, expressed* (*LOC_Os03g59040*) and *cysteine synthase, putative, expressed* (*LOC_Os12g42980*), that could be applied to rice heat-tolerant breeding programs (Wei et al. 2021).

Cold damage is a global issue that particularly affects rice-producing countries like Japan, Korea, and China. Through QTL-Seq, QTL mapping and RNA-Seq, six candidate genes with single nucleotide polymorphisms (SNPs) and InDels in qLTG9 region were identified including two *Heat shock factors* (*HSF*; *LOC_Os09g28200* and *LOC_Os09g28354*), *bHLH* (*LOC_Os09g28210*), lysine motif-containing protein (*LYP4*; *LOC_Os09g27890*), emp24/gp25L/p24 family protein (*LOC_Os09g27980*) and retrotransposon protein (*LOC_Os09g28370*), which can be used as targets for cold tolerance functional characterization (Yang et al. 2023). The direct RNA-Seq technique recently developed by ONT provides an alternative way to characterize the transcriptome (Garalde et al. 2018). RNA-Seq profiling using nanopore technology and genome re-sequencing between the two near-isogenic lines (NILs), Towada and ZL31 with different cold sensitivity, revealed that a potential candidate gene *OsWRKY24* (*LOC_Os01g61080*), contained sequence variations in the promoter, 5′ untranslated region (UTR), and the exon regions, which may change its transcript abundance in the two varieties (Wu et al. 2023).

In Asia, rice has been identified as one of the major sources of heavy metal intake for humans (Fangmin et al. 2006). The toxic effect of cadmium (Cd) has been a worldwide concern since the outbreak of Itai-Itai disease in the mid-20th century in Japan caused by the daily consumption of Cd-contaminated rice (Horiguchi et al. 1994). It was reported a comprehensive transcriptome analysis by RNA-Seq following Cd exposure in rice roots and demonstrated that Cd exposure upregulates metal transporter genes including Heavy metal transporting ATPase (HMA) family, multi-antimicrobial extrusion protein (MatE), Zrt (zinc-regulated transporter) / IRT-like protein (Zip), Cation_efflux, natural resistance-associated macrophage protein

(NRAMP) family transporter, pleiotropic drug resistance-type ATP-binding protein (PDR family), and low-affinity cation transporter (LCT) (Oono et al. 2014). The nodes play a key role in grain Cd accumulation by regulating the redistribution of Cd in nodes (Wang et al. 2020). Plants have evolved various molecular mechanisms to respond to heavy metal stress, including microRNA (miRNA)-guided gene regulation at the posttranscriptional level (Ding et al. 2017). An increasing number of miRNAs have been shown to play important regulatory roles in plant responses to various metal stresses. Liu et al. investigated the genotypic differences in gene and miRNA expression patterns in panicle node and first node in response to Cd using two rice genotypes differing in grain Cd accumulation. Distinct expression patterns of various transporter genes such as *nitrate transporter 1.5A* (*OsNRT1.5A*; *LOC_Os02g46460*), *OsNramp5* (*LOC_Os07g15370*), *Iron-Regulated Transporter 1* (*OsIRT1*; *LOC_Os03g46470*), *Vacuolar Iron Transporter 2* (*OsVIT2*; *LOC_Os09g23300*) and *ATP-binding cassette transporter multidrug resistance protein 1* (*OsABCC1*; *LOC_Os04g52900*) resulted in differential Cd accumulation in different nodes (Liu et al. 2020a). Little is known about the miRNA-mediated post-transcriptional regulatory mechanisms involved in arsenic (As) responses and tolerance in plants. In a study with cv. Nipponbare, it has been shown that under As stress, specific genes were identified related to heavy metal transportation, jasmonate (JA) biosynthesis and signaling. Furthermore, it has been reported that 14 As-responsive miRNAs in rice regulated gene expression in metabolism, transport, and signaling pathways (Yu et al. 2012). Selenium (Se), a trace element, is a known antagonist of As toxicity. In an integrated study of RNA-Seq and proteome profiling, the expression of As transporter genes *viz., Nodulin like Intrinsic Protein* (*NIP1;1*; *LOC_Os02g13870*), *NIP2;1* (*LOC_Os02g51110*), *ATP-binding cassette (ABC) subfamily G protein (ABCG5*; *LOC_Os03g17350*), *NRAMP1* (*LOC_Os07g15460*), *NRAMP5* (*LOC_Os07g15370*), *Delta-tonoplast intrinsic protein* (*TIP2;2*; *LOC_Os06g22960*) as well as sulfate transporters, *SULTR3;1* (*LOC_Os10g28440*) and *SULTR3;6* (*LOC_Os01g52130*), were higher in As + Se compared to As alone exposure, which resulted in reduced As accumulation and toxicity (Chauhan et al. 2020).

6.3 Rice Pan-Genome

In the past decade, owing to the high degree of genomic variation, research has demonstrated that a single reference genome does not fully represent the diversity of a species (Bayer et al. 2020). The concept of pan-genomes was first proposed by Tettelin in 2005 (Tettelin et al. 2005). Pan-genomes represent the genomic diversity of a species and include core genes, found in all individuals of the species, as well as variable genes, which are absent in some individuals (Danilevicz et al. 2020; Lei et al. 2021). Abiotic stress responsive genes are frequently enriched within variable gene groups (Petereit et al. 2022). Thus, pan-genomes aid in breeding varieties better adapted to diverse environmental conditions, thereby promoting sustainable agriculture (Jayakodi et al. 2020).

A Rice Pan-genome Browser (RPAN; https://cgm.sjtu.edu.cn/3kricedb/ or https://www.rmbreeding.cn/pan3k) was developed. It provides multiple search and visualization functions of genomic sequences, gene annotations, gene presence–absence variation (PAV) information of 3010 accessions and gene expression data of the rice pan-genome (Sun et al. 2017). On the other hand, a tremendous amount of transcriptomic data also continues to accumulate, which could be used to identify genes with potential for plant breeding applications. Recently, a study performed a transcriptome analysis of the Global Mini-Core Rice collection consisting of 202 accessions after salt stress treatment of plants at the seedling stage. Using rice super pan-genome mapping and by combining GWAS, eQTL, and DEG datasets, the authors swiftly pinpointed the most likely causal candidate gene *salt tolerance gene 5* (*STG5*; *LOC_Os05g49700*). Further, the authors clarified that *STG5* conferred salt tolerance through the regulation of Na^+ and K^+ homeostasis (Wei et al. 2024). It has been shown that building a pan-transcriptome for diverse rice subspecies will improve understanding of their cold resistance system (Zhong et al. 2024). In a pan-transcriptome study combining transcriptome data from 11 rice cultivars under cold stress, *CATALASE C* (*OsCATC*; *LOC_Os03g03910*) and *LOC_Os03g49440* encoding a protein similar to a sugar-starvation induced protein (fragment), were identified as candidate genes that could be used to enhance cold tolerance in rice by genetic engineering (Zhong et al. 2024). Another study reported that heat stress tolerance genes have been identified by combining data from pan-genomes, pan-transcriptomes, and QTLs (Woldegiorgis et al. 2022). The genomic variation analysis detected 76,435 SNPs as the heat-tolerance-related SNPs and mapped 1677 DEGs to 46 heat stress (HS)-tolerant QTL regions. Two DEGs, *calmodulin-binding protein 60 A-like* (*maker_00000041*) and *cysteine-rich receptor-like protein kinase 6* (*maker_00001878*), were up-regulated and harbored SNPs linked to HS tolerance in the non-reference rice sequences (Woldegiorgis et al. 2022).

6.4 RNA-Seq Profiling Under Abiotic Stress in Barley

A staple food of ancient civilizations, today barley (*Hordeum vulgare* L.) with a diploid genome of 5.1 gigabases (Gb), is more adaptable to severe environmental conditions than its close relative wheat, and maintains an important role in human nutrition in harsh climatic regions (Dawson et al. 2015). These unique properties make barley a model crop for the study of the effect of abiotic stresses in general. Suberin enhancement upon water deficit could be an effective acclimation mechanism against drought by sealing the central vascular cylinder and preventing passive water loss. A multifaceted RNA-Seq study showed that water deficit, mimicked by different osmotic potentials through polyethylene glycol (PEG) treatment, markedly upregulated the genes related to suberin biosynthetic pathway (such as

HORVU3Hr1G085020, *HORVU1Hr1G042910*) in barley seminal roots (Kreszies et al. 2019), which is shown in Table 6.2. The increasing of wax under drought stress improves water preservation, and reduces the cuticular transpiration, which helps plants to endure drought stress via avoidance mechanism (Ebercon et al. 1977; Surwenshi et al. 2010). *ECERIFERUM 1* (*CER1*) plays a pivotal role in the drought stress-induced biosynthesis of cuticular wax (Ahmad et al. 2022). A transcriptomic profile was compared between drought-tolerant (cv. Otis) and drought-susceptible (cv. Baronesse) barley plants subjected to drought stress and potential drought tolerance genes such as wax biosynthesis gene *ECERIFERUM1* (*CER1*, *HORVU1Hr1G039830*), and two NAC TFs (*HORVU0Hr1G017490*, *HORVU2Hr1G080460*) were uniquely induced in Otis leaves under drought stress (Harb et al. 2020).

Barley has strong salt tolerance (Xu et al. 2023); however, the underlying genetic basis is not fully clear. Recent studies have revealed that alternative splicing (AS) is significantly altered in response to diverse stresses to rapidly reprogram gene expression, that is optimal for plant survival under these unfavorable growth conditions (Ganie and Reddy. 2021). A comparative transcriptome profiling between rice (cv. Nipponbare) and barley (cv. XZ26) showed that a higher level of AS in the barley genes related to ion-transporters and TFs contribute to the higher salt tolerance in barley, due to the superior K^+/Na^+ homeostasis (Fu et al. 2019). Quertani et al. performed transcriptomic analysis of salt-stress-responsive genes in barley and a number of candidate genes involved in hormone and kinase signaling pathways, as well as several TF families and transporters, were identified (Nefissi Ouertani et al. 2021). Another comparison of leaf transcriptomes between salt-tolerant and susceptible genotypes following salt exposure allowed the identification of key candidate genes enhanced exclusively in the tolerant cv. Boulifa including *sucrose synthase 4*, *catalase 3*, *oxidative stress 3* (*OXS3*), and *salt overly sensitive 1* (*SOS1*) as well as several TFs such as *MYB20*, *MYB41* and growth-regulating factor 4 (*GRF4*) (Nefissi Ouertani et al. 2022). Combining GWAS and RNA-Seq analysis between tolerant (CM72) and susceptible (Gairdner) genotypes revealed five promising salt-responding candidate genes including *2-phosphoglycerate kinase* (*PGK2*, *HORVU4Hr1G088190*), *BASS3* (*HORVU6Hr1G016070*), *SINAT2* (*HORVU6Hr1G016120*), *Aquaporin* (*AQP*, *HORVU6Hr1G064140*), and *SYT3* (*HORVU7Hr1G101310*) from the hot spot regions (Xu et al. 2023).

Signal transduction links the sensing mechanism with genetic responses, which is important for signal transmission to the cellular machinery to initiate adaptive responses upon an environmental change (Huang et al. 2012). For example, comparative transcriptome profiles of two hulless barley (*H. vuglare*) genotypes (H8 as cold tolerant and T148 as susceptible) under cold acclimation and freezing treatments showed that the G-protein-coupled receptor signaling pathway participates in cold tolerance (Yuan et al. 2017). RNA-Seq analysis using two barley genotypes differing in freezing tolerance (cv. Nure, tolerant and cv. Tremois, susceptible) found that the *C-repeat binding factors* (*HvCBF9*; *MLOC_41994*) and *HvCBF14* (*MLOC_59224*) TFs that localized under the *Frost resistance-H2* (*Fr-H2*) QTL have a major role in freezing tolerance in barley (Wang et al. 2016).

Table 6.2 Transcriptomic studies on analyzing abiotic stress responses in barley

Stress	Keywords	Tissue	Stage	Cultivars	References
Drought	Suberin biosynthetic pathway genes (such as *HORVU3Hr1G085020*, *HORVU1Hr1G042910*)	Seminal roots	12-day-old plants	1. cv. Scarlett, (S)[a]	Kreszies et al. (2019)
Drought	*ECERIFERUM1* (*CER1*, *HORVU1Hr1G039830*), and two *NAC* TFs (*HORVU0Hr1G017490*, *HORVU2Hr1G080460*)	Leaves	Tillering stage Z21 (Zadoks scale) (Zadoks et al. 1974)	1. cv. Otis, a two-rowed, spring feed barley with high growth and yield in drylands (Mornhinweg et al. 2009), released by Colorado State University, (T)[b] 2. cv. Baronesse, a two-rowed, spring, feed barley cultivar, released by Peterson Seed Company Incorporation, (S)	Harb et al. (2020)
Salt	*Sucrose synthase 4*, *Catalase 3*, *Oxidative stress 3* (*OXS3*), *Salt Overly Sensitive 1* (*SOS1*), TFs such as *MYB20*, *MYB41* and *GRF*4	Leaves	Eight-day-old seedlings	1. cv. Boulifa, both cultivars are Tunisian barley landraces, (T) 2. cv. Testour (S)	Nefissi Ouertani et al. (2022)
Salt	*2-phosphoglycerate kinase* (*PGK2*, *HORVU4Hr1G088190*), *BASS3* (*HORVU6Hr1G016070*), *SINAT2* (*HORVU6Hr1G016120*), *Aquaporin* (*AQP*, *HORVU6Hr1G064140*), and *SYT3* (*HORVU7Hr1G101310*)	Leaves	Seedling stage	1. cv. CM72 (T) 2. cv. Gairdner (S)	Xu et al. (2023)
Cold	*C-repeat binding factors* (*HvCBF9*; *MLOC_41994*) and	Third fully expanded leaves	Three-leaf-stage seedlings	1. cv. Nure, winter barley, originated from Italy (T)	Wang et al. (2016)

(continued)

Table 6.2 (continued)

Stress	Keywords	Tissue	Stage	Cultivars	References
	HvCBF14 (*MLOC_59224*) TFs			2. cv. Tremois, spring barley, originated from France (S)	
Cd	*PLANT CADMIUM RESISTACE 2* (*HvPCR2*)	Roots & shoots	Seven-day-old seedlings	1. cv. Morex	Kintlová et al. (2021)
Cd	*ABC transporter* with PDR_assoc domain (*HORVU6Hr1G019080* and *HORVU6Hr1G036800*)	Shoots	14-day-old seedlings	1. cv. Rec, parental cultivars of Oregon Wolfe Barley (OWB) mapping population, developed at Oregon State University, (T) 2. cv. Dom (S)	Derakhshani et al. (2020b)

[a]*S* susceptible, [b]*T* tolerant

A genome-wide transcriptome analysis on barley plants identified the four different copies of *PLANT CADMIUM RESISTACE 2* (*HvPCR2*) gene, which is known to be responsible for heavy metal detoxification, as one of the most responsive genes to a Cd stimulus (Kintlová et al. 2021). Heterologous expression of all five barley copies in a Cd-susceptible yeast mutant restored cadmium resistance. RNA sequencing has made it possible to discover genetic variants that may underlie various phenotypic differences. Derakhshani et al. by combining the data from QTL mapping with RNA-Seq analysis, identified 16 Cd-tolerance-related candidate genes; 9 of these genes were metal ion transporters. Notably, the gene *HORVU6Hr1G019080* an *ABC transporter* with PDR_assoc domain showed 56 missense variants in Cd susceptible genotype (cv. Dom) and six frameshift variants in Cd tolerant genotype (cv. Rec), which may correspond to its different function in Cd exclusion between cv. Rec and cv. Dom (Derakhshani et al. 2020b).

6.5 RNA-Seq Profiling Under Abiotic Stress in Wheat

Ranked as the third most important crop, bread wheat (*Triticum aestivum* L.) is the staple food in many parts of the world. Wheat has a very large and complex genome with approximately 17 Gb, that is about 37 times the size of the rice genome (Smith and Flavell 1975). The polyploid and complex genetic background of wheat strongly improved wheat plants' adaptability to diverse environments and accelerated its spread in the world (Dubcovsky and Dvorak. 2007; Feldman et al. 2012). RNA-Seq shows significant differences in transcriptome response to stress in wheat cultivars from different geographic regions. For instance, drought-tolerant wheat cv. Saratovskaya-29 showed a rise in expression of ATPases associated proteins (AAA 16 family), signal and regulatory proteins with F-box domain, proteins of α/β hydrolase superfamily, and ABC transporters in response to water deficit. At the same time, drought-susceptible wheat cv. Yanetzkis Probat, showed a reduced expression of transmembrane carriers of amino acids (Konstantinov et al. 2021). According to a comparative transcriptomics survey in a set of drought tolerant and susceptible genotypes, the genes encoding enzymes from the phenylalanine ammonia-lyase (PAL), cytochrome P450 (CYP450), glutathione S-transferase (GST), TaABA8OH2, and TFs *AP2/ERF-ERF, MYB-related, B3, WRKY, Threonine, Isoleucine, Phenylalanine, Tyrosine* (*Tify*), and *NAC* families were important for drought tolerance (Rahimi et al. 2021). The summary of their study is illustrated in Table 6.3. Comparative analysis of root transcriptome revealed that TFs, i.e., *NAC* (*TraesCS2A01G102000*), *WRKY* (*TraesCS7D01G497400*), *MYB* (e.g., *TraesCS7D01G540700*), *bHLH* (*TraesCS6A01G373500*), *pyroline-5-carboxylate reductase* (*P5CR*, *TraesCS3B01G538100*), and LEA protein (e.g., *TraesCS4D01G177500*), were upregulated in the tolerant cv. Colotana, which are important for enhanced water-stress tolerance and root growth under stress condition. Furthermore, among upregulated genes in cv. Colotana, the authors detected one potential candidate gene *TraesCS2A01G102000* encoding NAC TF, harboring one missense variant, which may influence the drastic root length of cv. Colotana under water stress condition (Derakhshani et al. 2020a). Another study revealed that drought significantly upregulated auxin receptor *Auxin related F-box proteins* (*AFB2*) and ABA responsive TFs (*MYB78*, *WRKY18* and *G-BOX BINDING FACTOR 3* (*GBF3*)), *1-aminocyclopropane-1-carboxylic acid (ACC) oxidase* and *2-oxoglutarate/Fe(II)-dependent oxygenase* (*2OG oxygenase*) in roots of tolerant genotype Raj3765 (Dalal et al. 2018).

Most wheat varieties are susceptible to salt stress, which is a major limitation for wheat production. The plant response to salinity comprises two phases. The first phase is the osmotic phase in which the root senses salt stress and second messengers as ROS and Ca^{2+} are spread as signals to the aerial parts. Second, the ionic phase continues because of salt accumulation in leaves and takes days or weeks to manifest. In a transcriptome profiling at osmotic stage (cv. Zentos-tolerant and cv. Syn86-susceptible) and ionic stage (cv. Altay2000-tolerant and cv. Bobur-susceptible) of salt stress response in wheat, the expression of genes from four genotypes showed upregulation of some ABC transporters and Na^+/Ca^{2+} exchangers in the tolerant genotypes, indicating their involvement in the mechanisms of sodium exclusion and ion homeostasis. Furthermore, *TraesCS2A02G395000*, which codes for an *oxoglutarate/iron-dependent dioxygenase*, was reported to be central in controlling salinity stress in wheat (Duarte-Delgado et al. 2020). F-type H^+/Na^+-transporting ATPase subunit beta, and Ca^{2+} signal transduction sensors and key regulatory genes, including *zinc finger*, *NAC* and *WRKY* showed higher expression in cv. JM22 under salinity stress (Dugasa et al. 2021). Zhang et al. introduced some genes such as a *NAC* TF (homologous to *Arabidopsis AtNAC025*), a histone-lysine N-methyltransferase (homologous to *Arabidopsis SET Domain Group 16*

Table 6.3 Transcriptomic studies on analyzing abiotic stress responses in wheat

Stress	Keywords	Tissue	Stage	Cultivars	References
Drought	*PAL*, *CYP450*, *GST*, *TaABA8OH2* and TFs *AP2/ERF-ERF*, *MYB-related*, *B3*, *WRKY*, *Tify*, and *NAC*	Roots	Tillering (Zadoks 20–25) stages	1. cv. PI627299, (all cultivars are Iranian wheat landraces), (T)[a] 2. cv. PI627038, (T) 3. cv. PI624837, (S)[b]	Rahimi et al. (2021)
Drought	NAC TF (*TraesCS2A01G102000*), harboring one missense variant, *P5CR* and LEA proteins	Roots	Two-week-old seedlings	1. cv. Colotana (AGG20118WHEA1), (T) 2. cv. Tincurrin (AGG20578WHEA1), (S)	Derakhshani et al. (2020a)
Drought	*Auxin related F-box proteins* (*AFB2*), ABA responsive TFs (*MYB78*, *WRKY18*, *G-BOX BINDING FACTOR 3* (*GBF3*)), *1-aminocyclopropane-1-carboxylic acid (ACC) oxidase* and *2-oxoglutarate/Fe (II)-dependent oxygenase* (*2OG oxygenase*)	Roots	25-day-old plants, vegetative stage	1. cv. Raj3765, (T) 2. cv. HD2329, (S)	Dalal et al. (2018)
Salt	*Oxoglutarate/iron-dependent dioxygenase* (*TraesCS2A02G395000*)	Leaves	Seedling	1. cv. Zentos, an elite German winter wheat cultivar, (T); 2. cv. Syn86, synthetic genotype, (S); 3. cv. Altay2000, winter wheat Turkish cultivar, (T); 4. cv. Bobur, winter wheat Uzbek cultivar, (S)	Duarte-Delgado et al. (2020)
Salt	F-type H^+/Na^+-transporting ATPase, Ca^{2+} signal transduction sensor, Zinc finger, *NAC* and *WRKY*	Leaves	Four-leaf stage	1. cv. Jimai22, (T) 2. cv. Yangmai20, (S)	Dugasa et al. (2021)
Salt	*NAC* (homologous to *Arabidopsis AtNAC025*), Histone-lysine N-methyltransferase (homologous to *Arabidopsis AtSDG16*), MYB (homologous to *Arabidopsis AtMYB33*) and *TaRSL4*	Roots	10-day-old seedlings	1. cv. Qing Mai 6, released by Qingdao Agricultural University in 2007, (T) 2. cv. Chinese Spring, (S)	Zhang et al. (2016)

Salt	Aquaporin (*TaAQP*), Zinc finger (*TaZFP*) and Glutathione S-transfer (*TaGST*)	Roots	Three-leaf stage seedlings	1. cv. CH7034 breeding line, (T)	Chen et al. (2024)
Salt	Expansin, Xyloglucan endotransglucosylase/hydrolase (*XTH*), Dehydrins, Peroxidases, *WRKY*, *MYB*, *NAC*, *bHLH* and *AP2/ERF*	Roots	Anthesis stage	1. cv. Kharchia Local, bread wheat landrace, native to saline-sodic soils of India (T)	Mahajan et al. (2020)
Heat	*HsfA-6e* (*TraesCS1A02G375600*), *HsfB-2b* (*TraesCS7A02G270100.1*), and *HsfC-1b* (*TraesCS3A02G280800.1*)	Leaves	19-day-old seedlings	1. cv. Jopum (accession no. 2002-874), (T) 2. cv. Keumkang (accession no. 1997-4435), moderately (T) 3. cv. Olgeuru (accession no. 1997-4432), (S)	Lee et al. (2022)
Heat	*6-phosphogluconate dehydrogenase*, *S6 RPS6-2 ribosomal protein*, *Peptidylprolyl isomerase*, *Plasma membrane proton ATPase*, *Heat shock cognate-70*, *FtsH protease*, *RuBisCO activase B*, *Methionine synthase*, *Cytochrome C (class I)*, and *HMW-glutenin*	Spikes	Early- (14-dpa (days post anthesis)) and late-grain filling (30-dpa) stages	1. cv. Banks (AUS20599), all three cultivars were originally procured from Australian gene bank, (S) 2. cv. EGA Gregory (AUS34283), (S) 3. cv. Fang-60 (AUS24511), (T)	Rangan et al. (2020)
Heat	*Rubisco activase B* (*Rca1β*)	Awn	5–10 days post-anthesis	1. cv. PBW343, Indian winter wheat, (S)	Chaudhary et al. (2021)
Cold	*bHLH*, *MYB*, *NAC*, *WRKY* and *ERF*	–	Seedlings after cold stress treatment for 0, 10, 20 and 30 days	1. cv. Jing 841, winter wheat (Duan et al. 2015)	Lei et al. (2023)
Cold	*AP2/ERF*, *bZIP*, *NAC*, *WRKY*, *bHLH* and *MYB*	Tillering nodes	Seedling	1. cv. Dongnongdongmai1, winter wheat, (T)	Tian et al. (2022)
Cold	Sucrose synthase, Trehalose-6-phosphate synthase (*TPS*) and Trehalose-6-phosphate phosphatase (*TPP*)	Spikes	Anther connective tissue formation phase	1. cv. Yannong19, winter wheat, (T) 2. cv. Xinmai26, winter wheat, (S)	Jiang et al. (2022)
Cd			–	–	

(continued)

Table 6.3 (continued)

Stress	Keywords	Tissue	Stage	Cultivars	References
	Copper transporter 3 (*COPT3*) and *Zzinc transporter 1* (*ZnT1*)	Roots, stems, and leaves			Zheng et al. (2023)

[a]*T* tolerant, [b]*S* susceptible

(*AtSDG16*), a MYB TF (homologous to *Arabidopsis AtMYB33*) and *ROOT HAIR DEFECTIVE6-LIKE 4* (*TaRSL4*) gene (positively associated with root hair development) as necessary genes for salt stress tolerance in bread wheat root (Zhang et al. 2016). Upregulation of genes associated with *LEA*, *dehydrin*, and *potassium transportation*, and *sodium/cation exchange* and *aquaporin*, enhanced long-term salinity stress tolerance (Bhanbhro et al. 2020). Aquaporins located on the cell plasma membrane, mediate the regulation of root hydraulic conductivity in response to osmotic and salt stress (Boursiac et al. 2005; Domec et al. 2021). Zinc finger protein (ZFP), located in the nucleus, is a type of well-studied TF related to salt stress responsiveness (Mukhopadhyay et al. 2004; Sakamoto et al. 2004). In RNA-Seq profiling on the roots of salt-tolerant wheat breeding line CH7034 after NaCl treatment, an aquaporin gene (*TaAQP*), a zinc finger gene (*TaZFP*) and a *TaGST* gene were associated with the root salt-tolerance phenotype (Chen et al. 2024). Furthermore, genes encoding expansin, xyloglucan endotransglucosylase/hydrolase (XTH), dehydrins, peroxidases, and TFs, i.e., *WRKY*, *MYB*, *NAC*, *bHLH*, *AP2/ERF* also found to be upregulated under salt stress (Mahajan et al. 2020).

In particular, wheat is highly susceptible to heat stress, and it is estimated that global wheat production will decrease by 4–8.5%/°C increase in temperature (Asseng et al. 2015; Asseng et al. 2011; Shew et al. 2020). Analyzing the transcript expression in three wheat varieties with different susceptibilities to heat stress caused identification of three new candidate genes that might be involved in the heat-stress response of wheat seedlings including *HsfA-6e* (*TraesCS1A02G375600*), *HsfB-2b* (*TraesCS7A02G270100*), and *HsfC-1b* (*TraesCS3A02G280800*) (Lee et al. 2022). Anthesis and post-anthesis developmental stages are more susceptible to heat stress, with the grain-filling stage being crucial for sustained grain yield and quality. Rangan et al. reported the cluster of upregulated genes comprising *6-phosphogluconate dehydrogenase*, *S6 RPS6-2 ribosomal protein*, *peptidylprolyl isomerase*, *plasma membrane proton ATPase*, *Heat shock cognate-70*, *filamentation temperature sensitive* (*FtsH*) *protease*, *RuBisCO activase B*, *methionine synthase*, *cytochrome C (class I)*, and *HMW-glutenin* as key candidate genes for heat stress tolerance during grain filling stage in the tolerant genotype Fang-60 (Rangan et al. 2020). Wheat awns act as a compensation source of assimilate supply under stress conditions (Grundbacher. 1963). Transcriptome analysis of thermo-susceptible Indian wheat cv. PBW343 awn tissue under heat stress caused identification of candidate gene *Rubisco activase B* (*Rca1β*), which is involved in the activation of enzyme Rubisco and thus enhances the CO_2 fixation rate and photosynthetic efficiency. Interestingly, ectopic expression of Ta*Rca1β* in rice transgenics showed enhanced photosynthetic efficiency and thermotolerance under heat stress condition (Chaudhary et al. 2021).

Periods of cold are common in wheat-growing regions throughout the world, including Australia, China, Europe, and the United States (Crimp et al. 2016; Holman et al. 2011; Trnka et al. 2014). TFs play important regulatory functions in plant stress response, for example, genes encoding TFs such as *bHLH*, *MYB*, *NAC*, *WRKY* and *ERF* have been found to be up-regulated in wheat cv. Jing 841 under cold stress conditions (Lei et al. 2023). A transcriptome analysis of winter wheat cv. Dongnongdongmai1 (Dn1) identified major TFs (*AP2/ERF*, *bZIP*, *NAC*, *WRKY*, *bHLH* and *MYB*) participating in the cv. Dn1 cold stress response (Tian et al. 2022). The accumulation of sucrose and soluble sugar in plants under cold stress plays an important role in Osmo protection, maintenance of membrane stability, protein stability, carbohydrate homeostasis and intracellular expansion (Nuccio et al. 2015; Ramazan et al. 2021). Jiang et al. found that the upregulation of the sucrose synthase gene and the differential expression of trehalose-6-phosphate synthase (TPS) and trehalose-6-phosphate phosphatase (TPP) under low temperature treatments may play a key role in the cold tolerance of wheat (Jiang et al. 2022). Chilling conditions coinciding with the early stage of flowering cause significant yield losses in temperate climate zone wheat crops. Onyemaobi et al. compared transcriptomes of a chilling-tolerant (cv. Young) and a chilling-susceptible (cv. Wyalkatchem) wheat varieties to understand cold acclimation during the young microspore stage of spike development (Onyemaobi et al. 2022). The authors suggested differential regulation of auxin metabolism and signaling, including genes of the *Small Auxin Up-Regulated* (*SAUR*) gene family, as well as genes involved in lipid metabolism and cuticular wax synthesis may contribute to chilling tolerance in wheat varieties.

Wheat comprises approximately 20% of the human diet worldwide; therefore, reducing the concentrations of Cd in wheat grain will have significant impacts on the intake of Cd in food products. Cd accumulation ability differs among varieties in wheat, but the underlying molecular mechanism is largely unknown. It was shown that DEGs related to antioxidant defense mechanisms, ion binding and sulfotransferase activity were more accumulated in roots of the low-Cd-accumulating wheat (cv. L17) genotype (Zhou et al. 2019). Understanding the Cd resistance mechanisms is crucial for combating Cd phytotoxicity and meeting the increasing daily food demand. The ascorbic acid-glutathione (ASA-GSH) cycle is also an important component of the plant antioxidant system and is involved in countering oxidative damage induced by Cd stress (Zhang et al. 2021). In wheat plants exposed to gradient concentrations of Cd in soil, copper transporter 3 (COPT3) and zinc transporter 1 (ZnT1) were identified for the first time as the major transporters. Overexpression of the *nicotianamine synthase* and *pectinesterase* genes suggested that nicotianamine and pectin are the key chelators in Cd detoxification (Zheng et al. 2023). It has been reported that five miRNAs and their targeted genes were differentially expressed in roots and leaves of wheat seedlings under Cd stress. Among them, miR398 may participate in the tolerance to oxidative stress by modulating its target Cu–Zn superoxide dismutase to involve in Cd stress in wheat (Qiu et al. 2016).

6.6 Crosstalk Between Abiotic Stresses in Transcriptome Profile of Crops

Unraveling the molecular mechanisms of combined abiotic stress responses is essential to secure crop productivity under unfavorable environmental conditions. Drought and salinity are the two important and commonly co-occurring abiotic stresses affecting plant growth and productivity worldwide. The simultaneous occurrence of these stresses enhances their negative effects, and previous studies have revealed how these stresses induce plant responses at molecular levels. For example, a larger number of transcripts encoding for members of *NAC* and *D-Box Binding PAR BZIP* (*DBP*) TF families in drought-tolerant [cv. Nagina 22 (N22)] and members of *bHLH* and *Cys2/His2-type* (*C2H2*) TF families in salinity-tolerant (cv. Pokkali) rice cultivar exhibited differential regulation under desiccation and salinity stresses, respectively (Shankar et al. 2016). While transcripts involved in wax and terpenoid metabolism were upregulated in cv. Pokkali. Transcripts encoding thioredoxin and those related to phenylpropanoid metabolism were upregulated in cv. N22. This demonstrates how rice plants respond to varying levels of abiotic stress through a variety of TFs and metabolic pathways. Global climate change is predicted to result in increased yield losses of agricultural crops caused by environmental conditions. Heat and drought stress are major factors that negatively affect plant development and reproduction (Sato et al. 2024). A *de novo* transcriptome assembly using PacBio Iso-Seq for novel transcript discovery led to the identification of a new gene *B12288*, which was upregulated in drought/heat tolerant cv. N22 during combined heat and drought stress (Schaarschmidt et al. 2020). This gene is a homolog of *responsive to abscisic acid 21*(*RAB21*), a member of the dehydrin family of LEA proteins controlling cellular membrane traffic. The identified gene in N22 was more closely related to *RAB21* isoforms from wild rice species than to the homolog from cv. Nipponbare. These results indicate that the newly generated rice transcriptomes were a cost-effective alternative to identify candidate genes for stress tolerance breeding not present in the reference transcriptomes/genomes (Schaarschmidt et al. 2020). Cd and As contaminated plants share a broad range of stress response genes. Root transcriptome analysis of the rice after Cd and As treatment revealed that the genes shared by Cd and As stress were involved in signal transduction, transmembrane transport, redox control, stress response, transcriptional regulation and the biosynthesis and metabolism of macromolecules and Sulphur compounds. Furthermore, *LOC_Os09g20220*, annotated as *OsGSTU5*, was found to be highly expressed under As and Cd exposure, which is vital for ROS scavenging (Huang et al. 2019).

The effects of water deficit, salinity stress, and combined stress through transcriptome sequencing of barley seminal roots after 6 h and 24 h stress treatment was revealed (Osthoff et al. 2019). Combined salt and water deficit stress application results in overrepresentation of *bHLH*, *ethylene response factors* (*ERF*), and *HSF* TFs at each time point (6 h and 24 h). In addition, the *bZIP* was enriched after both water deficits and combined stress at both time points. In recent years, Weighted

Gene Co-Expression Network Analysis (WGCNA) has been performed to identify hub genes in response to abiotic stress in crops, such as rice (Wang et al. 2022), and wheat (Luo et al. 2019). It was reported two barley genotypes, stress tolerant (cv. Otis) and stress-susceptible (cv. Golden Promise), exhibiting contrasting responses to heat and drought, implied both as singly and in combination, during the heading stage (Mahalingam et al. 2022). WGCNA identified hub genes linked with RNA metabolism and Hsp70 chaperone components, suggesting potential for enhanced multiple abiotic stresses tolerance. Chemical similarity between As and phosphate (P) indicates possible antagonism between them in uptake and transportation. A comparative RNA-Seq study was conducted on the two barley genotypes differing in As tolerance (cv. ZDB160, As tolerant; and cv. ZDB475, As susceptible). It was suggested that cv. ZDB160 achieves its tolerance to As under P by up-regulating P transporters, resulting in more P uptake and less As translocation (Zvobgo et al. 2018).

In a genome-wide transcriptome analysis of the two contrasting wheat cultivars (cv. JM22, drought/salt tolerant; cv. YM20, salt susceptible) in response to drought (10% soil moisture) and salinity (100 mM NaCl) stresses, 10 DEGs, annotated to cellular process, metabolic process, osmotic regulation, and MAPK signaling pathway, were co-identified as drought and salinity tolerance associated DEGs (Dugasa et al. 2021). The gene expression and regulatory network analysis under combined drought and heat stress in wheat implicated that *HSFs* and *DREBs* were involved in the regulation of combined stresses, indicating a crosstalk between heat and drought stress (Liu et al. 2015). Bread wheat contains three copies of most genes, one on each of the A, B, and D homoeologous chromosomes, and these genes share over 95% identity in coding sequences (Krasileva et al. 2013). A fraction of expressed homoeologs in allopolyploids are likely to respond differently when subjected to stresses. It is hypothesized that a gene that is responsive to certain stress may show differences in homoeologous gene expression in response to those conditions. For example, a large proportion (68.4%) of wheat homoeologous genes exhibited differential expression patterns under drought, heat, or heat-drought condition. In addition, the calcium-dependent protein kinase *TaCPK2* has differential responses to stress between homeologs, with the D homoeolog upregulated in response to cold stress and the A homeolog upregulated in response to powdery mildew infection (Geng et al. 2013).

6.7 RNA-Seq Databases for Abiotic Stress Responsive Genes

The proliferation in the use of next-generation sequencing technologies by the plant research community has led to the accumulation of large volumes of expression data. To understand precisely abiotic stress-responsive transcriptome and regulatory networks, it is essential to investigate large-scale transcriptome data under various abiotic stress conditions using a single, uniform analysis platform. Transcriptomic resources have made major contributions to plant functional genomics, determining that the gene expression level often constitutes one of the first steps towards its functional characterization. For researchers, such large-scale resources provide further opportunities to re-analyze the data with new hypotheses.

At present, many gene expression databases for rice have been established, including Transcriptome ENcyclopedia Of Rice (TENOR, https://tenor.dna.affrc.go.jp/), Plant-Specific Alternative Splicing Database (PlaASDB; http://zzdlab.com/PlaASDB/ASDB/contact.html), EBI Gene Expression Atlas (https://www.ebi.ac.uk/gxa/experiments?species=oryza+sativa), Rice Expression Database (RED, https://ngdc.cncb.ac.cn/red/index), Network-based Rice Expression Analysis server (NetREx, https://bioinf.iiit.ac.in/netrex/index.html), and Plant public RNA-Seq database (PPRD, https://plantrnadb.com/), which are represented in Table 6.4.

TENOR was constructed based on RNA-Seq data from root and shoot of cv. Nipponbare using 470 samples under environmental stresses and hormone treatments. It contains the information of expression profiles, co-expressed genes and *cis*-regulatory elements of transcripts (Kawahara et al. 2016). PlaASDB provides extensive details of AS, gene expression and gene co-expression events by analyzing 3255 RNA-Seq data of rice and *Arabidopsis* under environmental (abiotic and biotic) stresses (Guo et al. 2023). The authors collected 410 abiotic stress samples for rice. The EBI Gene Expression Atlas includes gene expression profiles inferred from RNA-Seq and microarray data in plant and animal samples of different cell types, developmental stages, diseases, and other conditions (Petryszak et al. 2016). At the time of writing, EBI contained data from 296 and 32 abiotic stress samples for rice and wheat, respectively. RED contains 284 high-quality RNA-Seq data, from a wide range of rice growth stages and treatments. It also provides a list of housekeeping and tissue-specific genes and constructs co-expression networks for gene of interest (Xia et al. 2017). NetREx provides information of 192 samples for the expression and interaction of rice genes under various environmental stress (drought, flood, cold stresses) and hormonal (abscisic and jasmonic acid) treatment (Sircar et al. 2022). This web server also allows searches for orthologous genes from *A. thaliana*, *T. aestivum*, *Z. mays*, *H. vulgare*, and *S. bicolor* plants. PPRD consists of 11,726 RNA-Seq libraries of rice, wheat, maize, soybean and cotton from Gene Expression Omnibus (GEO), Sequence Read Archive (SRA), European Nucleotide Archive (ENA) and DNA Data Bank of Japan (DDBJ) databases. It shows the expression patterns from different tissues, developmental stages, and mutants under abiotic or biotic stresses (Yu et al. 2022).

The comprehensive RNA-Seq based Barley Expression Database (BarleyExpDB; http://barleyexp.com/common.html) uses cv. Morex V2, V3, cv. B1K-04–12 and cv. Zangqing320 gene models, contains a total of 3492 RNA-Seq samples, and provides the expression levels for nearly all barley tissues, developmental stages, many stress experiments (812 abiotic stress samples), as well as differential expression of mutants and populations. It also implements several practical utilities for sequence homology searching, visualization, and functional annotation of genes (Li et al. 2023b).

Table 6.4 List of RNA-Seq databases of rice, wheat and barley

Crop	Database	Keywords	RNA-Seq Samples No.	URL	References
Rice	TENOR	Expression profiles, co-expressed genes and cis -regulatory elements	470 samples, from a time-course study of shoots and roots under high salinity, high and low phosphate, high, low and extremely low cadmium, drought, osmotic, cold, flood, ABA, and JA in cv. Nipponbare	https://tenor.dna.affrc.go.jp/	Kawahara et al. (2016)
Rice, *Arabidopsis*	PlaASDB	Plant-Specific AS events, gene expression and gene co-expression	2280 samples for *Arabidopsis* under abiotic stress (total 2703 samples adding with 423 samples under biotic stress) and 410 samples for rice under abiotic stress (total 552 samples adding with 142 samples under biotic stress)	http://zzdlab.com/PlaASDB/ASDB/contact.html	Guo et al. (2023)
Rice, Wheat	EBI Gene Expression Atlas	Gene expression of different cell types, developmental stages	296 samples for rice under abiotic stresses, 32 samples for wheat under limited phosphate, high and cold temperature conditions	https://www.ebi.ac.uk/gxa/experiments?species=oryza+sativa https://www.ebi.ac.uk/gxa/experiments?species=triticum+aestivum	Petryszak et al. (2016)
Rice	RED	Rice gene expression profiles, list of housekeeping and tissue-specific genes, constructs	174 samples for abiotic stress from root, shoot and leaves (284 total samples from a	https://ngdc.cncb.ac.cn/red/index	Xia et al. (2017)

(continued)

Table 6.4 (continued)

Crop	Database	Keywords	RNA-Seq Samples No.	URL	References
		co-expression networks	diversity of rice growth stages and tissues)		
Rice	NetREx	Expression and interaction of rice genes, ortholog search	192 samples from drought, cold, flood, osmotic stress, ABA and JA hormone	https://bioinf.iiit.ac.in/netrex/index.html	Sircar et al. (2022)
Rice, Wheat, etc.	PPRD	Expression from different tissues, developmental stages, and mutants	11,726 samples covering different mutants, tissues, developmental stages and abiotic or biotic stresses	https://plantrnadb.com/	Yu et al. (2022)
Barley	BarleyExpDB	Expression levels for nearly all barley tissues, developmental stages, stress experiments, differential expression, functional annotation, homology search	812 samples for abiotic stress conditions (total 3492 samples)	http://barleyexp.com/introduction.html	Li et al. (2023b)
Wheat	expVIP	Wheat gene expression as bar graphs for single genes or as heatmaps for multiple genes, homoeolog gene expression	104 samples under abiotic stress (Total 977 samples)	http://www.wheat-expression.com/	Borrill et al. (2016)
Wheat	Wheat abiotic stress eFP	Visualization of flag leaf transcriptome	32 samples under heat, drought, salinity and their combinations in cv. Stettler	https://bar.utoronto.ca/efp_wheat/cgi-bin/efpWeb.cgi?dataSource=Wheat_Abiotic_Stress	Da Ros et al. (2023)
Wheat	WheatExp	Homoeologue-specific database of gene	14 samples under drought, heat, and their	https://wheat.pw.usda.gov/WheatExp/	Pearce et al. (2015)

(continued)

Table 6.4 (continued)

Crop	Database	Keywords	RNA-Seq Samples No.	URL	References
		expression profiles	combination in cv. TAM107		

For wheat, five expression browsers are currently available: EBI Gene Expression Atlas (https://www.ebi.ac.uk/gxa/experiments?species=triticum+aestivum), and PPRD (https://plantrnadb.com/), Wheat Expression Browser (expVIP; http://www.wheat-expression.com), wheat abiotic stress electronic Fluorescent Pictograph (eFP browser; https://bar.utoronto.ca/efp_wheat/cgi-bin/efpWeb.cgi?dataSource=Wheat_Abiotic_Stress), and WheatExp (https://wheat.pw.usda.gov/WheatExp), which are depicted in Table 6.4.

The wheat expVIP browser uses RefSeq v1.0, v1.1, IWGSC v2.26, and TGAC v1 gene models and includes expression data from 977 samples across a diverse range of wheat tissues, developmental stages, cultivars, and environmental conditions including various abiotic (104 samples) and biotic stress treatments (Borrill et al. 2016). It can display expression values for gene(s) and homeoeolog gene(s) of interest either as bar graphs or as heatmaps. Users can also analyze public and user-specified data sets with minimal bioinformatics knowledge using the expVIP virtual machine. Wheat eFP browser hosted by the Bio-Analytic Resource (BAR), contains gene expression data from a developmental time course and different tissue types from grain, root, leaf, and spike samples of the cv. Azhurnaya (Winter et al. 2007). Wheat abiotic stress eFP browser was created and integrated into the Wheat eFP Browser. It includes 32 RNA-Seq samples and allows for rapid visualization of flag leaf transcriptome of wheat genotype (cv. Stettler) subjected to salinity, drought and heat stress treatments and their combinations (Da Ros et al. 2023). WheatExp is a publicly accessible expression database and visualization tool, hosted on the GrainGenes website, to analyze and compare homoeologue-specific transcript profiles of different tissues over development, senescence, photomorphogenesis, as well as drought and heat stress in wheat (Pearce et al. 2015).

6.8 Conclusion and Future Perspectives

Modest projections of a 2 °C increase in the global mean temperature suggest forthcoming alterations in the timing and severity of abiotic stresses (Varshney et al. 2024). To prevent further environmental damage, it is imperative to boost the productivity of staple crops without expanding agricultural land area. Recent developments in NGS-based transcriptome sequencing provide a potential solution to identify novel functional and regulatory genes involved in stress adaptation that have been lost through crops domestication. In addition, in-depth studies on post-transcriptional regulation by nc-RNAs and miRNAs could be useful for the genetic improvement of abiotic stress tolerance in crops. The wealth of information generated through transcriptomics provides invaluable insights into potential targets for crop improvement. In addition to unveiling the novel candidate genes, the combined analysis of transcriptomics with other omics technologies will help in narrowing down the number of candidate genes to be subjected to functional validation by utilizing the mutant resource freely available to plant scientists. The utilization of such information would greatly facilitate the ability to address the challenges posed by climate change and speed up the efforts to breed crop plants that can maintain high yields under limiting growing conditions. Plant transcriptomics is an integral component of systems biology and is widely applied in plant sciences in both model as well as crop plants. Future development in plant molecular biology, computational biology and systems biology is based on plant transcriptomics for several aspects, thus, this field is expected to play a major role in future research on plant sciences.

Acknowledgments Financial assistance from National Agriculture and Food Research Organization (NARO), Tsukuba, Ibaraki, Japan and JSPS KAKENHI (grant number JP16F16396).

References

Ahmad HM, Wang X, Mahmood Ur R et al (2022) Morphological and physiological response of Helianthus annuus L. to drought stress and correlation of wax contents for drought tolerance traits. Arab J Sci Eng 47:6747–6761

Asseng S, Foster IN, Turner NC (2011) The impact of temperature variability on wheat yields. Glob Chang Biol 17:997–1012

Asseng S, Ewert F, Martre P et al (2015) Rising temperatures reduce global wheat production. Nat Clim Chang 5:143–147

Bayer PE, Golicz AA, Scheben A et al (2020) Plant pan-genomes are the new reference. Nat Plants 6:914–920

Bhanbhro N, Xiao B, Han L et al (2020) Adaptive strategy of allohexaploid wheat to long-term salinity stress. BMC Plant Biol 20:210

Borrill P, Ramirez-Gonzalez R, Uauy C (2016) expVIP: a customizable RNA-seq data analysis and visualization platform. Plant Physiol 170:2172–2186

Boursiac Y, Chen S, Luu D-T et al (2005) Early effects of salinity on water transport in Arabidopsis roots. Molecular and cellular features of aquaporin expression. Plant Physiol 139:790–805

Cartagena JA, Yao Y, Mitsuya S et al (2021) Comparative transcriptome analysis of root types in salt tolerant and sensitive rice varieties in response to salinity stress. Physiol Plant 173:1629–1642

Chaudhary C, Sharma N, Khurana P (2021) Decoding the wheat awn transcriptome and overexpressing Ta Rca1β in rice for heat stress tolerance. Plant Mol Biol 105:133–146

Chauhan R, Awasthi S, Indoliya Y et al (2020) Transcriptome and proteome analyses reveal selenium mediated amelioration of arsenic toxicity in rice (Oryza sativa L.). J Hazard Mater 390:122122

Chen J, Zhang L, Liu Y et al (2024) RNA-Seq-based WGCNA and association analysis reveal the key regulatory module and genes responding to salt stress in wheat roots. Plants [Online] 13
Crimp SJ, Zheng B, Khimashia N et al (2016) Recent changes in southern Australian frost occurrence: implications for wheat production risk. Crop Pasture Sci 67:801–811
Da Ros L, Bollina V, Soolanayakanahally R et al (2023) Multi-omics atlas of combinatorial abiotic stress responses in wheat. Plant J 116:1118–1135
Dalal M, Sahu S, Tiwari S et al (2018) Transcriptome analysis reveals interplay between hormones, ROS metabolism and cell wall biosynthesis for drought-induced root growth in wheat. Plant Physiol Biochem 130:482–492
Danilevicz MF, Fernandez CGT, Marsh JI et al (2020) Plant pangenomics: approaches, applications and advancements. Curr Opin Plant Biol 54:18–25
Dawson IK, Russell J, Powell W et al (2015) Barley: a translational model for adaptation to climate change. New Phytol 206:913–931
Derakhshani B, Ayalew H, Mishina K et al (2020a) Comparative analysis of root transcriptome reveals candidate genes and expression divergence of homoeologous genes in response to water stress in wheat. Plants [Online] 9
Derakhshani B, Jafary H, Maleki Zanjani B et al (2020b) Combined QTL mapping and RNA-Seq profiling reveals candidate genes associated with cadmium tolerance in barley. PLoS One 15: e0230820
Dilday RH, Yan WG, Moldenhauer KA et al (2001) Chinese and other foreign germplasm evaluation
Ding Y, Wang Y, Jiang Z et al (2017) MicroRNA268 overexpression affects rice seedling growth under cadmium stress. J Agric Food Chem 65:5860–5867
Domec J-C, King JS, Carmichael MJ et al (2021) Aquaporins, and not changes in root structure, provide new insights into physiological responses to drought, flooding, and salinity. J Exp Bot 72:4489–4501
Duan Z-B, Shen C-C, Li Q-Y et al (2015) Identification of a novel male sterile wheat mutant dms conferring dwarf status and multi-pistils. J Integr Agric 14:1706–1714
Duarte-Delgado D, Dadshani S, Schoof H et al (2020) Transcriptome profiling at osmotic and ionic phases of salt stress response in bread wheat uncovers trait-specific candidate genes. BMC Plant Biol 20:428
Dubcovsky J, Dvorak J (2007) Genome plasticity a key factor in the success of polyploid wheat under domestication. Science 316:1862–1866
Dugasa MT, Feng X, Wang N-H et al (2021) Comparative transcriptome and tolerance mechanism analysis in the two contrasting wheat (Triticum aestivum L.) cultivars in response to drought and salinity stresses. Plant Growth Regul 94:101–114
Ebercon A, Blum A, Jordan WR (1977) A rapid colorimetric method for epicuticular wax contest of sorghum leaves 1. Crop Sci 17:179–180
Fangmin C, Ningchun Z, Haiming X et al (2006) Cadmium and lead contamination in japonica rice grains and its variation among the different locations in southeast China. Sci Total Environ 359: 156–166
Feldman M, Levy AA, Fahima T et al (2012) Genomic asymmetry in allopolyploid plants: wheat as a model. J Exp Bot 63:5045–5059
Fu L, Shen Q, Kuang L et al (2019) Transcriptomic and alternative splicing analyses reveal mechanisms of the difference in salt tolerance between barley and rice. Environ Exp Bot 166: 103810
Ganie SA, Reddy ASN (2021) Stress-induced changes in alternative splicing landscape in rice: functional significance of splice isoforms in stress tolerance. Biology (Basel):10
Garalde DR, Snell EA, Jachimowicz D et al (2018) Highly parallel direct RNA sequencing on an array of nanopores. Nat Methods 15:201–206
Geng S, Li A, Tang L et al (2013) TaCPK2-A, a calcium-dependent protein kinase gene that is required for wheat powdery mildew resistance enhances bacterial blight resistance in transgenic rice. J Exp Bot 64:3125–3136

Grundbacher FJ (1963) The physiological function of the cereal awn. Bot Rev 29:366–381
Guo X, Wang T, Jiang L et al (2023) PlaASDB: a comprehensive database of plant alternative splicing events in response to stress. BMC Plant Biol 23:225
Harb A, Simpson C, Guo W et al (2020) The effect of drought on transcriptome and hormonal profiles in barley genotypes with contrasting drought tolerance. Front Plant Sci 11:618491
Heikema AP, Horst-Kreft D, Boers SA et al (2020) Comparison of illumina versus nanopore 16S rRNA gene sequencing of the human nasal microbiota. Genes [Online] 11
Holman JD, Schlegel AJ, Thompson CR et al (2011) Influence of precipitation, temperature, and 56 years on winter wheat yields in western Kansas. Crop Manag 10:1–10
Horiguchi H, Teranishi H, Niiya K et al (1994) Hypoproduction of erythropoietin contributes to anemia in chronic cadmium intoxication: clinical study on Itai-itai disease in Japan. Arch Toxicol 68:632–636
Huang G-T, Ma S-L, Bai L-P et al (2012) Signal transduction during cold, salt, and drought stresses in plants. Mol Biol Rep 39:969–987
Huang Y, Chen H, Reinfelder JR et al (2019) A transcriptomic (RNA-seq) analysis of genes responsive to both cadmium and arsenic stress in rice root. Sci Total Environ 666:445–460
Jayakodi M, Padmarasu S, Haberer G et al (2020) The barley pan-genome reveals the hidden legacy of mutation breeding. Nature 588:284–289
Jiang G, Hassan MA, Muhammad N et al (2022) Comparative physiology and transcriptome analysis of young spikes in response to late spring coldness in wheat (Triticum aestivum L.). Front Plant Sci 13
Kawahara Y, Oono Y, Wakimoto H et al (2016) TENOR: database for comprehensive mRNA-Seq experiments in rice. Plant Cell Physiol 57:e7
Kintlová M, Vrána J, Hobza R et al (2021) Transcriptome response to cadmium exposure in barley (Hordeum vulgare L.). Front Plant Sci 12
Konstantinov DK, Zubairova US, Ermakov AA et al (2021) Comparative transcriptome profiling of a resistant vs susceptible bread wheat (Triticum aestivum L.) cultivar in response to water deficit and cold stress. PeerJ 9:e11428
Krasileva KV, Buffalo V, Bailey P et al (2013) Separating homeologs by phasing in the tetraploid wheat transcriptome. Genome Biol 14:R66
Kreszies T, Shellakkutti N, Osthoff A et al (2019) Osmotic stress enhances suberization of apoplastic barriers in barley seminal roots: analysis of chemical, transcriptomic and physiological responses. New Phytol 221:180–194
Lee MH, Kim K-M, Sang W-G et al (2022) Comparison of gene expression changes in three wheat varieties with different susceptibilities to heat stress using RNA-Seq analysis. Int J Mol Sci [Online] 23
Lei L, Zheng H, Bi Y et al (2020) Identification of a major QTL and candidate gene analysis of salt tolerance at the bud burst stage in rice (Oryza sativa L.) using QTL-Seq and RNA-Seq. Rice 13:55
Lei L, Goltsman E, Goodstein D et al (2021) Plant pan-genomics comes of age. Annu Rev Plant Biol 72:411–435
Lei L, Han C, Yang J et al (2023) Comprehensive transcriptome analysis of responses during cold stress in wheat (Triticum aestivum L.). Gene 14:844
Li P, Jiang J, Zhang G et al (2023a) Integrating GWAS and transcriptomics to identify candidate genes conferring heat tolerance in rice. Front Plant Sci 13
Li T, Li Y, Shangguan H et al (2023b) BarleyExpDB: an integrative gene expression database for barley. BMC Plant Biol 23:170
Liang Y, Tabien RE, Tarpley L et al (2021) Transcriptome profiling of two rice genotypes under mild field drought stress during grain-filling stage. AoB Plants 13:plab043
Liu Z, Xin M, Qin J et al (2015) Temporal transcriptome profiling reveals expression partitioning of homeologous genes contributing to heat and drought acclimation in wheat (Triticum aestivum L.). BMC Plant Biol 15:152

Liu A, Zhou Z, Yi Y et al (2020a) Transcriptome analysis reveals the roles of stem nodes in cadmium transport to rice grain. BMC Genomics 21:127
Liu G, Zha Z, Cai H et al (2020b) Dynamic transcriptome analysis of anther response to heat stress during anthesis in thermotolerant rice (Oryza sativa L.). Int J Mol Sci 21
Luo Y, Pang D, Jin M et al (2019) Identification of plant hormones and candidate hub genes regulating flag leaf senescence in wheat response to water deficit stress at the grain-filling stage. Plant Direct 3:e00152
Mahajan MM, Goyal E, Singh AK et al (2020) Shedding light on response of Triticum aestivum cv. Kharchia Local roots to long-term salinity stress through transcriptome profiling. Plant Growth Regul 90:369–381
Mahalingam R, Duhan N, Kaundal R et al (2022) Heat and drought induced transcriptomic changes in barley varieties with contrasting stress response phenotypes. Front Plant Sci 13
Matsui A, Ishida J, Morosawa T et al (2008) Arabidopsis transcriptome analysis under drought, cold, high-salinity and ABA treatment conditions using a tiling array. Plant Cell Physiol 49: 1135–1149
Mehta S, James D, Reddy M (2019) Omics technologies for abiotic stress tolerance in plants: current status and prospects:1–34
Mohanty B, Kitazumi A, Cheung CYM et al (2016) Identification of candidate network hubs involved in metabolic adjustments of rice under drought stress by integrating transcriptome data and genome-scale metabolic network. Plant Sci 242:224–239
Mornhinweg DW, Bregitzer PP, Porter DR et al (2009) Registration of 'Sidney' spring feed barley resistant to Russian wheat aphid. J Plant Registrations 3:214–218
Mosa KA, Ismail A, Helmy M (2017) Omics and system biology approaches in plant stress research. In: Mosa KA, Ismail A, Helmy M (eds) Plant stress tolerance: an integrated omics approach. Springer, Cham, pp 21–34
Mukhopadhyay A, Vij S, Tyagi AK (2004) Overexpression of a zinc-finger protein gene from rice confers tolerance to cold, dehydration, and salt stress in transgenic tobacco. Proc Natl Acad Sci USA 101:6309–6314
Muthurajan R, Rahman H, Manoharan M et al (2018) Drought responsive transcriptome profiling in roots of contrasting rice genotypes. Indian J Plant Physiol 23:393–407
Nefissi Ouertani R, Arasappan D, Abid G et al (2021) Transcriptomic analysis of salt-stress-responsive genes in barley roots and leaves. Int J Mol Sci 22
Nefissi Ouertani R, Arasappan D, Ruhlman TA et al (2022) Effects of salt stress on transcriptional and physiological responses in barley leaves with contrasting salt tolerance. Int J Mol Sci [Online] 23
Nuccio ML, Wu J, Mowers R et al (2015) Expression of trehalose-6-phosphate phosphatase in maize ears improves yield in well-watered and drought conditions. Nat Biotechnol 33:862–869
Onyemaobi O, Sangma H, Garg G et al (2022) Transcriptome profiling of the chilling response in wheat spikes: I, acclimation response to long-term chilling treatment. Curr Plant Biol 31:100255
Oono Y, Yazawa T, Kawahara Y et al (2014) Genome-wide transcriptome analysis reveals that cadmium stress signaling controls the expression of genes in drought stress signal pathways in rice. PLoS ONE 9:e96946
Osthoff A, Donà Dalle Rose P, Baldauf JA et al (2019) Transcriptomic reprogramming of barley seminal roots by combined water deficit and salt stress. BMC Genomics 20:1–14
Pearce S, Vazquez-Gross H, Herin SY et al (2015) WheatExp: an RNA-seq expression database for polyploid wheat. BMC Plant Biol 15:299
Petereit J, Bayer PE, Thomas WJW et al (2022) Pangenomics and crop genome adaptation in a changing climate. Plants 11:1949
Petryszak R, Keays M, Tang YA et al (2016) Expression Atlas update—an integrated database of gene and protein expression in humans, animals and plants. Nucleic Acids Res 44:D746–D752
Qiu Z, Hai B, Guo J et al (2016) Characterization of wheat miRNAs and their target genes responsive to cadmium stress. Plant Physiol Biochem 101:60–67

Rahimi Y, Ingvarsson PK, Bihamta MR et al (2021) Characterization of dynamic regulatory gene and protein networks in wheat roots upon perceiving water deficit through comparative transcriptomics survey. Front Plant Sci 12

Ramazan S, Qazi HA, Dar ZA et al (2021) Low temperature elicits differential biochemical and antioxidant responses in maize (Zea mays) genotypes with different susceptibility to low temperature stress. Physiol Mol Biol Plants 27:1395–1412

Rangan P, Furtado A, Henry R (2020) Transcriptome profiling of wheat genotypes under heat stress during grain-filling. J Cereal Sci 91:102895

Rasheed S, Bashir K, Matsui A et al (2016) Transcriptomic analysis of soil-grown Arabidopsis thaliana roots and shoots in response to a drought stress. Front Plant Sci 7:183475

Sakamoto H, Maruyama K, Sakuma Y et al (2004) Arabidopsis Cys2/His2-type zinc-finger proteins function as transcription repressors under drought, cold, and high-salinity stress conditions. Plant Physiol 136:2734–2746

Sasaki T, International Rice Genome Sequencing P (2005) The map-based sequence of the rice genome. Nature 436:793–800

Sato H, Mizoi J, Shinozaki K et al (2024) Complex plant responses to drought and heat stress under climate change. Plant J 117:1873–1892

Schaarschmidt S, Fischer A, Lawas LMF et al (2020) Utilizing PacBio Iso-Seq for novel transcript and gene discovery of abiotic stress responses in Oryza sativa L. Int J Mol Sci [Online] 21

Shahid M, Khalid S, Abbas G et al (2015) Heavy metal stress and crop productivity:1–25

Shan X, Li Y, Jiang Y et al (2013) Transcriptome profile analysis of maize seedlings in response to high-salinity, drought and cold stresses by deep sequencing. Plant Mol Biol Report 31:1485–1491

Shankar R, Bhattacharjee A, Jain M (2016) Transcriptome analysis in different rice cultivars provides novel insights into desiccation and salinity stress responses. Sci Rep 6:23719

Shew AM, Tack JB, Nalley LL et al (2020) Yield reduction under climate warming varies among wheat cultivars in South Africa. Nat Commun 11:4408

Sircar S, Musaddi M, Parekh N (2022) NetREx: Network-based Rice Expression Analysis Server for abiotic stress conditions. Database 2022:baac060

Smith DB, Flavell RB (1975) Characterisation of the wheat genome by renaturation kinetics. Chromosoma 50:223–242

Song Y, Wang C, Linderholm HW et al (2022) The negative impact of increasing temperatures on rice yields in southern China. Sci Total Environ 820:153262

Sun C, Hu Z, Zheng T et al (2017) RPAN: rice pan-genome browser for ~3000 rice genomes. Nucleic Acids Res 45:597–605

Surwenshi A, Chimmad VP, Jalageri BR et al (2010) Characterization of sorghum genotypes for physiological parameters and yield under receding soil moisture conditions. Res J Agric Sci 1: 242–244

Tettelin H, Masignani V, Cieslewicz MJ et al (2005) Genome analysis of multiple pathogenic isolates of Streptococcus agalactiae: implications for the microbial "pan-genome". Proc Natl Acad Sci USA 102:13950–13955

Tian Y, Peng K, Lou G et al (2022) Transcriptome analysis of the winter wheat Dn1 in response to cold stress. BMC Plant Biol 22:277

Tiwari P, Srivastava D, Chauhan AS et al (2021) Root system architecture, physiological analysis and dynamic transcriptomics unravel the drought-responsive traits in rice genotypes. Ecotoxicol Environ Saf 207:111252

Trnka M, Rötter RP, Ruiz-Ramos M et al (2014) Adverse weather conditions for European wheat production will become more frequent with climate change. Nat Clim Chang 4:637–643

Varshney RK, Barmukh R, Bentley A et al (2024) Exploring the genomics of abiotic stress tolerance and crop resilience to climate change. Plant Genome 17:e20445

Wang X, Wu D, Yang Q et al (2016) Identification of mild freezing shock response pathways in barley based on transcriptome profiling. Front Plant Sci 7:106

Wang Q, Zeng X, Song Q et al (2020) Identification of key genes and modules in response to Cadmium stress in different rice varieties and stem nodes by weighted gene co-expression network analysis. Sci Rep 10:9525

Wang X, Liu H, Zhang D et al (2022) Photosynthetic carbon fixation and sucrose metabolism supplemented by weighted gene co-expression network analysis in response to water stress in rice with overlapping growth stages. Front Plant Sci 13:864605

Wei Z, Yuan Q, Lin H et al (2021) Linkage analysis, GWAS, transcriptome analysis to identify candidate genes for rice seedlings in response to high temperature stress. BMC Plant Biol 21:85

Wei H, Wang X, Zhang Z et al (2024) Uncovering key salt-tolerant regulators through a combined eQTL and GWAS analysis using the super pan-genome in rice. Natl Sci Rev 11:nwae043

Winter D, Vinegar B, Nahal H et al (2007) An "electronic fluorescent pictograph" browser for exploring and analyzing large-scale biological data sets. PLoS ONE 2:e718

Woldegiorgis ST, Wu T, Gao L et al (2022) Identification of heat-tolerant genes in non-reference sequences in rice by integrating pan-genome, transcriptomics, and QTLs. Genes (Basel) 13

Wu B, Chen S, Cheng S et al (2023) Transcriptome analysis revealed the dynamic and rapid transcriptional reprogramming involved in cold stress and related core genes in the rice seedling stage. Int J Mol Sci 24

Xia L, Zou D, Sang J et al (2017) Rice Expression Database (RED): an integrated RNA-Seq-derived gene expression database for rice. J Genet Genomics 44:235–241

Xie Z, Wang J, Wang W et al (2021) Integrated analysis of the transcriptome and metabolome revealed the molecular mechanisms underlying the enhanced salt tolerance of rice due to the application of exogenous melatonin. Front Plant Sci 11

Xu T, Meng S, Zhu X et al (2023) Integrated GWAS and transcriptomic analysis reveal the candidate salt-responding genes regulating Na+/K+ balance in barley (Hordeum vulgare L.). Front Plant Sci 13

Yang Y, Zhang C, Zhu D et al (2022) Identifying candidate genes and patterns of heat-stress response in rice using a genome-wide association study and transcriptome analyses. Crop J 10: 1633–1643

Yang L, Liu H, Lei L et al (2023) Combined QTL-sequencing, linkage mapping, and RNA-sequencing identify candidate genes and KASP markers for low-temperature germination in Oryza sativa L. ssp. Japonica. Planta 257:122

Yu L-J, Luo Y-F, Liao B et al (2012) Comparative transcriptome analysis of transporters, phytohormone and lipid metabolism pathways in response to arsenic stress in rice (Oryza sativa). New Phytol 195:97–112

Yu Y, Zhang H, Long Y et al (2022) Plant public RNA-seq database: a comprehensive online database for expression analysis of ~45,000 plant public RNA-Seq libraries. Plant Biotechnol J 20:806–808

Yuan H, Zeng X, Ling Z et al (2017) Transcriptome profiles reveal cold acclimation and freezing tolerance of susceptible and tolerant hulless barley genotypes. Acta Physiol Plant 39:1–13

Zadoks JC, Chang TT, Konzak CF (1974) A decimal code for the growth stages of cereals. Weed Res 14:415–421

Zhang T, Zhao X, Wang W et al (2012) Comparative transcriptome profiling of chilling stress responsiveness in two contrasting rice genotypes. PLoS ONE 7:e43274

Zhang Y, Cheng Y, Guo J et al (2014) Comparative transcriptome analysis to reveal genes involved in wheat hybrid necrosis. Int J Mol Sci [Online]:15

Zhang Y, Liu Z, Khan AA et al (2016) Expression partitioning of homeologs and tandem duplications contribute to salt tolerance in wheat (Triticum aestivum L.). Sci Rep 6:21476

Zhang J, Su L, Wang Y et al (2020) Improved high-throughput sequencing of the human oral microbiome: from illumina to PacBio. Can J Infect Dis Med Microbiol 2020:6678872

Zhang T, Xiao J, Zhao Y et al (2021) Comparative physiological and transcriptomic analyses reveal ascorbate and glutathione coregulation of cadmium toxicity resistance in wheat genotypes. BMC Plant Biol 21:459

Zheng S, Qi J, Fu T et al (2023) Novel mechanisms of cadmium tolerance and Cd-induced fungal stress in wheat: transcriptomic and metagenomic insights. Ecotoxicol Environ Saf 256:114842

Zhong Y, Luo Y, Sun J et al (2024) Pan-transcriptomic analysis reveals alternative splicing control of cold tolerance in rice. Plant Cell:koae039

Zhou M, Zheng S, Liu R et al (2019) Comparative analysis of root transcriptome profiles between low-and high-cadmium-accumulating genotypes of wheat in response to cadmium stress. Funct Integr Genomics 19:281–294

Zvobgo G, Sagonda T, Lwalaba JLW et al (2018) Transcriptomic comparison of two barley genotypes differing in arsenic tolerance exposed to arsenate and phosphate treatments. Plant Physiol Biochem 130:589–603

Chapter 7
Advancing Crop Resilience: Integrating Multi-Omics Approaches for Abiotic Stress Tolerance

Mather A. Khan, Divya Gupta, Zahid H. Siddiqui, Pravej Alam, and Sanjib Kumar Panda

Abstract Abiotic stresses are among the most significant challenges limiting crop production. These stresses adversely affect plant growth and development, leading to devastating economic losses and threatening global food security. To meet the demand of growing population in a changing climate, it is estimated that crop yield must increase by about 70%. In the challenging condition of increased food insecurity, reduction in agricultural land, and deleterious effects of climate change on agricultural productivity, developing crops with enhanced resilience to abiotic stress is crucial to for sustainable agricultural practices and food security. Plant under abiotic stress conditions often suffer from multiple stress factors simultaneously, such as drought and salinity or drought along with nutrient and heat stress. These combined stresses can have interactive effects, worsening the plant growth and productivity. To decipher the global mechanisms of plant response to stress, multi-omics approach is necessary. This chapter explores the potential of integrating multi-omics to develop crops that are resilient to abiotic stress. It covers the current state of research, highlight successful studies, and outlines future directions for utilizing these advanced technologies in crop improvement. By harnessing the power of

M. A. Khan (✉)
Bond Life Sciences Center, Division of Plant Science and Technology, University of Missouri, Columbia, MO, USA
e-mail: khanma@missouri.edu

D. Gupta · S. K. Panda
Department of Biochemistry, Central University of Rajasthan, Ajmer, India

Z. H. Siddiqui
Department of Biology, Faculty of Science, University of Tabuk, Tabuk, Kingdom of Saudi Arabia

P. Alam
Department of Biology, Prince Sattam bin Abdulaziz University, Alkharj, Kingdom of Saudi Arabia

S. K. Panda, M. A. Khan (eds.), *Plant Functional Genomics for Abiotic Stress Resilience*, Plant in Challenging Environments 7,
https://doi.org/10.1007/978-3-032-01704-8_7

multi-omics, we can create innovative strategies to boost crop resilience and productivity, thereby contributing to global food security amidst increasing environmental challenges.

Keywords Abiotic stress · Food security · Crop yield · Regulatory network · Tolerance · Omics

7.1 Introduction

Abiotic stresses are among the most significant challenges limiting crop yield worldwide. These stresses adversely affect plant growth and development, leading to devastating economic losses and threatening global food security. According to the FAO report (2023), hunger affected 9.2% of the global population, about 760 million people, in 2022. Additionally, 900 million people (11.3% of the world population) were exposed to severe food insecurity, and an estimated 18.3% (over 1.4 billion people) experienced food insecurity at moderate level. This amounts to about 2.4 billion people facing severe to moderate food insecurity. Furthermore, the global population is expected to exceed 10 billion in the next 35 years. On the other hand, about 86 million hectares of agricultural land declined between 2000 and 2021, which is almost equal to the total land area of France and Germany combined. The increasing incidence of drought, heat waves, flooding and water scarcity due to climate change negatively affects the crop productivity. To meet the demand of growing population, it is estimated that the crop production must increase by about 70% (2023). In this adverse scenario of increased food insecurity, reduction in agricultural land, and the deleterious effects of climate change on agricultural productivity, developing crops with enhanced resilience to abiotic stress is crucial for sustainable agricultural practices and food security.

Abiotic stresses, such as salinity, drought, extreme temperature, nutrient deficiencies and heavy metal contamination (Fig. 7.1), significantly reduces crop growth and production. These stresses can cause osmotic stress, ion toxicity, and reduced photosynthesis, ultimately leading to significant lower yields and economic losses (Bailey-Serres et al. 2019; Zhang et al. 2023). It is estimated that abiotic stress can cause crop yield losses ranging from 50% to 70% annually (Bailey-Serres et al. 2019; Rahman et al. 2022). Though all abiotic stresses negatively affect crop growth and development leading to reduction in productivity, but their mechanism and genes/protein involved in providing protection/tolerance/resistance are different and varied from crops to crops based on their stages of life cycle and geographical location. For instance, cold stress severely affects the membrane fluidity and metabolic homeostasis leading to significant morphological and physiological changes in growing plants, such as reduced growth and development, photosynthesis (Wu et al. 2022). During drought conditions, plants reduced stomatal conductance to minimize water loss while under heat stress, plants keep stomata open and transpire more water to maintain the temperature (Bheemanahalli et al. 2022). The multi-omics

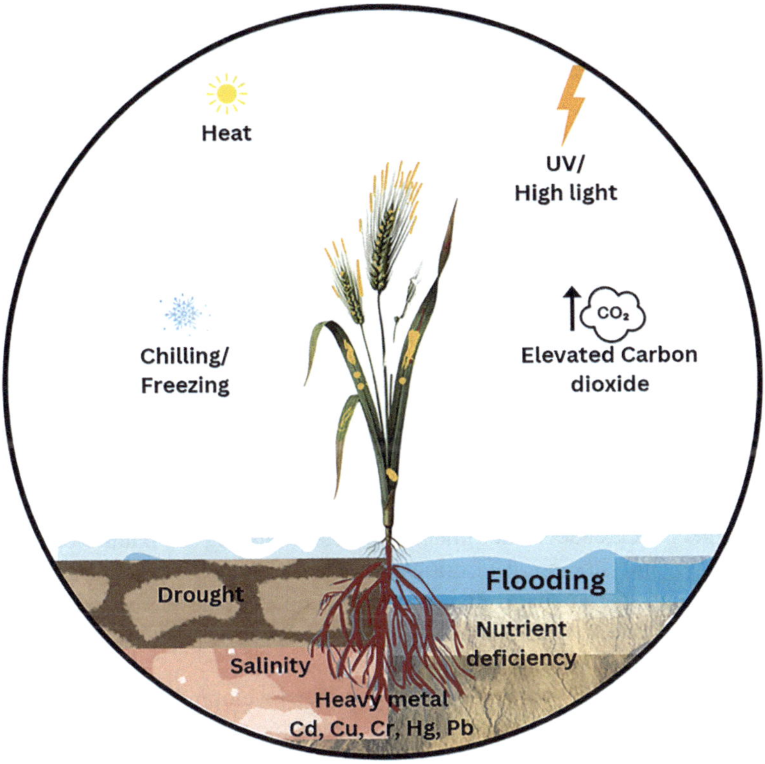

Fig. 7.1 Different abiotic stress in plants: All these stresses affect plant growth and development, leading to devastating economic losses and threatening global food security

technologies including genomics, transcriptomics, proteomics, metabolomics and ionomics, have been used to identify and discover the mechanisms of different stress response in various crops including wheat, maize, millet and soybean (Zhang et al. 2007; Deshmukh et al. 2014; Lešková et al. 2017; Sharma et al. 2020; Godoy et al. 2021; Yang et al. 2021; Wu et al. 2022; Bheemanahalli et al. 2022). By providing a comprehensive understanding of the molecular mechanisms underlying stress responses, these integrative approaches enable the identification of key genes, pathways, and biomarkers associated with abiotic stress tolerance.

This chapter explores the potential of integrating multi-omics to develop crops that are resilient to abiotic stress. It covers the current state of research, highlights successful studies, and outlines future directions for utilizing these advanced technologies in crop improvement. By harnessing the power of multi-omics, we can create innovative strategies to boost crop resilience and productivity, thereby contributing to global food security amidst increasing environmental challenges.

7.2 Omics Approaches to Deciphering Drought Stress Responses in Crop Plants: A Case Study

Drought stress is one of the most devastating abiotic stresses, significantly impairing crop productivity worldwide, with far-reaching implications for food security and economic stability. As the predominant abiotic stressor affecting crop yields worldwide, drought conditions are responsible for significant agricultural losses, particularly in rain-fed farming systems that support the livelihoods of millions. According to the Food and Agriculture Organization (FAO), nearly half (45%) of the world's arable land is affected by drought. In a recent report of FAO, drought results into economic losses of $37 billion in crop and livestock production between 2008 and 2018 (FAO 2021). Climate models predict an alarming increase in both the frequency and severity of drought events, potentially worsening the current challenges and threatening the stability of global food systems.

Drought hits plants hard. The physiological damage of drought stress shows up in multiple ways, affecting nearly every system of plants. At the cellular level, water deficit triggers a cascade of detrimental effects including stomatal closure, membrane damage, and disruption of critical enzymatic processes including ATP synthesis and energy metabolism. These primary impacts lead to secondary damages such as inhibited photosynthetic activity, accumulation of reactive oxygen species (ROS), and extensive cellular damage.

The impact of drought varies depending on the developmental stage of the plant, the degree and duration of the stress, and the genotypic types of plants variety. Short-duration drought stress typically reduces grain yield, while prolonged drought stress can lead to complete plant death. For instance, drought stress in wheat can result in significant yield reductions, with severe cases leading to total crop loss.

7.2.1 Transcriptomics: Decoding the Genetic Blueprint of Drought Response

Recent advances in transcriptomic technologies have significantly enhanced our understanding of how plants respond to drought stress. High-throughput transcriptomics tools has identified various transcription factor families, such as WRKY, MYB, DREB, AP2/ERF, bZIP, and NAC, which play crucial roles in drought adaptation (Li et al. 2022; Mahalingam et al. 2022; Roychowdhury et al. 2023; Pereira et al. 2025). These proteins act as molecular switches, controling the expression of numerous drought-responsive genes and coordinating various protective mechanisms.

The ABA-dependent signaling pathway is central to drought response networks. ABA-responsive element binding factors (ABFs) regulate genes involved in stomatal regulation, osmotic adjustment, and cellular protection. Comparative transcriptomic studies have provided valuable insights into drought tolerance. For example, research on barley (Mahalingam et al. 2022) revealed that tolerant genotypes upregulate protective genes like *HvP5CS* (proline biosynthesis), *HvNAC6* (stress-associated transcription factor), and *HvCER* (cuticular wax formation), while sensitive cultivars inadequately activate these mechanisms and overexpress oxidative stress related genes. Similarly, studies on soybean (Li et al. 2022), identified *NCED1* and *SDR* as crucial ABA biosynthesis genes consistently upregulated during drought, working with ABF transcription factors to mediate stress signaling.

The advent of single-cell RNA sequencing (scRNA-seq) has added unprecedented resolution to our understanding of abiotic stress responses. A groundbreaking study on *Arabidopsis* leaves identified distinct mesophyll cell populations with opposing reactions to water deficit. Cells near the leaf margin and vasculature showed typical drought response, upregulating ABA, *COLD-REGULATED* (*COR*) genes, and *EARLY-RESPONSIVE TO DEHYDRATION* (*ERD*) genes (Tenorio Berrío et al. 2025). This study provides a detailed scRNA-seq atlas of young Arabidopsis leaves, capturing transcriptomic profiles from both well-watered and drought-stressed seedlings. The platform includes a gene expression visualization tool integrated into a schematic leaf representation, available at: http://www.single-cell.be/plant/leaf-drought. This spatial heterogeneity in drought responses suggests an intricate level of tissue-specific regulation that was previously difficult to study.

7.2.2 *Proteomics: Unraveling the Protein Networks Underlying Drought Tolerance*

Proteomic approaches have provided valuable insights into how plants respond to drought by examining changes in protein abundance, modification, and interaction networks. Proteomics techniques like tandem mass tags (TMT) and liquid chromatography-tandem mass spectrometry (LC-MS/MS), have identified many drought-responsive proteins in various crops.

Drought stress severely impacts photosynthetic proteins. Research on wheat (Michaletti et al. 2018) showed significant differences between tolerant (Kavir) and susceptible (Bahar) cultivars. The tolerant wheat maintained stable levels of photosynthetic proteins like RuBisCO and ATP synthase during drought, while the susceptible cultivar experienced severe declines in these proteins. Additionally, proteins involved in nitrogen metabolism (glutamine synthetase) and detoxification (GST, glyoxalase I) were greatly reduced in susceptible plants indicating a breakdown of essential metabolic processes. These findings suggest that protein stability and efficient redox homeostasis are crucial for drought tolerance. RuBisCO activase, GST, and energy metabolism proteins could be potential targets for breeding drought resilient wheat varieties. In tomato, researchers used a combination of TMT (Tandem Mass Tag)-based quantitative proteomics and RNA-seq to investigate drought responses (Liu et al. 2023). This multi-omics approach demonstrates the effectiveness of TMT proteomics for high-throughput, reproducible protein quantification, helping to uncover drought adaptation.

In another most comprehensive proteomic studies, Ren et al. (2022) used iTRAQ-based proteomics in maize seedlings. This quantitative proteomics analysis involved redox homeostasis, cell wall organization, photosynthesis, and ABA biosynthesis were differently abundant under drought conditions (Ren et al. 2022). Molecular chaperones, particularly the HSP20 family, and dehydrins like COR410 were consistently upregulated during short- and long-term drought exposure.

7.2.3 *Metabolomics: Mapping Metabolic Reprogramming During Drought Stress*

Metabolomic studies have provided crucial understanding of how plants adapt biochemically to combat water deficit. Advanced techniques like GC-MS and LC-MS have helped researchers to identify various drought-responsive metabolites involved in osmotic adjustment, antioxidant defense, stress signaling and regulatory mechanisms.

The work of Ghorbanzadeh et al. (2023) offers an interesting example of metabolic adaptation in rice under drought conditions. Their untargeted GC-MS metabolomics approach showed that drought-tolerant rice varieties promoted root elongation through metabolic shifts in the TCA cycle and amino acid biosynthesis pathways. These drought tolerant genotypes accumulated higher levels of protective osmolytes like proline, trehalose, and sucrose compared to susceptible cultivars. In contrast, drought susceptible genotypes exhibited limited root plasticity, relying instead on cell wall thickening and less efficient metabolic adjustments, including elevated GABA levels, which were linked to restricted root growth (Ghorbanzadeh et al. 2023).

Phytohormonal regulation also plays a critical role in drought response. Wang et al. (2024) demonstrated that drought stress in in alfalfa varieties triggered significant change in hormone profiles, with ABA levels increasing while IAA, GA_3, and ZT declined. Notably, rhizomatous and tap-rooted varieties showed stronger drought tolerance, maintaining more stable hormone levels and requiring fewer metabolic changes than compared to creeping-rooted varieties (Wang et al. 2024).

7.2.4 *Ionomics: Understanding ion Homoeostasis Under Drought Conditions*

The emerging field of ionomics offers novel insights into how drought disrupts mineral nutrient homeostasis in plants. Using advanced analytical tools like high-resolution inductively coupled plasma mass spectrometry (HR-ICP-MS) and capillary electrophoresis, researchers can detect even subtle shifts in nutrient uptake and assimilation triggered by water deficit.

In an interesting and comprehensive study, D'Oria et al. (2022) showed that how *Brassica napus* and *Triticum aestivum* ionome profile changes under different drought conditions. Under moderate to severe water deficits (40%–25% field capacity), these plants experienced a significant decline in microcronutrient uptake. Essential micronutrients like Fe, Zn, Mn and Mo were severely declined. Most significantly, Fe uptake declined by 83% under severe stress conditions. Transcriptomic analyses of these plants provided further understanding and regulation patterns of nutrient transporter genes. The transporter of some nutrients (N/P/B/Mo/Ni) being consistently downregulated. Others, such as those for S, Mg, K, Ca, Fe, Zn, and Mn, displayed mixed regulations patterns. Interestingly, transporters for Cl, and Co were upregulated (D'Oria et al. 2022). Recent work from another research group (Maghrebi et al. 2024) extended this understanding in durum wheat. Drought tolerant cultivars maintained or even enhanced their accumulation of (SO_4^{2-}, NO_3^-, PO_4^{3-}, K^+, Ca^{2+}, Mg^{2+}). These responses were accompanied by the upregulation of stress related genes *TdP5CS*, and *TdWRKY2*. In contrast, susceptible cultivars showed reduced ion uptake and elevated oxidative stress markers.

The true power of modern plant stress biology lies in the integration of multiple omics datasets. This multi-layered approach enables researchers to build more broad models of drought response networks. Several important insights have emerged through such integrative approaches:

Transcriptome-Proteome Concordance: In many cases, drought-induced transcripts corelate with increased protein expression- like dehydrins and HSPs. Yet, not all responses are linear. For various metabolic enzymes and signaling molecules, post-transcriptional plays a key role. These layers add complexity and nuance to gene-protein relationships.

Metabolome-Ionome Interactions: The maintenance of osmotic balance under drought is not solely organic. Plants rely on both organic osmolytes and inorganic ions to maintain cellular water balance. The coordination between these systems is crucial, disruptions in either can compromise drought resilience.

Tissue-Specific Responses: Leaves, roots, and reproductive organs each face distinct challenges during drought environment. Interestingly, they do not act in isolation. Each tissues adapts in a tailored way, yet they remain functionally coordinated across the plant to cope with water deficit.

Predictive Modeling: Sophisticated models that combine transcriptomics, proteomics, ionomics, and metabolomics to phenomics are emerging. These frameworks can simulate plant behavior across varying drought conditions- informing both research and crop management.

As climate change accelerates, drought is becoming more intense, more frequent and more damaging to agriculture worldwide. Tackling this threat demands a shift in how we understand plant stress. A multi-omics approach offers a distinctive toolkit that captures the complexity of drought tolerance across different biological layers.

By integrating transcriptomics, proteomics, metabolomics, and ionomics researchers are uncovering an intricate network of responses. Each layer provides unique insight. Together they offer a system-level view of how plants perceive and respond to drought. This integrative approach has already helped in identifying key genes and metabolites critical for improving drought resilience. Given the complexity of abiotic stress, and need of designing crop plants for better future, research must focus on translating these fundamental discoveries into practical solutions through interdisciplinary collaborations.

Future progress in plant abiotic stress biology depends on more than just individual genes or traits. It requires holistic thinking. A system level approach that recognizes the importance of physiological, molecular and environmental factors. By embracing this complexity, we stand a better chance of developing climate resilient crops. The challenge is enormous. But the scientific tools and techniques we now have are equally incredible and more ready than ever to meet the giant challenges.

7.3 Integration of Omics Approaches

7.3.1 Multi-Omics Integration

Most of the time, plants under abiotic stress conditions are likely suffering from multiple stress factors simultaneously, such as salinity and drought, or drought along with nutrient and heat stress. Sometimes, a combination of abiotic and biotic stress is also prevalent. These combined stresses can have synergistic effects, exacerbating the overall impact on plant growth and productivity. For instance, drought stress can reduce a plant's ability to uptake nutrients, while salinity can cause ion toxicity and osmotic stress, further complicating the plant's physiological responses (Haak et al. 2017; González-Morales et al. 2021; Atta et al. 2023). The mechanisms adapted by plants under drought stress alone will differ from those under multiple stresses at the same time. Under drought stress, plants may close their stomata to reduce water loss, activate osmoprotectants, and alter root architecture to enhance water uptake (Sallam et al. 2019; Zahra et al. 2021; Mahreen et al. 2022; Bheemanahalli et al. 2022). However, when faced with multiple stresses, such as drought and salinity, plants must balance these responses with mechanisms to mitigate ion toxicity and maintain cellular homeostasis(Wu et al. 2022; Kumar et al. 2023; Atta et al. 2023). This complexity necessitates a universal understanding of how plants integrate signals from different stressors and coordinate their responses. Therefore, to develop crop plants with tolerance to abiotic stress, it is crucial to understand the mechanisms behind each specific stress and how plants respond and adapt to them. This understanding can inform breeding programs and genetic engineering efforts aimed at enhancing stress tolerance. For example, identifying key regulatory genes involved in stress responses can lead to the development of crops that are more resilient to multiple stress factors (Fig. 7.2).

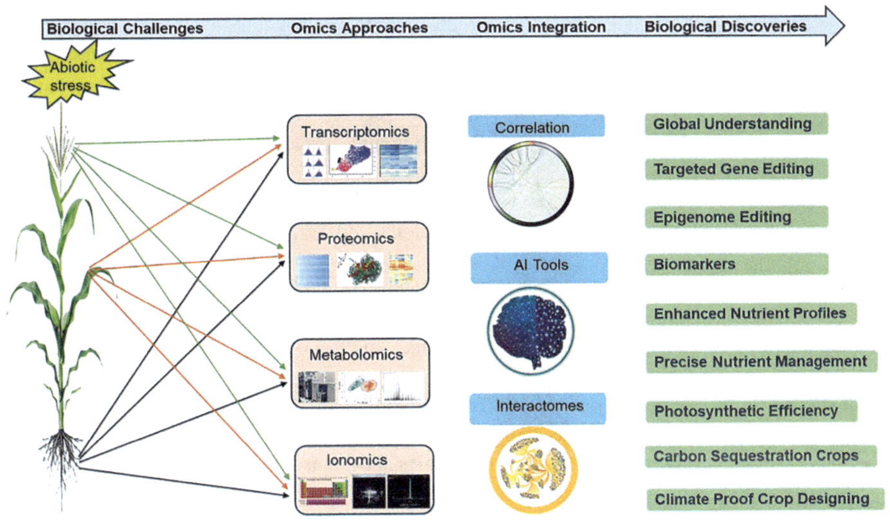

Fig. 7.2 Multi-omics approaches for enhancing abiotic stress tolerance in crop plants: Integrated multi-omics analysis involves merging multiple omics approaches within a single study under the same conditions. This holistic strategy employs advanced computational tools, including artificial intelligence (AI), to drive diverse biological discoveries

To understand the global mechanisms of plant response to stress, a multi-lateral and integrative approach involving multiple omics tools is necessary (Table 7.1). Each omics layer contributes a unique dimension to biological information. Genomics can identify stress-responsive genes and regulatory sequences. Transcriptomics can reveal dynamic changes in gene expression in response to stress. Proteomics can provide insights into protein abundance, modifications, stability, and interactions-offering a closer look at the functional molecules executing cellular responses. Metabolomics can profile the metabolites that reflect the plant's physiological and biochemical state under stress conditions (Haak et al. 2017; Sharma et al. 2020; Scossa et al. 2021; Rahman et al. 2022; Wu et al. 2022; Roychowdhury et al. 2023; Kamali and Singh 2023; Mmbando 2024; Tenorio Berrío and Dubois 2024; Zheng et al. 2024). Integrating these omics data can uncover complex regulatory networks and molecular pathways that single-layer studies might miss. When three or more omics tools are combined—for example, transcriptomics, proteomics, and metabolomics, it becomes possible to link transcriptional changes to protein functions and resulting metabolites adjustments. Such integration offers a holistic view of the plant's stress response mechanism and improves the chance of identifying candidate genes, enzymes, or metabolites for stress tolerance. Moreover, the inclusion of ionomics and phenomics adds further depth. Ionomics captures elemental profiles and nutrient dynamics, enabling them to corelate molecular data with nutrient accumulation, uptake and assimilation. When combined with phenomics, data on plant morphology, physiology and performance helps in linking genotype to phenotype under stress conditions. Importantly, the choice of omics tools should be

Table 7.1 Recent studies on multi-omics approaches against abiotic stress in plants

Abiotic stress	Plants	Omics approaches	References
Heat stress	Maize	Phenomics and Genomics	Longmei et al. (2021)
Heat stress	Maize	Transcriptomics, proteomics, and Metabolomics	Guo et al. (2021)
Heat stress	Maize	Transcriptomics and DAP-Seq	Li et al. (2024)
Low phosphorous	Wheat	Metabolome and Transcriptome	Li et al. (2023b)
Drought	Tomato	Transcriptomics and Proteomics	Liu et al. (2023)
Drought and salinity	Barley (*Hordeum vulgare*)	Phenomics and Genomics	Sabouri et al. (2024)
Salt stress	Wheat	Transcriptomics and Metabolomics	Wang et al. (2023b)
Salinity	Rice	Genomics and Transcriptomics	Sun et al. (2019)
Salinity	Brassica	Transcriptomics, Proteomics, and Metabolomics	Shu et al. (2022)
Salinity	Rapeseed (*Brassica napus*)	whole-genome re-sequencing and Transcriptomics	Zhou et al. (2021)
Heavy metal (Chromium)	*Canna indica*	Transcriptomics and Metabolomics	Zhu et al. (2024)
Multiple abiotic stress	Rice	Meta analysis of Transcriptomics data	Azad et al. (2024)
Arsenic	Rice	Ionomics and Transcriptomics	Xiao et al. (2021)
Arsenic	Rice	Genomics and Transcriptomics	Lee et al. (2022)

guided by nature of the abiotic stress being studied or applied and the goals of the experiments.

Integrating multiple omics layers such as genomics, transcriptomics, proteomics, metabolomics, and ionomics provides a comprehensive understanding of how organisms respond to abiotic stress. Ultimately, such integrative strategies are important in advancing the development of climate-resilient crop varieties and ensuring agriculture sustainability in the face of increasing environment pressures. Integration of these tools, help to:

1. **Identify Key Regulatory Networks**: Understanding how genes, proteins, and metabolites interact helps in identifying crucial regulatory networks and pathways involved in stress responses. For instance, integrating transcriptomic and proteomic data can reveal post-transcriptional modifications and protein-protein interactions that are critical under stress conditions (Scossa et al. 2021; Roychowdhury et al. 2023; Chao et al. 2023). In a recent study in tomato (Liu et al. 2023), the integrative analysis of transcriptomic and proteomic data was employed to understand the molecular mechanisms underlying drought tolerance

in tomato plants. In this study, comprehensive approach involving post-transcriptional processes and the correlation between mRNA expression and protein abundance was analyzed and identified several genes/proteins including TAS14, WSIP-3, ARS4, GSH-Px, APX, Hsp20, and Hsp70 which were differentially expressed at both transcript and protein levels, highlighting their roles in stress-related pathways and regulatory networks. In rice, genomics and transcriptomics approach were used to investigate the molecular basis of salinity tolerance in a novel strong salt-tolerant rice landrace, Changmaogu (Sun et al. 2019). This study led to the identification of candidate gene, OsPP2C8 and pathways, enhancing our understanding of plant responses to abiotic stress and providing valuable insights for future breeding efforts.

2. **Discover Mechanistic Insights**: Multi-omics integration allows for the discovery of novel mechanisms and interactions that single-layer studies might miss. For example, combining Genomics, transcriptomics, ionomics, proteomics, and metabolomics, can help to discover the interplay of cellular functions at different levels. In wheat, integrating metabolomic and transcriptomic data has revealed that salt treatment caused stress induced alterations in both transcriptomes and metabolomes (Wang et al. 2023a, b), leading to the identification of key pathways linked to glycerolipid and phenylalanine metabolism involved in stress tolerance. In *Brassica napus*, use of transcriptomics, proteomics, and metabolomics approaches helps to understand led to understand the role of the jasmonic acid and abscisic acid working jointly to reduce the damage caused by salt stress (Shu et al. 2022). In this study, several key metabolites were identified which involved in stress response and maintaining balance of reactive oxygen species such as 1-Decanol, 4-Pyridoxic acid, N α -Acetyl-L-glutamine, Trigonelline, 10-Formyl-THF N-acetyl-5-hydroxytryptamine, L-Cysteine and L-(+)-Arginine.
3. **Enhance Predictive Models and Facilitate Targeted Interventions**: Combining data from different omics layers, such as genomics, transcriptomics, proteomics, ionomics, and metabolomics, significantly improves the accuracy of predictive models and helps to gain a comprehensive understanding of the molecular mechanisms underlying stress responses. A holistic view enables the identification of potential targets for genetic engineering or breeding programs aimed at enhancing stress tolerance. For instance, integrating genomics and proteomics can identify stress-responsive genes and their protein products, which can be targeted for genetic modification. High-throughput sequencing technologies have revolutionized crop breeding by enabling the identification of beneficial quantitative trait loci (QTL), genes, and alleles for crop improvement. For example, in barley, integrating genomic and phenotypic data has led to the identification of QTLs associated with drought and salinity tolerance (Makhtoum et al. 2022; Sabouri et al. 2024). Heat stress is one of the major abiotic stresses that hampers crop production and yields with its impact varying extensively across growing seasons due to its complex inheritance (Thistlethwaite et al. 2020; Longmei et al. 2021; Djalovic et al. 2024). Recently, the transcription factor *ZmHSF20* was identified in maize for heat shock response (Z. Li et al. 2024). In this study, RNa-Seq was performed on B73 seedlings exposed to higher

temperature at varying durations and them weighted gene co-expression network analysis (WCGNA) was performed to identify this transcription factor *ZmHsf20*. These findings contribute to a better understanding of the molecular mechanism underlying heat tolerance in maize. Overall, the integration of multi-omics data not only enhances the predictive power of models for stress tolerance but also accelerates the breeding process, making it possible to develop crops that are better equipped to withstand the challenges posed by changing environmental conditions.

7.3.2 Bioinformatics Tools

Integrating data from different omics analysis is essential for gaining a comprehensive understanding of biological systems and their dynamic responses to environmental stimuli (Picard et al. 2021). While individual omics platform offers valuable insights into specific molecular layers, they represent only partial views of the overall biological story. To uncover the complexity of cellular process and regulatory networks, it is crucial to analyze and integrate information across these layers.

A wide range of computational algorithms and analytical methods have been developed to facilitate multi-omics integration. These methods differ in complexity, assumptions and the type of biological questions they address, each with its unique approach and advantages.

- **Correlation-Based Methods:** These methods identify correlations between different omics datasets, helping to uncover relationships and interactions between various biological molecules. For example, WGCNA utilizes correlation-based methods to construct co-expression networks (Li et al. 2024), identifying modules of highly correlated genes that may be involved in similar biological processes. Recently, AI-enhanced correlation tools have been developed to improve the accuracy and efficiency of these analyses. For instance, DeepInsight (Sharma et al. 2019) uses deep learning to transform high-dimensional omics data into low-dimensional images, which can then be analyzed for correlations using image recognition techniques. Though DeepInsight has not been used in crops studies, but with advent of AI and machine learning, this kind of approach can be integrated into crop omics analysis.
- **Network-Based Approaches:** These approaches involve constructing and analyzing biological networks to integrate omics data. By representing biological entities (genes, proteins, and metabolites) as nodes and their interactions as edges, network-based methods can reveal complex relationships and regulatory pathways. Cytoscape (https://cytoscape.org) is one of the widely used tool for visualizing and analyzing biological networks, allowing to integrate data from multiple omics studies and identify key regulatory hubs. Recently, AI-driven tools like GraphSAGE (Ferebee and Buckler 2023) have been employed to enhance network-based analyses by learning representations of nodes in large graphs, enabling more accurate predictions of biological interactions. In recent

years, graph neural networks (GNNs) are also widely used (Zhang et al. 2022), which apply deep learning to biological networks including protein-protein interactions. Correlation AnalyzeR (Miller and Bishop 2021) is an another tool recently developed for studying co-expression correlations and predicting gene functions, gene-gene relationships, and gene set topology.

- **Machine Learning and AI:** The application of machine learning (ML) and artificial intelligence (AI) in plant science is rapidly growing. ML algorithms are increasingly used for data integration and pattern recognition in multi-omics studies (Wekesa and Kimwele 2023). These techniques can handle large, complex datasets and uncover hidden patterns that may not be apparent through traditional methods. Random forests, support vector machines (SVMs), and neural networks are some of the machine learning algorithms applied to integrate omics data. For instance, DNNGP is a deep neural network-based framework designed for genomics prediction using multi-omics data in plants (Wang et al. 2023a). In this study, the newly developed deep learning tool, deep neural network genomic prediction (DNNGP) was compared with other popular prediction methods including GBLUP, SVR, DLGWAS, LightGBM, and deepGS on wheat, maize, and tomato to evaluate efficiency and performance.

7.3.3 Visualization Tools

Effective visualization of integrated omics data is crucial for interpreting complex biological information and deriving meaningful insights (Fukushima et al. 2014). As multi-omics studies often involve the simultaneous analysis of diverse data types, ranging from gene expression, protein abundance to metabolite profile and phenomics traits, visualization tools serves as essential platforms for data exploration, hypothesis generation, and effective communication of findings. A growing number of specialized software tools and platforms have been developed to facilitate the visualization of integrated omics data. These tools vary widely in their functionality, input requirements, and visualization formats, each offering unique features and capabilities depending on the research objective and type of analysis.

- **Cytoscape:** Cytoscape is a widely used open-source platform for visualizing complex networks and integrating these with any type of attribute data. It is particularly useful for constructing and analyzing biological networks, such as protein-protein interaction networks, gene regulatory networks, and metabolic pathways. Cytoscape handles diverse types of biological networks, including protein-protein interaction networks, gene co-expression and gene regulatory networks and signaling cascades. Users can construct their own networks or imports them from different biological databases like STRING, BioGRID, KEGG. A major feature of Cytoscape is its plugin based architecture, which enables seamless extension of its functionalities. One such plugin is BINGO (Biological Networks Gene Ontology) which is used for GO enrichment analysis

(Maere et al. 2005; Vilarrasa-Blasi et al. 2024) . Other widely used plugins include ClueGo, for pathway and functional analysis; cytoHubba, for identifying key regulatory nodes and hub genes within networks.

- **Integrative Genomics Viewer (IGV):** IGV (Robinson et al. 2023) is a high-performance visualization tool (https://igv.org) for interactive exploration of largescale genomic datasets. IGV provides an intuitive platform for the real-time analysis of diverse types of genomic, transcriptomics, and epigenomic data. It is particularly well-suited for navigating large datasets generated from next-generation sequencing platforms. This platform allows users to examine different features like gene expression levels, alternative splicing events, single nucleotide polymorphisms (SNPs) and DNA methylation patterns. In a study on a comparative flower transcriptome network in chickpea, IGV were used to shows upregulation of genes in tolerant chickpea genotype compared to sensitive genotype in response to salt stress (Kaashyap et al. 2022).
- **MiBiOmics:** MiBiOmics (Zoppi et al. 2021) is a web-based application that facilitates multi-omics data visualization, exploration, integration, and analysis. It provides easy access to interactive protocols for ordination techniques and network-based approaches. It incorporates weighted-gene correlation network analysis and can be implemented as an easy-to-use R shiny application but is currently limited to a maximum of three omics datasets (Munk et al. 2024). MiBiOmics is currently available as a Shiny app at https://shiny-bird.univ-nantes.fr/app/Mibiomics and as a standalone application at https://gitlab.univ-nantes.fr/combi-ls2n/mibiomics.
- **Visual Omics:** Visual Omics is web-based platform for omics data analysis and visualization (Li et al. 2023a). It integrates multiple omics analyses, including differential expression analysis, enrichment analysis, and protein-protein interaction analysis, with extensive graph presentations. It can also independently plot and customize basic charts that are involved in omics analysis including PCA plot, bar plot, box plots, heat maps, set intersection diagrams, bubble charts, and volcano plots. Visual Omics can be used at http://bioinfo.ihb.ac.cn/visomics. Source code can be downloaded at http://bioinfo.ihb.ac.cn/software/visomics/visomics-1.1.tar.gz
- **VANTED (Visualization and Analysis of Networks containing Experimental Data):** VANTED is a tool for the integrative visualization and analysis of omics data (Junker et al. 2006). By combining data, it enables users to visualize and evaluate metabolic and regulatory networks, offering insights into biological processes. One of the main strength of VANTED is its ability to support the simultaneous visualization of multiple types of quantitative data on network diagrams. This feature facilitates the interpretation of dynamic changes in molecular abundance across different experimental conditions or time points. It supports network construction and editing, statistical analysis, and dynamic simulations. This platform can also connect with external databases and pathway repositories, such as KEGG, enhancing the biological relevance of network models.

- **Gene Expression Plotter (GXP):** GXP (Eiteneuer et al. 2022) is web-based tool for visualizing and analyzing RNA-seq and metabolomics data. It allows for the easy plotting and analysis of gene expression data without the need for custom installations or programming. GXP supports a wide range of visualization features for effective data interpretation. Its main tools are interactive heatmaps, volcano plots, boxplots, scatter plots. These plots allow users to identify differentially expresses genes, analysis overall expression patterns, and compare expression levels across conditions/treatments. Its also provides data filtering and statistical testing directly through its interface. Gene Expression Plotter is freely available on GitHub (https://usadellab.github.io/GeneExpressionPlots/; accession date 11 January 2022).

7.4 Challenges and Future Directions in Omics Research for Advancing Crop Resilience

Omics data is inherently complex due to its high dimensionality and variability. Large volumes of data are generated by all omics techniques which can be fascinating to handle and evaluate. Sophisticated computational tools that can handle large-scale data while preserving accuracy and efficiency are needed for the integration of these distinct datasets. The integration of multi-omics data presents a notable barrier because of variations in data formats, sizes, and types. Advance bioinformatics tools and techniques are required to integrate these datasets and enable significant biological interpretations. Moreover, multi-omics integration demands a significant amount of computing power. High-performance computing infrastructure, enormous storage capabilities, and specialized software are necessary for processing and evaluating omics data. Additionally, the expertise required to run these systems and understand the outcomes is frequently a limiting factor, necessitating interdisciplinary collaboration between plant biologist, bioinformaticians and data scientists.

The advent of artificial intelligence (AI) and machine learning (ML) is transforming how omics data is analyzed and interpreted. These technologies are increasingly being applied to omics data to enhance data integration, pattern recognition, and predictive modelling.AI and ML algorithms, such as deep learning, ensemble methods, and unsupervised clustering, have demonstrated promise in identifying key regulatory elements, gene modules, and genotype to phenotype associations across omics layers. In the near future, it is expected that a new generation of AI-powered tools will enhance the resolution of multiple omics data analysis, improved visualization of biological networks, and prediction modeling of stress dynamics. These developments are necessary for enhancing crop yield, nutrient quality, and resilience to abiotic stresses.

In conclusion, omics tools and technologies are advancing our understanding of abiotic stress tolerance in crops. However, the key challenges for data integration, computational infrastructure, and need for interdisciplinary expertise remain crucial. Strategic investments and developments in computational biology, AI driven analytics are required. Together these efforts will help to unlock the full potential of omics for sustainable agriculture and global food security.

References

Atta K, Mondal S, Gorai S, Singh AP, Kumari A, Ghosh T, Roy A, Hembram S, Gaikwad DJ, Mondal S, Bhattacharya S, Jha UC, Jespersen D (2023) Impacts of salinity stress on crop plants: improving salt tolerance through genetic and molecular dissection. Front Plant Sci 14:1241736. https://doi.org/10.3389/FPLS.2023.1241736/BIBTEX

Azad M, Tohidfar M, Ghanbari Moheb Seraj R, Mehralian M, Esmaeilzadeh-Salestani K (2024) Identification of responsive genes to multiple abiotic stresses in rice (Oryza sativa): a meta-analysis of transcriptomics data. Sci Rep 14. https://doi.org/10.1038/s41598-024-54623-7

Bailey-Serres J, Parker JE, Ainsworth EA, Oldroyd GED, Schroeder JI (2019) Genetic strategies for improving crop yields. Nature 575. https://doi.org/10.1038/s41586-019-1679-0

Bheemanahalli R, Vennam RR, Ramamoorthy P, Reddy KR (2022) Effects of post-flowering heat and drought stresses on physiology, yield, and quality in maize (Zea mays L.). Plant Stress 6: 100106. https://doi.org/10.1016/j.stress.2022.100106

Chao H, Zhang S, Hu Y, Ni Q, Xin S, Zhao L, Ivanisenko VA, Orlov YL, Chen M (2023) Integrating omics databases for enhanced crop breeding. J Integr Bioinf 20

D'Oria A, Courbet G, Billiot B, Jing L, Pluchon S, Arkoun M, Maillard A, Roux CPL, Trouverie J, Etienne P, Diquélou S, Ourry A (2022) Drought specifically downregulates mineral nutrition: plant ionomic content and associated gene expression. Plant Direct 6. https://doi.org/10.1002/pld3.402

Deshmukh R, Sonah H, Patil G, Chen W, Prince S, Mutava R, Vuong T, Valliyodan B, Nguyen HT (2014) Integrating omic approaches for abiotic stress tolerance in soybean. Front Plant Sci 5

Djalovic I, Kundu S, Bahuguna RN, Pareek A, Raza A, Singla-Pareek SL, Prasad PVV, Varshney RK (2024) Maize and heat stress: physiological, genetic, and molecular insights. Plant Genome 17

Eiteneuer C, Velasco D, Atemia J, Wang D, Schwacke R, Wahl V, Schrader A, Reimer JJ, Fahrner S, Pieruschka R, Schurr U, Usadel B, Hallab A (2022) GXP: analyze and plot plant omics data in web browsers. Plants 11. https://doi.org/10.3390/PLANTS11060745/S1

FAO (2021) The impact of disasters and crises on agriculture and food security: 2021. FAO

FAO (2023) World Food and Agriculture—Statistical Yearbook 2023. FAO

Ferebee TH, Buckler ES (2023) Exploring the utility of regulatory network-based machine learning for gene expression prediction in maize. bioRxiv 05/14/2023. https://doi.org/10.1101/2023.05.11.540406

Fukushima A, Kanaya S, Nishida K (2014) Integrated network analysis and effective tools in plant systems biology. Front Plant Sci 5

Ghorbanzadeh Z, Hamid R, Jacob F, Zeinalabedini M, Salekdeh GH, Ghaffari MR (2023) Comparative metabolomics of root-tips reveals distinct metabolic pathways conferring drought tolerance in contrasting genotypes of rice. BMC Genomics 24. https://doi.org/10.1186/s12864-023-09246-z

Godoy F, Olivos-Hernández K, Stange C, Handford M (2021) Abiotic stress in crop species: improving tolerance by applying plant metabolites. Plants 10:1–19

González-Morales S, Solís-Gaona S, Valdés-Caballero MV, Juárez-Maldonado A, Loredo-Treviño A, Benavides-Mendoza A (2021) Transcriptomics of biostimulation of plants under abiotic stress. Front Genet 12. https://doi.org/10.3389/FGENE.2021.583888/FULL

Guo J, Gu X, Lu W, Lu D (2021) Multiomics analysis of kernel development in response to short-term heat stress at the grain formation stage in waxy maize. J Exp Bot 72:6291–6304. https://doi.org/10.1093/jxb/erab286

Haak DC, Fukao T, Grene R, Hua Z, Ivanov R, Perrella G, Li S (2017) Multilevel regulation of abiotic stress responses in plants. Front Plant Sci 8

Junker BH, Klukas C, Schreiber F (2006) Vanted: a system for advanced data analysis and visualization in the context of biological networks. BMC Bioinf 7:1–13. https://doi.org/10.1186/1471-2105-7-109/FIGURES/6

Kaashyap M, Ford R, Mann A, Varshney RK, Siddique KHM, Mantri N (2022) Comparative flower transcriptome network analysis reveals DEGS involved in chickpea reproductive success during salinity. Plants 11. https://doi.org/10.3390/plants11030434

Kamali S, Singh A (2023) Genomic and transcriptomic approaches to developing abiotic stress-resilient crops. Agronomy 13

Kumar R, Sharma VK, Rangari SK, Jha UC, Sahu A, Paul PJ, Gupta S, Gangurde SS, Kudapa H, Mir RR, Gaur PM, Varshney RK, Elango D, Thudi M (2023) High confidence QTLs and key genes identified using Meta-QTL analysis for enhancing heat tolerance in chickpea (Cicer arietinum L.). Front Plant Sci 14. https://doi.org/10.3389/FPLS.2023.1274759/FULL

Lee SB, Kim GJ, Shin JD, Chung W, Park SK, Choi GH, Park SW, Park YJ (2022) Genome-scale profiling and high-throughput analyses unravel the genetic basis of arsenic content variation in rice. Front Plant Sci 13. https://doi.org/10.3389/fpls.2022.905842

Lešková A, Giehl RFH, Hartmann A, Fargašová A, von Wirén N (2017) Heavy metals induce iron deficiency responses at different hierarchic and regulatory levels. Plant Physiol 174:1648–1668. https://doi.org/10.1104/pp.16.01916

Li M, Li H, Sun A, Wang L, Ren C, Liu J, Gao X (2022) Transcriptome analysis reveals key drought-stress-responsive genes in soybean. Front Genet 13. https://doi.org/10.3389/fgene.2022.1060529

Li H, Shi M, Ren K, Zhang L, Ye W, Zhang W, Cheng Y, Xia X-Q (2023a) Visual Omics: a web-based platform for omics data analysis and visualization with rich graph-tuning capabilities. https://doi.org/10.1093/bioinformatics/btac777

Li P, Ma X, Wang J, Yao L, Li B, Meng Y, Si E, Yang K, Shang X, Zhang X, Wang H (2023b) Integrated analysis of metabolome and transcriptome reveals insights for low phosphorus tolerance in wheat seedling. Int J Mol Sci 24. https://doi.org/10.3390/ijms241914840

Li Z, Li Z, Ji Y, Wang C, Wang S, Shi Y, Le J, Zhang M (2024) The heat shock factor 20-HSF4-cellulose synthase A2 module regulates heat stress tolerance in maize. Plant Cell 36. https://doi.org/10.1093/plcell/koae106

Liu M, Zhao G, Huang X, Pan T, Chen W, Qu M, Ouyang B, Yu M, Shabala S (2023) Candidate regulators of drought stress in tomato revealed by comparative transcriptomic and proteomic analyses. Front Plant Sci 14. https://doi.org/10.3389/fpls.2023.1282718

Longmei N, Gill GK, Zaidi PH, Kumar R, Nair SK, Hindu V, Vinayan MT, Vikal Y (2021) Genome wide association mapping for heat tolerance in sub-tropical maize. BMC Genomics 22. https://doi.org/10.1186/s12864-021-07463-y

Maere S, Heymans K, Kuiper M (2005) BiNGO: a Cytoscape plugin to assess overrepresentation of Gene Ontology categories in Biological Networks. Bioinformatics 21:3448–3449. https://doi.org/10.1093/BIOINFORMATICS/BTI551

Maghrebi M, Marín-Sanz M, Miras Moreno MB, Quagliata G, Caldo F, Gatti N, Mannino G, Pesenti M, D'Alessandro S, Nocito FF, Lucini L, Sestili F, Astolfi S, Barro F, Vigani G (2024) The drought-induced plasticity of mineral nutrients contributes to drought tolerance discrimination in durum wheat. Plant Physiol Biochem 215. https://doi.org/10.1016/j.plaphy.2024.109077

Mahalingam R, Duhan N, Kaundal R, Smertenko A, Nazarov T, Bregitzer P (2022) Heat and drought induced transcriptomic changes in barley varieties with contrasting stress response phenotypes. Front Plant Sci 13. https://doi.org/10.3389/fpls.2022.1066421

Mahreen N, Yasmin S, Asif M, Yousaf S, Yahya M, Ejaz K, Shahid Hussain H, Sajjid ZI, Arif M (2022) Integrated analysis of osmotic stress and infrared thermal imaging for the selection of resilient rice under water scarcity. Front Plant Sci 13. https://doi.org/10.3389/FPLS.2022.834520/FULL

Makhtoum S, Sabouri H, Gholizadeh A, Ahangar L, Katouzi M (2022) QTLs controlling physiological and morphological traits of barley (Hordeum vulgare L.) seedlings under salinity, drought, and normal conditions. Biotech 11. https://doi.org/10.3390/biotech11030026

Michaletti A, Naghavi MR, Toorchi M, Zolla L, Rinalducci S (2018) Metabolomics and proteomics reveal drought-stress responses of leaf tissues from spring-wheat. Sci Rep 8. https://doi.org/10.1038/s41598-018-24012-y

Miller HE, Bishop AJR (2021) Correlation AnalyzeR: functional predictions from gene co-expression correlations. BMC Bioinf 22. https://doi.org/10.1186/s12859-021-04130-7

Mmbando GS (2024) Omics: a new, promising technologies for boosting crop yield and stress resilience in African agriculture. Plant Stress 11

Munk K, Ilina D, Ziemba L, Brader G, Molin EM (2024) Holomics—a user-friendly R shiny application for multi-omics data integration and analysis. BMC Bioinf 25. https://doi.org/10.1186/s12859-024-05719-4

Pereira HD, Nonato JVA, Duarte RCRM, Gerhardt IR, Dante RA, Arruda P, Yassitepe JEDCT (2025) Temporal field phenomics of transgenic maize events subjected to drought stress: cross-validation scenarios and machine learning models. Plant Phenome J 8. https://doi.org/10.1002/ppj2.70015

Picard M, Scott-Boyer M-P, Bodein A, Périn O, Droit A (2021) Integration strategies of multi-omics data for machine learning analysis. Comput Struct Biotechnol J 19. https://doi.org/10.1016/j.csbj.2021.06.030

Rahman MU, Zulfiqar S, Raza MA, Ahmad N, Zhang B (2022) Engineering abiotic stress tolerance in crop plants through CRISPR genome editing. Cells 11

Ren W, Shi Z, Zhou M, Zhao B, Li H, Wang J, Liu Y, Zhao J (2022) iTRAQ-based quantitative proteomic analysis provides insight into the drought-stress response in maize seedlings. Sci Rep 12. https://doi.org/10.1038/s41598-022-13110-7

Robinson JT, Thorvaldsdottir H, Turner D, Mesirov JP (2023) igv.js: an embeddable JavaScript implementation of the Integrative Genomics Viewer (IGV). Bioinformatics 39. https://doi.org/10.1093/BIOINFORMATICS/BTAC830

Roychowdhury R, Das SP, Gupta A, Parihar P, Chandrasekhar K, Sarker U, Kumar A, Ramrao DP, Sudhakar C (2023) Multi-omics pipeline and omics-integration approach to decipher plant's abiotic stress tolerance responses. Genes (Basel) 14

Sabouri H, Pezeshkian Z, Taliei F, Akbari M, Kazerani B (2024) Detection of closely linked QTLs and candidate genes controlling germination indices in response to drought and salinity stresses in barley. Sci Rep 14. https://doi.org/10.1038/s41598-024-66452-9

Sallam A, Alqudah AM, Dawood MFA, Baenziger PS, Börner A (2019) Drought stress tolerance in wheat and barley: advances in physiology, breeding and genetics research. Int J Mol Sci 20. https://doi.org/10.3390/IJMS20133137

Scossa F, Alseekh S, Fernie AR (2021) Integrating multi-omics data for crop improvement. J Plant Physiol 257. https://doi.org/10.1016/j.jplph.2020.153352

Sharma A, Vans E, Shigemizu D, Boroevich KA, Tsunoda T (2019) DeepInsight: a methodology to transform a non-image data to an image for convolution neural network architecture. Sci Rep 9. https://doi.org/10.1038/s41598-019-47765-6

Sharma A, Kumar V, Shahzad B, Ramakrishnan M, Preet Singh Sidhu G, Shreeya Bali A, Handa N, Kapoor D, Yadav P, Khanna K, Bakshi P, Rehman A, Kaur Kohli S, Khan EA, Daman Parihar R, Yuan H, Kumar Thukral A, Bhardwaj R, Zheng B (2020) Photosynthetic response of plants under different abiotic stresses: a review. 39:509–531. https://doi.org/10.1007/s00344-019-10018-x

Shu J, Ma X, Ma H, Huang Q, Zhang Y, Guan M, Guan C (2022) Transcriptomic, proteomic, metabolomic, and functional genomic approaches of Brassica napus L. during salt stress. PLoS One 17. https://doi.org/10.1371/journal.pone.0262587
Sun BR, Fu CY, Fan ZL, Chen Y, Chen WF, Zhang J, Jiang LQ, Lv S, Pan DJ, Li C (2019) Genomic and transcriptomic analysis reveal molecular basis of salinity tolerance in a novel strong salt-tolerant rice landrace Changmaogu. Rice 12. https://doi.org/10.1186/s12284-019-0360-4
Tenorio Berrío R, Dubois M (2024) Single-cell transcriptomics reveals heterogeneity in plant responses to the environment: a focus on biotic and abiotic interactions. J Exp Bot. https://doi.org/10.1093/jxb/erae107
Tenorio Berrío R, Verhelst E, Eekhout T, Grones C, De Veylder L, De Rybel B, Dubois M (2025) Dual and spatially resolved drought responses in the Arabidopsis leaf mesophyll revealed by single-cell transcriptomics. New Phytol 246:840–858. https://doi.org/10.1111/nph.20446
Thistlethwaite RJ, Tan DKY, Bokshi AI, Ullah S, Trethowan RM (2020) A phenotyping strategy for evaluating the high-temperature tolerance of wheat. Field Crop Res 255. https://doi.org/10.1016/j.fcr.2020.107905
Vilarrasa-Blasi J, Vellosillo T, Jinkerson RE, Fauser F, Xiang T, Minkoff BB, Wang L, Kniazev K, Guzman M, Osaki J, Barrett-Wilt GA, Sussman MR, Jonikas MC, Dinneny JR (2024) Multi-omics analysis of green lineage osmotic stress pathways unveils crucial roles of different cellular compartments. Nat Commun 15. https://doi.org/10.1038/s41467-024-49844-3
Wang K, Ali Abid M, Rasheed A, Crossa J, Hearne S, Li H (2023a) DNNGP, a deep neural network-based method for genomic prediction using multi-omics data in plants. Mol Plant 16: 279–293. https://doi.org/10.1016/j.molp.2022.11.004
Wang W, Cao J, Huang S, Wang Z, Wang W, Zou J, Wang F, Luo M, Zhang J (2023b) Integrated transcriptomics and metabolomics analyses provide insights into salt-stress response in germination and seedling stage of wheat (Triticum aestivum L.). Curr Plant Biol 33. https://doi.org/10.1016/j.cpb.2023.100274
Wang K, Nan LL, Xia J, Wu SW, Yang LL (2024) Metabolomics reveal root differential metabolites of different root-type alfalfa under drought stress. Front Plant Sci 15. https://doi.org/10.3389/fpls.2024.1341826
Wekesa JS, Kimwele M (2023) A review of multi-omics data integration through deep learning approaches for disease diagnosis, prognosis, and treatment. Front Genet 14
Wu J, Nadeem M, Galagedara L, Thomas R, Cheema M (2022) Recent insights into cell responses to cold stress in plants: signaling, defence, and potential functions of phosphatidic acid. Environ Exp Bot 203
Xiao W, Liu P, Wang K, Yang Z, Wang L (2021) Relationship between ionomics and transcriptomics of rice plant in response to arsenite stress. Environ Exp Bot 189:104565. https://doi.org/10.1016/J.ENVEXPBOT.2021.104565
Yang Y, Saand MA, Huang L, Abdelaal WB, Zhang J, Wu Y, Li J, Sirohi MH, Wang F (2021) Applications of multi-omics technologies for crop improvement. Front Plant Sci 12
Zahra N, Wahid A, Hafeez MB, Ullah A, Siddique KHM, Farooq M (2021) Grain development in wheat under combined heat and drought stress: plant responses and management. Environ Exp Bot 188:104517. https://doi.org/10.1016/J.ENVEXPBOT.2021.104517
Zhang F, Wang Y, Lou Z (2007) Effect of heavy metal stress on antioxidative enzymes and lipid peroxidation in leaves and roots of two mangrove plant seedlings (Kandelia candel and Bruguiera gymnorrhiza). 67:44–50. https://doi.org/10.1016/j.chemosphere.2006.10.007
Zhang Q, Li B, Zhang Y, Wang S (2022) Suitability evaluation of crop variety via graph neural network. https://doi.org/10.1155/2022/5614974
Zhang Q, Teng R, Yuan Z, Sheng S, Xiao Y, Deng H, Tang W, Wang F (2023) Integrative transcriptomic analysis deciphering the role of rice bHLH transcription factor Os04g0301500 in mediating responses to biotic and abiotic stresses. Front Plant Sci 14. https://doi.org/10.3389/fpls.2023.1266242

Zheng X, Wang C, Xu Y, Kang Z, Cao P, Zhang H, Chen Q (2024) Multi-omics analysis of Nicotiana tabacum unveils tissue-specific regulatory expression strategies for cadmium accumulation. Ind Crop Prod 222. https://doi.org/10.1016/j.indcrop.2024.119661

Zhou T, Yue C-P, Liu Y, Zhang T-Y, Huang J-Y, Hua Y-P (2021) Multiomics reveal pivotal roles of sodium translocation and compartmentation in regulating salinity resistance in allotetraploid rapeseed. J Exp Bot 72:5687–5708. https://doi.org/10.1093/jxb/erab215

Zhu S, Zhao W, Sheng L, Yang X, Mao H, Sun S, Chen Z (2024) Integrated transcriptome and metabolomics analyses revealed key functional genes in Canna indica under Cr stress. Sci Rep 14. https://doi.org/10.1038/s41598-024-64877-w

Zoppi J, Guillaume JF, Neunlist M, Chaffron S (2021) MiBiOmics: an interactive web application for multi-omics data exploration and integration. BMC Bioinf 22:1–14. https://doi.org/10.1186/S12859-020-03921-8/FIGURES/5

Chapter 8
Mechanism of Temperature Perception, Signalling and Acclimation in Plants

Kishore C. S. Panigrahi, Adwitiya Khalua, Aman Kumar, Ananthapadmanabhan Ajith, Anish Dash, Joshua Joseph Julison, Meera Vijayan, Soubhagya Ranjan Bastia, Soumyaranjan Barik, and Madhusmita Panigrahy

Abstract Numerous studies have demonstrated the substantial influence of temperature on plant growth, quality, and productivity. This chapter elucidates the different biological phenomena in plants involved in the perception and response to temperature changes in the environment of the organism. The chapter sheds light on the role of various components in temperature sensing, signalling and acclimation. Response to various ranges of temperature—ambient and stressful—is discussed. Temperature is an essential cue used to entrain plant growth cycles, and in the context of modern challenges faced by plants with regard to rapid climate change due to anthropogenic interventions, understanding the mechanisms in plants related to temperature becomes crucial to address sustainable agricultural practices, food security issues and preservation of ecological niches.

Keywords Temperature · Thermomorphogenesis · Vernalisation · Circadian · Phytohormones · Stress · Photoreceptors · Calcium · Epigenetic

K. C. S. Panigrahi (✉) · A. Khalua · A. Kumar · A. Ajith · A. Dash · J. J. Julison · M. Vijayan · S. R. Bastia · S. Barik
School of Biological Sciences, NISER, Bhubaneswar, Odisha, India

Homi Bhabha National Institute (HBNI), Mumbai, India
e-mail: panigrahi@niser.ac.in; adwitiya.khalua@niser.ac.in; aman.kumar@niser.ac.in; ananthapadmanabhan.a@niser.ac.in; anish.dash@niser.ac.in; joshuaj.julison@niser.ac.in; meera.vijayan@niser.ac.in; soubhagyaranjan.b@niser.ac.in; soumyaranjan.barik@niser.ac.in

M. Panigrahy
Department of Molecular Biology & Biotechnology, Institute of Agricultural Sciences, Siksha 'O' Anusandhan University, Mumbai, India
e-mail: madhusmitapanigrahy@soa.ac.in

S. K. Panda, M. A. Khan (eds.), *Plant Functional Genomics for Abiotic Stress Resilience*, Plant in Challenging Environments 7,
https://doi.org/10.1007/978-3-032-01704-8_8

Abbreviations

ABA	Abscisic Acid
ABI5	Abscisic Acid Insensitive 5
ADC1	Arginine decarboxylase1
ADT	Average Daily Temperature
AFP	Anti-Freeze Protein
AHK	ARABIDOPSIS HISTIDINE KINASE
AHP	ARABIDOPSIS HISTIDINE PHOSPHOTRANSFER PROTEIN
AOS	Allene Oxide Synthase
AOX	alternative oxidases
AP2	Apetala 2
APX	ascorbate peroxidase
ARF	Auxin Response Factor
ARR	ARABIDOPSIS RESPONSE REGULATOR
ASA	ascorbate
ASF	ANTI-SILENCING FACTOR
ATBH2	*A. thaliana* homeobox protein 2
AtOP3	Arabidopsis thaliana Oligopeptide Transporter3
AUX1	Auxin Influx Transporter
BES1	Bri1-EMS Supressor1
bHLH	basic helix-loop-helix
BIN2	Brassinosteroids-Insensitive
BRs	Brassinosteroids
BZR1	Brassinazole-Resistant 1
CaM	Calmodulin
CAMTA	calmodulin binding transcription activator
CAT	catalase
CBF	C-REPEAT BINDING FACTOR
CBK3	CaM-binding protein kinase 3
CCA1	CIRCADIAN CLOCK ASSOCIATED 1
CCD	Carotenoid cleavage dioxygenase
CCS1	Copper Chaperone for Superoxide Dismutase 1
CDK	Cyclin Dependent Kinase
CK	Cytokinin
CKX	cytokinin dehydrogenase/oxidase
CNGC	Cyclic Nucleotide Gated Calcium Channel
COI1	Coronatine-insensitive 1
COP1	Constitutive Photomorphogenesis 1
COR	Cold-Responsive or Cold-regulated genes
CRF	CYTOKININ RESPONSE FACTORS
CRY	Cryptochrome
CSD	Copper/Zinc Superoxide Dismutase

CTR1	Copper Transporter Protein
DAG	diacylglycerol
DGDG	digalactosyldiacylglycerol
DIF	Difference between Day Temperature and Night Temperature
DNA	Deoxyribonucleic Acid
DRE	DEHYDRATION RESPONSIVE ELEMENT
DREB	DEHYDRATION RESPONSIVE ELEMENT BINDING
DT	Day Temperature
DWF4	Dwarf4
EBF	EIN-3 BINDING F-BOX
EC	Evening Complex
EIN	ETHYLENE INSENSITIVE
EL	electrolyte leakage
ELF3	Early Flowering 3
ELF4	Early Flowering 4
ER	Endoplasmic reticulum
ERD	early dehydration-inducible
ERF	ETHYLENE RESPONSE FACTOR
ETR1	Ethylene Receptor1
ETS	Electron Transport System
FAD2	fatty acid desaturase 2
FBH1	FLOWERING BASIC HELIX-LOOP-HELIX 1
FCA	FLOWERING CONTROL LOCUS A
FER	FERONIA
FLC	FLOWERING LOCUS C
FMN	FLAVIN MONONUCLEOTIDE
FT	FLOWERING LOCUS T
GA	Gibberellic Acid
GABA	Gamma-aminobutyric acid
GAD	L-glutamate decarboxylase
GAPDH	glyceraldehyde-3-phosphate dehydrogenase
GCN5	General Control Non-derepressible 5
GhAOC2	Gossypium hirsutum Allene Oxide Cyclase2
GI	GIGANTEA
GPAT9	glycerol-3-phosphate (G3P) acyltransferase 9
GPX	guaiacol peroxidase
GR	glutathione reductase
GSH	glutathione (reduced)
GSSG	glutathione disulphide (oxidised glutathione)
HAMK	heat-associated MAP kinase
HOS1	High Expression of Osmotically Responsive Gene 1
HSE	Heat Shock Element
HSF	Heat Shock Factor

HSP	Heat Shock Protein
HSR	heat stress-responsive genes
HY5	LONG HYPOCOTYL 5
IAA	Indole-3-Acetic Acid
IAA29	Indole-3-acetic acid inducible 29
ICE1	Inducer of CBF Expression 1
IP3	inositol-1,4,5-trisphosphate
IPT	isopentyl transferase
JA	Jasmonic Acid
JOX	Jasmonic acid Oxidase
JUB1	JUNGBRUNNEN 1
LBS	LUX Binding Site
LEA	late-embryogenesis-abundant
LHY	LATE ELONGATED HYPOCOTYL
lncRNA	long non-coding RNA
LOV	LIGHT-OXYGEN-VOLTAGE
LTI	low-temperature induced
LUX	LUX ARRYTHMO
MAPK	mitogen-activated protein kinase
MAX1	More axillary growth
MBF1c	MULTIPROTEIN BRIDGING FACTOR 1C
MDA	malondialdehyde
MEcPP	2-C-methyl-D-erythritol-2,4-cyclopyrophosphate
MGDG	monogalactosyldiacylglycerol
miR	micro RNA
MKK	MAP kinase kinase
mRNA	messenger RNA
NAD	nicotinamide adenine dinucleotide
NCED	9-cis-epoxycarotenoid dioxygenase
NO	Nitric Oxide
NT	Night Temperature
OEC	Oxygen evolving complex
OsAOC	Oryza sativa Allene Oxide Cyclase
OsD14	Oryza sativa Dwarf 14
OsLOX2	Oryza sativa Lipoxygenase 2
PAP	30-PHOSPHOADENOSINE 50-PHOSPHATE
PC	phosphatidylcholine
PCD	programmed cell death
PCNA	proliferating cell nuclear antigen
PE	phosphatidylethanolamine
phyB	Phytochrome B
PI	phosphatidylinositol
PIF	Phytochrome Interacting Factor

PIN	PIN-Formed proteins
PIP2	phosphatidylinositol-4,5-bisphosphate
PLD	Prion-Like Domain
PP2CD	2C Protein phosphatases
PP7	Protein Phosphatase 7
PRR	PSEUDO-RESPONSE REGULATOR
PTM	Post-translational Modifications
PYL/PYR	PYR1-like/Pyrabactin Resistance
RAB18	Responsive to Abscisic Acid 18
Rboh	Respiratory burst homology
RCB	REGULATOR OF CHLOROPLAST BIOGENESIS
RD	responsive to desiccation
RdDM	RNA-dependent DNA methylation
RG4	RNA G-quadruplex
RNA	Ribonucleic Acid
ROS	Reactive Oxygen Species
Rubisco	Ribulose-1,5-bisphosphate carboxylase/oxygenase
SA	Salicylic Acid
SAMK	Stress Activated MAP kinase
SAUR	SMALL AUXIN-UP RNA
SCRM	Scream
SDIR1	Salt- and Drought-induced Ring Finger1
SGT1	suppressor of the G2 allele of SKP1
SIZ1	Small Ubiquitin-related Modifier 1
SLs	Strigolactones
SOC1	SUPPRESSOR OF CONSTANS 1
SOD	Superoxide Dismutase
sRNA	small RNA
ST2A	sulfotransforase2
SUMO	small ubiquitin-related modifier
TAA1	TRYPTOPHAN AMINOTRANSFERASE OF ARABIDOPSIS 1
TAG	triacylglycerol
TAT	Tryptophan Amino Transferase
TE	transposable elements
TFs	Transcription factors
TGS	transcriptional gene silencing
TIR2	Transport Inhibitor Response2
TOC1	TIMING OF CAB EXPRESSION 1
TTL1	TETRATRICOPEPTIDE-REPEAT THIOREDOXIN-LIKE 1
UTR	Untranslated Region
XRN	exoribonuclease
YUC8	YUCCA8
ZTL	ZEITLUPE

8.1 Introduction

Plants are entrained by various environmental cues to optimise their growth and development. Temperature is one such essential cue that directs the biological fate of the organism. Some plants are sensitive enough to distinguish differences in temperature even by 1 °C (Penfield 2008). Such is the constancy of temperature patterns, which means that plants have mechanisms that anticipate certain temperature-related phenomena to complete their lifecycle. For instance, a cold winter is needed for a plant to become competent at flowering (vernalisation). The memory of winter leads to flowering in spring. (Michaels and Amasino 2000) The temperature patterns in different regions have led to the evolution of plants adapted to various climatic conditions (Nievola et al. 2017). The optimum temperature for growth varies from plant to plant. For instance, a plant like *Oryza sativa* found in tropical regions has an ambient growth temperature range of 25–35 °C (Sánchez et al. 2014). In contrast, *Arabidopsis thaliana*, generally found in temperate areas, has an ambient growth temperature range of 20–25 °C (Balasubramanian et al. 2006).

Nonetheless, an organism can accommodate a certain amount of fluctuation, beyond which survival becomes critical. Up to a specific temperature range, plants can acclimatise themselves, exhibiting thermomorphogenesis, in which the physiology of a plant is altered from what it would have been in optimum growth conditions to cope with the surroundings. Beyond such ranges, extreme temperature variation results in stressing out the plants to the extent of damage. Furthermore, alterations in expected temperature patterns affect a plant's life trajectory (Nievola et al. 2017; Penfield 2008).

Sessile plants cannot avoid the stresses in their environment; instead, they developed precise mechanisms to assess temperature-related phenomena. Plants sense changes in temperature, process the signal and respond accordingly. They can modulate their lifecycle according to variable environmental cues because of well-developed and established molecular machinery crafted through several years of evolution (Nievola et al. 2017; Penfield 2008). This chapter discusses the various physiological processes in plants based on temperature cues. Further, the roles of different components (Table 8.1) are highlighted that mediate temperature sensation, signals, and response.

8.2 Temperature-Related Phenomena Affecting Plant Growth and Development

8.2.1 Thermomorphogenesis and Temperature Stress

Thermomorphogenesis is a crucial process in plant physiology. It refers to the adaptive responses of plants to changes in ambient temperature, especially warmer conditions. This dynamic phenomenon profoundly impacts various aspects of plant

Table 8.1 Some key genes involved in temperature-related pathways in *A. thaliana*

Gene	Known function	References
PHYTOCHROME B	Temperature sensor; degradation of PIFs	Jung et al. (2016), Legris et al. (2016), and Yamashino et al. (2013)
PHYTOCHROME INTERACTING FACTORs (PIFs)	Thermoresponsiveness; hypocotyl elongation	Kim et al. (2020) and Box et al. (2015)
EVENING COMPLEX GENES: *EARLY FLOWERING 3 (ELF3)* *ELF4* *LUX ARRYTHMO (LUX)*	Transcriptional repressor for genes with LBS; regulates the expression of *PIFs*; temperature-based affinity for DNA binding	Silva et al. (2020), Mizuno et al. (2015), Ezer et al. (2017), and Nusinow et al. (2011)
HEAT SHOCK PROTEINs (HSPs)	Aid in protein folding (chaperone), aid in degradation of misfolded proteins and prevent aggregation of misfolded protein	Mondal et al. (2023) and Al-Whaibi (2011)
HEAT SHOCK FACTORs (HSFs)	Transcription factors that regulate the expression of *HSPs*	Ohama et al. (2017)
DEHYDRATION RESPONSIVE ELEMENT BINDING (DREBs)/ C-REPEAT BINDING FACTORs (CBFs)	Important transcription factors for cold acclimation; regulate the activity of COR	Gilmour et al. (2000) and Jaglo-Ottosen et al. (1998)
COLD REGULATION (COR) GENES*:* *RESPONSIVE TO ABSCISIC ACID* 18 *(RAB18)* *COR47* *COR15A* *COR78* *LOW TEMPERATURE INDUCED 29 (LTI29)* *LTI30*	Combined effects of these genes enhance cold tolerance	Dong et al. (2011) and Puhakainen et al. (2004)
INDUCER OF CBF EXPRESSION 1 (ICE1) *CALMODULIN BINDING TRANSCRIPTION ACTIVATOR 3 (CAMTA3)*	Key transcription factors that regulate the activity of CBFs	Doherty et al. (2009) and Chinnusamy et al. (2003)

development and growth, with significant implications for crop productivity and our understanding of plant adaptation to global climate change.

High-temperature stress significantly impacts plant lifecycle by causing changes in morphology, biochemistry, and physiology. It can damage proteins, disrupt enzyme function, harm membranes, and affect cell division, leading to stunted growth, oxidative damage, and lower quality and yield of seeds (Smertenko et al.

1997; Wahid et al. 2007). High temperatures can decrease seed germination potential, inducing poor germination and stand establishment. The impact of heat stress on cereal crops depends on the timing, duration, and intensity of the heat, with such stress reducing the number of spikes and florets per plant in rice and negatively affecting seed set in sorghum (Fahad et al. 2016, 2017; Prasad et al. 2006)

Thermomorphogenesis primarily impacts plant architecture by altering plant height, branching patterns, and tiller numbers, as seen in rice. These changes are thought to be driven by complex signalling networks involving hormones like gibberellin, salicylic acid (SA), and cytokinin (CK), helping plants adapt to temperature variations and enhance their survival and reproduction (Patel and Franklin 2009). Thermomorphogenesis not only alters plant morphology but also impacts physiological and biochemical functions. High-temperature stress can interfere with protein synthesis, inactivate enzymes, and damage cellular membranes, causing metabolic disturbances that can ultimately reduce crop productivity and pose a significant challenge to global food security as temperatures rise (Fahad et al. 2017).

Heat stress primarily affects plant germination, leading to reduced germination rates, poor seedling vigour, and abnormal seedling development, though the extent varies by crop species (Hasanuzzaman et al. 2013). Exposing pearl millet and maize seeds to constant heat stress (45 °C) for 8 days significantly reduced both the germination rate and final germination percentage for both species. (Ashraf and Hafeez 2004). A study showed that seeds possess a distinct mechanism to regulate Abscisic Acid (ABA) levels in response to increases in temperatures and employ a specific ABA-signalling pathway to inhibit germination (Toh et al. 2008).

High temperatures regulate GA levels independently of light conditions by stimulating ABA production and inhibiting GA biosynthesis and signalling in *Arabidopsis* seeds. Thermoinhibition may occur by keeping ABA levels above the germination threshold, not only through increased ABA biosynthesis but also by reducing its breakdown, regulated by GA in *Arabidopsis* seeds (Toh et al. 2008). Exposure to high temperatures, such as 45 °C during the day and 35 °C at night, slows plant growth due to stress, marked by increased electrolyte leakage (EL), indicating cell membrane damage. Research has proven EL to be a reliable indicator of heat stress in plants like wheat, rice, and strawberries. Higher temperature leads to greater EL in plants like legumes (Kumar et al. 2011; Almeselmani et al. 2009; Sohn and Back 2007; Gulen and Eris 2004; Ortiz and Cardemil 2001; Liu and Huang 2000).

Heat stress causes morphological symptoms such as scorching and sunburn on leaves and twigs, as well as damage to branches, stems, and fruits. In sugarcane, high temperatures lead to leaf-tip damage, leaf rolling and drying, and necrosis (Hasanuzzaman et al. 2013; Omae et al. 2012; Rodríguez et al. 2005). Flower initiation and development are highly sensitive to heat stress, which delays flowering and disrupts reproductive development in crops like wheat (Ullah et al. 2022; Aiqing et al. 2018). Heat stress can cause abnormalities in both male and female reproductive organs of a gametophyte, leading to poor grain quality due to inhibited protein and starch biosynthesis. Additionally, heat stress shortens the grain-filling period, reducing seed weight by hastening maturity and diminishing the flag leaf's assimilatory efficiency (Ullah et al. 2022).

8.2.2 Temperature Fluctuation in Seasonal Cycle

Vernalisation is a treatment method of exposing plants to cold temperatures to induce faster flowering. It controls the transition from the vegetative to the reproductive phase in the plant's life cycle (Thomson 1936). This phenomenon is especially crucial in temperate regions, where plants must synchronise their life cycle with seasonal changes to ensure successful reproduction and survival. Vernalisation ensures plants bloom in spring, avoiding hot summer and harsh winter conditions. A study says this delay between cold exposure and flowering, along with the need for vernalisation in each generation, suggests that an epigenetic mechanism and changes in chromatin structure at *FLOWERING LOCUS C (FLC)* gene are likely involved in this process (Sheldon et al. 2006; Finnegan et al. 2004, 2005; Bastow et al. 2004; Sung and Amasino 2004; Gendall et al. 2001). The FLC protein is a key flowering regulator in *Arabidopsis* and is crucial in vernalisation. It encodes a MADS-box protein that represses flowering by inhibiting the expression of the genes *SOC1* and *FT*, both of which promote flowering (Sheldon et al. 2006; Lee et al. 2000; Onouchi et al. 2000; Kardailsky et al. 1999; Kobayashi et al. 1999). In vernalisation-responsive *Arabidopsis* ecotypes, high *FLC* levels delay flowering without cold, but prolonged cold reduces *FLC* expression, leading to faster flowering during development (Sheldon et al. 1999, 2000, 2006; Michaels and Amasino 1999). Several genes are involved in mediating thermal responsive flowering (Blázquez et al. 2003).

8.2.3 Temperature Compensation in Circadian Clock

The plant circadian clock is responsive to environmental cues such as changes in temperature, yet the response is not so drastic as to alter its rhythmicity. In other words, the circadian clock is buffered against such physiologically relevant variations to maintain its functionality. Quantitative genetic approaches suggest that *GIGANTEA* (*GI*) and *ZEITLUPE* (*ZTL*) could have a role in this process (Edwards et al. 2005). At higher temperatures, *FLC* mediates temperature compensation, most probably through *LUX* (Edwards et al. 2006). PRR7 and PRR9 act in an FLC-independent manner for temperature compensation via modulation of LHY and CCA1 levels in response to changing temperature to above 12 °C (Salomé et al. 2010). The temperature-dependent negative transcriptional regulation of CCA1 via a bHLH transcription factor FBH1 also helps temperature compensation through a feedback mechanism. At elevated temperatures, overexpression of FBH1 shortens the period of CCA1, and CCA1 can also bind to FBH1 promoter to regulate its expression (Nagel et al. 2014). The mechanism of temperature compensation of the clocks in plants is still shrouded in mystery, with many components yet to be discovered and their activity yet to be understood.

8.2.4 Temperature Fluctuation in the Day-Night Cycle

The difference between day and night temperatures, known as the DIF, can significantly regulate a variety of plant characteristics. The growth rates of roots and shoots under negative DIF were significantly lesser than those observed under zero DIF and positive DIF conditions (Song et al. 2023).

Plants with the same average daily temperature (ADT) can grow differently based on the day and night temperatures. This effect is called 'thermoperiodism'. Generally, plants grow taller when it is warmer during the day than at night (positive DIF) compared to when it is warmer at night (negative DIF) (Thingnaes et al. 2003; Myster and Moe 1995; Went 1944). In *A. thaliana*, stem length increased with higher day temperatures (DT) and decreased with higher night temperatures (NT) within the 12–22 °C range. While leaf length decreased with higher NT, DT did not affect it. Similar effects on plant height were observed in *Lilium longiflorum* (Erwin et al. 1989). Additionally, the levels of IAA in flower stems were higher in samples exposed to colder nights than those exposed to colder days, suggesting that changes in IAA levels might influence temperature-dependent stem growth in *A. thaliana* (Thingnaes et al. 2003).

When DT was lower than NT, plants showed significantly lower fresh and dry weights and poorer root development, and their height was reduced by about half compared to those grown under constant or higher day temperatures. The number of flower buds dropped significantly when the ADT increased from 15 to 27 °C. NT had a stronger impact on reducing flower buds than DT, and a higher DIF further decreased the number of buds, especially when the difference was negative. Raising the ADT during propagation improved earliness, yield, and fruit quality, while temperature fluctuations between day and night did not impact earliness or yield but did enhance fruit quality (Grimstad and Frimanslund 1993).

8.2.5 Effect of Temperature on Seed Germination

Germination of the seed is vital for setting the foundation for a new generation to grow. Temperature plays a crucial role in this process, particularly soil temperature, which affects evaporation, aeration, and the chemical reactions necessary for seeds to sprout. How seeds respond to temperature is influenced by several factors, including their initial moisture content and how quickly they absorb water (Vertucci 1989). For instance, seeds with low moisture content might show reduced vigour if they absorb water too quickly at 25 °C. This underscores the importance of understanding how temperature, moisture, and seed traits interact during germination. Studies on various plant species have shown that germination speed and temperature thresholds are often linked, with plants adapting to specific conditions that match the environments in which they evolved (Dürr et al. 2015).

A study found that the germination rate increases with increasing stratification duration. The best germination results occurred with the longest stratification period, likely because it reduced the inhibitors that prevent seeds from sprouting. However, germination rates also depended on latitude, with seeds from lower latitudes generally germinating more slowly and less frequently (Cheng et al. 2022).

Temperature is the most crucial factor influencing seed germination, with the ideal range for many species being 10–20 °C. Within this range, seeds germinate most quickly, but each species has specific temperature limits beyond which germination will not occur (Milbau et al. 2009; Finch-Savage and Leubner-Metzger 2006; Pritchard 2000; Roberts 1988). If climate change causes temperatures to exceed these limits, it could prevent seeds from germinating, threatening the species' survival; temperature also affects germination by influencing moisture, hormone production, and enzyme activity, with warmer climates potentially reducing the moisture seeds need to sprout. Cold stratification breaks seed dormancy and boosts germination in many species, such as *Spartina alterniflora, Primula beesiana, Aster tripolium, Parthenium hysterophorus, Guizotia scabra*, etc. (Cheng et al. 2022; Yang et al. 2020; Al-Hawija et al. 2012; Karlsson et al. 2008). Additionally, moderate thermal stratification can positively impact seed germination in *Cirsium arvense* (Cheng et al. 2022; Bochenek et al. 2009).

ABA and gibberellins are two hormones which regulate seed germination. ABA keeps seeds dormant, while gibberellins trigger germination (Finch-Savage and Leubner-Metzger 2006). Temperature plays a key role by activating the genes that produce gibberellins, which release dormancy and allow the germination of the seed (Finch-Savage and Leubner-Metzger 2006). During germination, the embryo elongates, and the root emerges from the seed coat, a process that often requires enzymes to break down the seed's outer layers (Dove and Mangan 2010). These enzymes are regulated by temperature, and if the temperature is outside the ideal range, the enzymes might not work, preventing germination. Thus, significant temperature changes can greatly impact seed germination.

Cold stratification offers several benefits for seeds. It boosts superoxide dismutase (SOD) activity, helping to remove harmful oxygen-derived radicals and adapt seeds to winter conditions. It also reduces ABA levels, which increases germination (Feurtado et al. 2004). Additionally, cold stratification breaks seed dormancy in halophytes, allowing them to germinate at the optimal time to avoid salt stress (Cheng et al. 2022).

8.2.6 Stomatal Opening and Closure

The stomata are the outer openings of intercellular spaces within the plant, bounded by paired guard cells exhibiting high elasticity. Stomata are essential for controlling water vaporisation and carbon dioxide assimilation in plants. By adjusting the pressure in their guard cells, plants modify the size of the stomatal opening, which influences the gas exchange, with the efficiency also depending on the stomatal size

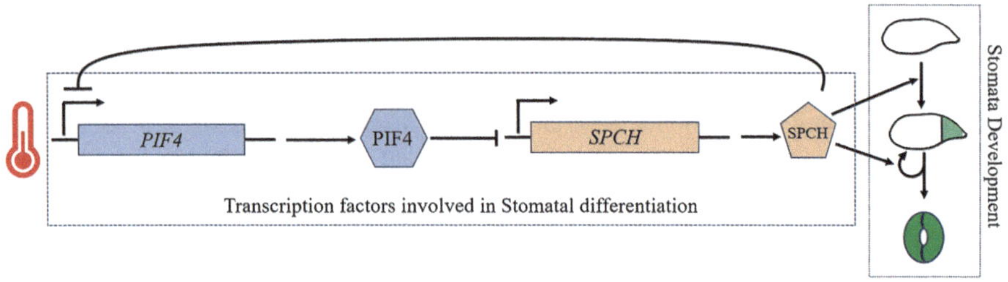

Fig. 8.1 PIF4 and SPCH regulate stomatal development in a temperature-dependent manner. PIF4, which is induced by high temperatures, accumulates in stomatal precursors and suppresses SPCH, hence preventing the induction of stomatal lineage. However, in certain cells, SPCH accumulates and negatively feeds back on PIF4 by binding to its promoter, facilitating SPCH accumulation and ensuring stomatal cell fate. (Adapted from Lau et al. 2018)

and density of the leaf (Driesen et al. 2020). High temperatures reduce plant stomata production by suppressing the expression of *SPPEECHLESS* (*SPCH*), a key transcription factor for initiating stomata development (Driesen et al. 2020; Lau et al. 2018; Zoulias et al. 2018). The transcription factor PIF4 accumulates in stomatal precursor cells and binds to the *SPCH* promoter, thus limiting its expression at elevated temperatures. As a result, fewer stomatal precursors are formed, reducing stomatal production (Driesen et al. 2020; Lau et al. 2018). Interestingly, there is a negative feedback loop where SPCH directly binds and suppresses *PIF4* expression, allowing SPCH to accumulate and ensure the development of guard cells. This feedback creates a balance that helps maintain stomatal production even under heat stress as illustrated in Fig. 8.1 (Driesen et al. 2020; Lau et al. 2018). Interestingly, *SCREAM* (or *ICE1*) and *SCREAM2* also regulate the stomatal development. They are bHLH proteins that can dimerise with bHLH proteins SPCH, MUTE and FAMA, which are the key-switch proteins that determine the stomatal differentiation state. This helps assert control over cell fate decisions. Furthermore, ICE1 is involved in cold stress responses, suggesting a link between developmental processes and environmental responses (Kanaoka et al. 2008).

Several factors affect stomatal opening and closing. There might be antagonistic signals arising from heat, which stimulates stomatal opening, and limited water availability, which triggers stomatal closure (Feller 2006). This becomes particularly crucial during heat waves, which are typically marked by both high temperatures and dry conditions.

High temperatures negatively impact photosynthesis by decreasing the activity of Rubisco, a key enzyme involved in carbon fixation. Rubisco activase, which helps reactivate Rubisco, is heat-sensitive, leading to slower reactivation during heat stress. A study suggests that under these conditions, stomatal opening is likely not the main factor limiting photosynthesis. However, stomatal regulation is crucial for leaf cooling and protecting the photosynthetic machinery, making the guard cells' response to heat and other factors, like drought, essential for plant stress tolerance. High temperatures also cause stomatal opening in *Arabidopsis*, *C. communis*, barley etc. (Kostaki et al. 2020) and allowing leaf cooling through evaporation but increasing water loss (Driesen et al. 2020; Reynolds-Henne et al. 2010; Feller 2006). Stomatal responses to heat are influenced by light-signalling components like phototropins, which are essential for full stomatal opening. Additionally, rapid temperature changes trigger reactive oxygen species (ROS) signals, which regulate stomatal opening in the phototropin-independent pathway manner (Driesen et al. 2020; Kostaki et al. 2020).

8.3 Components Involved in Temperature Response

8.3.1 Phytochrome B

The role of photoreceptor Phytochrome B (phyB) in temperature sensing was discovered recently. phyB exists as dimers in two forms: Pr and Pfr. The biologically inactive Pr converts into biologically active Pfr upon absorbing red light, and the reverse reaction occurs when Pfr absorbs far-red light. Thus, the red/far-red ratio received by the plants maintains the ratio of Pfr to Pr. However, Pfr can spontaneously convert into Pr with increased temperature, a process known as thermal reversion, also affecting the size of the phyB nuclear bodies formed by active Pfr with several other components (Fig. 8.2). Thus, the inactivation of phyB in the dark is temperature-dependent, implying the mechanism via which phytochromes convey temperature information to the plant over the course of the night. In conditions that reduced phytochrome activity, phyB targets were upregulated at night, but the same did not happen during the day. Furthermore, the relationship between phyB activity and warm temperature transcriptome is not detected during light conditions as seen

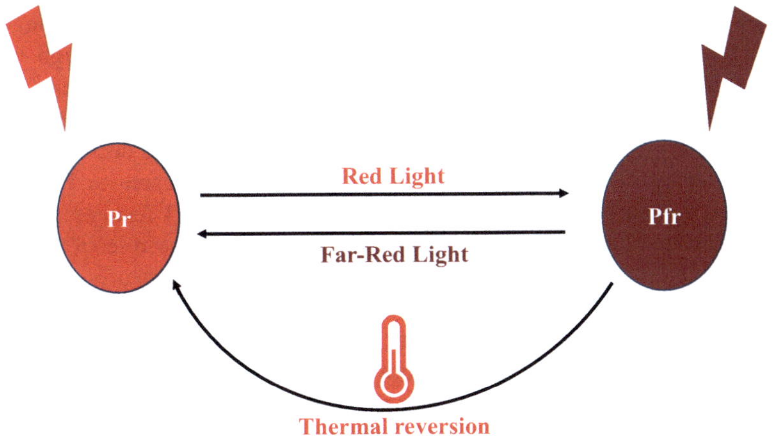

Fig. 8.2 In the presence of red light, PhyB transitions from its inactive Pr form to the active Pfr form. The active Pfr form can then be converted back to the inactive Pr form through exposure to either elevated temperatures or far-red light, a process known as thermal reversal or dark reversion

at night, consistent with the fact that most PhyB target genes are expressed at night, with phyB generally acting as a transcriptional repressor. The binding activity of phyB decreases with an increase in temperature without bringing about a change in the concentration of phyB protein or *PHYB* expression, thus affecting downstream processes, particularly the degradation of PIF4 and PIF7 proteins, which allow for the transcription of thermomorphogenesis genes. PhyB has temperature-dependent interactions with the promoters of several genes (Jung et al. 2016; Legris et al. 2016). Furthermore, the alteration of phytochrome activity affects the warm temperature transcriptome, with 79% of the temperature transcriptome being misregulated (Jung et al. 2016). Therefore, the reduction in the pool of active phyB dimers (Pfr-Pfr) in warmer temperatures mediates the temperature-related biological activity of phyB. phyB has a role in sensing temperatures both during the daytime and nighttime (Qiu et al. 2019).

8.3.2 Evening Complex

The Evening Complex (EC), a complex consisting of core clock components ELF3 (EARLY FLOWERING 3), ELF4 (EARLY FLOWERING 4) and DNA binding protein LUX (LUX ARRYTHMO), is a repressor complex expressed through the course of the night. The complex acts as a transcriptional repression for several genes, such as *PIF4*, *PIF5*, *GI*, *LUX* and *PRR7*, which have a LUX Binding Site (LBS), particularly in dark conditions. The EC maintains the circadian expression of its target genes. The expression of *PIFs* under normal growth conditions is repressed by the EC early at night. Evidently, this complex has a role in not only maintaining the circadian rhythm but also entraining the plant as per light and temperature cues. The complex acts as a thermosensor, its thermosensing mechanism being its affinity for DNA binding. This mechanism works in both sudden temperature changes and changes in steady-state temperature for growth. LUX alone has a high binding affinity for DNA, while this affinity lessens for a LUX-ELF3 complex (no such effect exists for the LUX-ELF4 complex). However, the binding affinity is restored in a LUX-ELF3-ELF4 complex. The binding affinity of the EC decreases with an increase in temperature and becomes stronger as temperature decreases; however, no such effect is seen for LUX alone. Overexpression of ELF4 restores the normal binding affinity of EC at higher temperatures, indicating the role of ELF4 as a modulator of the thermosensing activity of EC. At higher temperatures, the decrease in the DNA binding activity of EC results in *PIF4* gene activation, which leads to the expression of thermomorphogenesis genes. The EC may also have a role in regulating CBFs. Studies suggest that EC conveys the temperature information directly to its target genes at the start of the night, and the following cascade transmits it to a wider set of genes by dawn (Silva et al. 2020; Ezer et al. 2017; Mizuno et al. 2015; Nusinow et al. 2011).

The activity of PIF4 protein can also be regulated directly by ELF3, independent of the formation of the EC. This occurs through the interaction of the ELF3 with the bHLH domain of PIF4, thus not allowing PIF4 to activate its transcriptional targets. The levels of ELF3 accumulated in cells in light conditions are maintained by the antagonistic activity of PhyB and COP1, with COP1 degrading ELF3 and PhyB preventing this degradation by competing with COP1 for binding with ELF3 (Nieto et al. 2015).

ELF3 is also known to contain a prion-like domain (PLD) that contains polyglutamine repeats, whose length is suggested to be correlated to the temperature responsiveness of the protein. Increasing the length of the polyglutamine region increases the thermoresponsiveness of the plant. However, these effects are mild, and the plants without this region still show a response to changes in temperature. Plants adapted to hotter climates, such as *Brachypodium distachyon*, have no detectable PLD in their ELF3. Such ELF3 proteins are active at higher temperatures, leading the plants not to have thermal responsiveness like those with ELF3 with PLDs. The early flowering in high temperatures via liquid-liquid phase separation of *Arabidopsis* is lost when the ELF3 has no PLD. In response to high temperatures, ELF3-containing PLDs form liquid droplets via phase transition. When temperature increases, ELF3 binding to its target genes decreases, but this temperature response is not there for ELF3 without a PLD. Furthermore, ELF4 has a role in buffering the thermal response of ELF3 in elevated temperatures, but in lower temperatures, the same is not true. Higher levels of ELF4 also abolish the ELF3-mediated temperature responsiveness of the plants by stabilising ELF3. Thus, the PLD of ELF3 aids in its thermosensory activity (Jung et al. 2020).

8.3.3 Phytochrome Interacting Factors

Phytochrome Interacting Factors (PIFs) are transcription factors containing a basic-helix-loop-helix domain. Of the several PIFs found in plants, PIF4 is sufficient and necessary for thermoresponsiveness in *Arabidopsis thaliana* (Box et al. 2015). PIF5 acts redundantly with PIF4, while PIF7 is also involved in thermal responsiveness. The transcripts for these *PIFs* accumulate in the dark periods, with their regulation being maintained by the EC, ELF3 and PhyB, as mentioned earlier. Along with Phytochrome, Cryptochrome 1 (CRY1; a blue light photoreceptor) can also interact with the *PIF4* gene to repress its transcription as per the blue light signal and repress thermomorphogenesis (Ma et al. 2016).

PIF4/PIF5 are involved in the temperature-dependent hypocotyl elongation response via the activation of downstream genes such as *ATBH2* (*A. thaliana* homeobox protein 2) and *IAA29* (Indole-3-acetic acid inducible 29). PIF4 protein induces genes for auxin biosynthesis (*YUCCA8, TAA1, CYP79B2*), brassinosteroids' biosynthesis and growth-promoting factors (*ATBH2, PREs, LNGs*) (Kim et al. 2020). At normal temperatures (22°C for *Arabidopsis*), in long-day conditions, the circadian clock components (as discussed previously) repress the expression of the *PIF4* gene at night, allowing the gene to transcribe only at the end of night time. However, as the light period begins, light-activated phyB and DELLA protein degrade/inactivate PIF4 protein and thus stop its biological activity. In short-day conditions, on the other hand, just before dawn, the repression by phyB and DELLA does not occur, allowing the protein to accumulate, and consequently, elongation of hypocotyl is observed (Yamashino et al. 2013). The activity of phytochromes and circadian components together limits the period of temperature-responsive growth. As tissue-specific roles of PIF4 come to light, it is found that *PIF4* expression in epidermal cells (which increases in higher temperatures) causes hypocotyl elongation through auxin pathways (Kim et al. 2020). TOC1 is also involved in modulating thermoresponsiveness and suppressing thermomorphogenesis—overexpression of TOC1 causes thermoinsensitivity. PRR5 is also a close homolog of TOC1. Together, they ensure a circadian gating of thermal responses. The repression of thermal responsiveness in the evening and early night by negating the activity of PIF4 (the C-Terminal Domain of TOC1 interacts with the bHLH domain of PIF4; TOC1 can bind to *PIF4* promoter to repress gene expression; the G-box domain of target genes is a common target for both PIF4 and TOC1; TOC1, PRR5 and PIF4 may bind very close to each other on target genes) by high expression of *TOC1* is suggested. This repression is not present when the plant experiences warm temperatures at other times of the day (Zhu et al. 2016). *FCA*, a gene involved in thermal responsive flowering, is also involved in modulating the chromatin binding activity of PIF4 to attenuate stem growth in response to heat (Lee et al. 2014). HEMERA, in collaboration with REGULATOR OF CHLOROPLAST BIOGENESIS (RCB), interacts directly with PIF4 to promote its accumulation in warm temperatures and induction of its target genes (Qiu et al. 2019, 2021).

In conditions of heat stress, PIFs also promote thermal tolerance. PIF4 can activate the master regulator of heat shock response *HSFA2* via the G-box motif in the gene's promoter. This provides basal thermotolerance in normal conditions and resistance to heat in stressful situations. Unsurprisingly, PIF4 is accumulated in heat stress (Yang et al. 2022a, b). HSFA1d and its homologues play a critical role in thermomorphogenesis—they stabilise PIF4 by interfering with phyB activity, and PIF4 also directly interacts with HSFAs. HSFAs also accumulate in the nucleus in response to daytime warm temperatures, mediated by COP1 by suppressing BIN2-directed phosphorylation of HSFA1d (Tan et al. 2023).

The elongation of hypocotyl is also a characteristic feature of thermomorphogenesis and occurs in a PIF4-dependent manner. As discussed earlier, an increase in temperature causes a decrease in repression of the *PIF4/PIF5* genes at night time by clock components and phytochrome activity, allowing an accumulation of the corresponding proteins by dawn, even in long-day conditions. This leads to auxin-mediated hypocotyl growth. Thus, in ambient elevated temperature, the elongation of hypocotyl occurs at dawn by integration of temperature and light cues. In colder ambient temperatures, the repression is stronger, and elongation is not seen.

Elevated temperatures promote the accumulation of phosphorylated PIFs and their binding to target promoters. An increase in temperature results in the removal of H2A.Z nucleosomes from the *FT* gene locus, and PIF4 binding to the promoter of this gene increases. This suggests a mechanism for control of temperature-mediated flowering (Kumar et al. 2012). PIFs act antagonistically with HY5 (LONG

HYPOCOTYL 5) to integrate temperature cues. While PIFs respond to elevated temperatures, HY5 responds to lower temperatures by triggering pigment production, ROS production and cold-induced genes. In lower temperatures, there is an increase in stable HY5 as COP1 is depleted from the nucleus (Catalá et al. 2011). PIFs and HY5 regulate photosynthetic genes in a temperature-dependent manner through their ability to bind to G-box elements in DNA and the temperature-dependent abundance of these proteins (Toledo-Ortiz et al. 2014). Thus, PIFs play an important role in thermotolerance and thermoadaptation in a time-dependent manner throughout the day. The dynamics of the complete set of interactions involving PIFs are still being uncovered.

8.3.4 Phototropins

Phototropin is a blue light photoreceptor implicated in mediating chloroplast repositioning in response to temperature. Under normal conditions, chloroplasts are found at the periclinal cell wall, whereas at lower temperatures, they are at the anticlinal cell wall. In the fern *Adiantum capillus-veneris*, phot2 was found to mediate this response. Further studies in *Marchantia polymorpha* showed that the sensation of blue light by the single phototropin protein it has is essential for this response. Phototropins have two Light/Oxygen/Voltage domains (LOV1 and LOV2) at the N-terminal region. When the LOV domain is bound non-covalently to a flavin mononucleotide (FMN) molecule, it is inactive. A conserved cysteine residue in the LOV domain forms a covalent bond with FMN when it receives blue light, rendering the molecule active. The state of phototropin modulates the activity of the serine/threonine kinase at the C-terminal of the phototropin, which causes downstream processes such as chloroplast repositioning to occur. Temperature affects the transition between the active and inactive state, with a higher temperature making the LOV domains inactive. Active phototropin is higher at lower temperatures, and the increased lifetime of the active LOV2 domain of *M. polymorpha* phototropin at lower temperatures suggests a mechanism whereby plants perceive external temperature cues to mediate their physiological responses (Fujii et al. 2017). In *Arabidopsis*, phototropins have also been implicated in temperature-dependent stomatal opening (Kostaki et al. 2020).

8.3.5 Phytohormones

8.3.5.1 Auxin

Moderately high temperatures in *Arabidopsis* seedlings promote hypocotyl elongation by auxin biosynthesis through the *Tryptophan Amino Transferase* (*TAT*)/*Transport Inhibitor Response2* (*TIR2*) gene while the expression of the *YUCCA* gene reduces. In contrast, the level of auxin decreases endogenously in the anther of *Arabidopsis* during its development at high temperatures (Bianchimano et al. 2023; Zhu et al. 2015). Exogenous auxin can reverse high-temperature injury by restoring DNA replication-related genes inhibited by high-temperature stress.

In warm temperatures, the level of auxin increases. Due to the warm temperature, the affinity of Phytochrome interacting factor (PIF4) binding to the *YUC8* promotor increases, and the level of auxin synthesis consequently increases. PIF7 also binds to the *YUC8* promotor under warm temperature conditions. Warm temperature favours chromatin modification such that auxin synthesis gene expression can occur. During warm temperatures, the chromatin remodelling complex INO80 and HD9 work together and mediate the reduction of repressive Histone H2AZ in the nucleosome, so there will be a chance of binding of PIF4 to *YUC8* promotor for auxin synthesis to occur (Bianchimano et al. 2023; Zhu et al. 2015).

During warm temperatures, auxin synthesis occurs in cotyledon, then goes to the hypocotyl region and promotes growth. (Bellstaedt et al. 2019). Auxin binds to receptor TIR, inhibits AUX/IAA, and promotes Auxin Response Factor (ARF) activity. Due to this, activation of SMALL AUXIN UP RNA (SAUR) occurs. SAUR protein reduces type 2C Protein phosphatases (PP2CD) protein and activates H^+-ATPase in the plasma membrane, leading to acidification in the apoplast area. This loosens the cell wall, and cell elongation happens (Cosgrove 1997). A temperature rise affects the auxin level in roots. Warm temperature enhances primary root growth in an auxin-dependent manner. Low temperature hampers root growth. Low temperature reduces auxin accumulation in roots, thus affecting meristem activity and root growth. Furthermore, cold stress affects auxin transport by inhibiting auxin transport proteins Auxin Influx Transporter (AUX1) and PIN-Formed proteins (PIN) (Zhu et al. 2015).

8.3.5.2 Cytokinin (CK)

During heat stress, the CK level increases to produce antioxidant and heat shock proteins. However, prolonged exposure to heat stress deregulates the CK signalling pathway. CK is key in delaying senescence, mainly zeatin ribose and dihydro zeatin ribose. Overexpression of the SAG12:*IPT* construct maintains root formation and elongation under high-temperature stress. It also decreases the loss of chlorophyll and delays the senescence of the leaf. Studies in rice reveal that during heat stress, the CKX (cytokinin dehydrogenase) enzyme is activated, which reduces CK levels. The isopentyl transferase (IPT) is involved in CK biosynthesis. The overexpression of the *IPT* gene leads to the stomatal opening, which regulates gas exchange. Due to this, leaf transpiration happens, which is essential for cooling plants during heat stress (Srivastava and Pandey 2023; Liu et al. 2020).

CK has a crucial role at cold temperatures. A multistep phosphorylation system might relate CK and cold temperature response. This cascade activates various proteins needed for cold stress response. A-type *Arabidopsis* response regulators have been identified as positive regulators of cold tolerance that help plants cope with low temperatures. ARR22 (Arabidopsis response regulator 22) and C-type ARR maintain membrane integrity and prevent the membrane from cell damage. Cytokinin response Factors play an important role in cold stress. Cytokinin response factor 4 (CRF4) plays a crucial role in low-temperature stress, and a mutation in CRF4 makes the plant incapable of withstanding cold temperatures. CRF2 and CRF3 induce lateral root growth under cold stress. ARR1 is involved in low-temperature signals from two histidine kinases, AHK2 and AHK3, and signal mediated by AHP (*Arabidopsis* histidine phosphotransfer) proteins, which are part of a two-component signalling system. This AHP protein plays a role in enhancing the plant's resistance to cold stress by mediating the transmission of cold signals to ARR1, therefore activating downstream response that helps the plants withstand cold stress. In response to cold stress, ethylene and CK oppose each other. The C-binding factor (CBF) and DRE binding factor (DREB) play critical roles in response to cold stress. CBF activates cold-responsive genes during cold stress. Meanwhile, Ethylene inhibits CBF. This means that when ethylene levels are high, the plant's ability to respond to cold stress is diminished. A-type ARRs are important in integrating signals from both ethylene and CK. They act as regulator and balancing of both hormones (Srivastava and Pandey 2023; Liu et al. 2020).

CK plays an important role in response to hot ambient temperature. Floral damage may happen in hotter temperatures. CK enhances the survival of floral primordia in response to hot ambient temperatures. Evidence of CK cross-talk with gibberellin is emerging. In *Arabidopsis*, it was found that CK did not have any effect associated with gibberellin (Greenboim-Wainberg et al. 2005). This suggests that the presence of CK does not hamper the effect of GA in *Arabidopsis*. However, in tomatoes, CK inhibits the gibberellin expression (Fleishon et al. 2011). The GA2 Oxidase enzyme inhibits the effect of gibberellin, and CK enhances the activity of this enzyme (Sobol et al. 2014).

8.3.5.3 Ethylene

Mutation in the ethylene receptor Ethylene Receptor1 (ETR1) leads to plants showing basal or reduced thermal tolerance levels. Plants having defective Copper Transporter Protein (CTR1) proteins (a negative effector of ethylene signalling) show better adaptation to high-temperature stress. Ethylene insensitive 2 (EIN2) and Ethylene insensitive 3 (EIN3) are crucial in ethylene signalling and prevent high-temperature stress. EIN3-induced thermos-tolerance induced by root endophytes upstream of Heat shock factor A2 (HSFA2) and induction of transcription factors that regulate response to heat stress. During high-temperature stress, EIN3 becomes stabilised by inhibiting Salt- and Drought-induced Ring Finger1 (SDIR1). When an ethylene signal is received, EIN3 activates ERF transcription factors, which bind to DNA and produce proteins that prevent high-temperature stress. The Apetala 2 (AP2)/Ethylene Response Factor(ERF) family, particularly the ERF family, acts as a regulator in ethylene-mediated plant stress responses. ERF1 activates heat-responsive genes HSP70, HSP101, and HSFA3. These proteins protect plant cells from heat stress. ERF95 and ERF97 play important roles in heat stress. In response to heat stress, ERF74 and ERF 75 activate Respiratory burst homology D (RbohD), regulating the stress response. Ethylene-linked acetylation of H3 histone is associated with temperature stress response. General Control Non-derepressible 5 (GCN5) is a histone acetyltransferase that adds acetyl groups to histone H3K9 and H3K14 and provides resistance to plants against high temperatures. *At*ASF1A and *At*ASF1B are histone chaperones and cause histone acetylation (Huang et al. 2023a, b).

Cold stress is linked to a rapid increase in ethylene signalling components gene *ETR1*, *EBF1*, and *EBF2*. Ethylene signalling is suggested to activate the *Core Binding Factor* (*CBF*) gene and the Type-A Arabidopsis Response Regulator (ARR) gene. The main function of the *CBF* gene is to produce a cold-responsive gene expression in response to cold stress. Various signalling pathways like light signalling, circadian clock signalling, and ethylene regulate the *CBF* gene that provides resistance to cold stress (Huang et al. 2023b).

8.3.5.4 Abscisic Acid (ABA)

ABA is a hormone whose concentration dramatically increases during stress. In heat stress conditions, stomatal closure is mediated by ABA to check transpiration. ABA when bound to its receptor PYL/PYR (PYR1-like/Pyrabactin Resistance), activates the production of ROS, which leads to increased intracellular Ca^{2+} concentrations. Due to this, the inward flow of the K+ ion is blocked, and the outward flow of the K^+ ion is allowed. An increase in cytosolic Ca^{2+} causes the opening of the potassium ion channel in the plasma membrane. The efflux of K^+ ions reduces turgor pressure in guard cells and closes the stomata. During heat stress, ABA inhibits the ethylene genes in the root and enhances primary root growth and lateral root growth to meet water requirements. In this pathway, NAD and NO play crucial roles (Tao et al. 2022; Shi and Yang 2014).

In heat stress, ABA regulates sugar metabolism. Plants perform photosynthesis when carbon dioxide is available, but due to high temperatures, the photosynthesis process is inhibited. During high temperatures, many enzymes and pollen development are affected. In this case, ABA plays a critical role by enhancing sucrose transport and improving pollen fertility during high temperatures (Tao et al. 2022; Shi and Yang 2014).

High temperature accelerates ROS production and oxidative damage to cells, causing the degradation of Rubisco and lipid peroxidation. ABA promotes antioxidant production and prevents ROS-mediated oxidative damage. Due to Heat stress conditions, cellular proteins are degraded and sometimes misfolded; ABA helps the production of Heat-shock proteins, which helps in the proper folding of proteins and prevents proteins from degradation (Tao et al. 2022; Shi and Yang 2014).

During cold perception, ABA concentration levels increase. ABA binds to its receptor, and two pathways are generated: the ABA-dependent pathway and the ABA-independent pathway. In the ABA-dependent pathway, MYC-BZiP transcription factors are active. This Transcription factor (TF) binds to its target promotor region and changes its gene expression, generating cold stress tolerance factors. In the ABA-independent pathway, Dehydrin Responsive Element Binding (DREB) and NAM, ATAF1/2 and CUC2(NAC) transcription factors activate genes that help the plant tolerate stress responses (Tao et al. 2022; Shi and Yang 2014).

CBF and AREB are key TFs that regulate the expression of a *cold-responsive* (*COR*) gene, and are involved in ABA signalling, and help in coping with cold stress. Inducer of CBF Expression 1 (ICE1) and Calmodulin binding transcription activator (CAMTA) are upstream regulators that activate CBF in response to cold stress. MYB15 acts as a repressor of CBF. High Expression of Osmotically Responsive Gene 1 (HOS1) and SIZ1 regulate the stability of ICE. HOS1 causes the degradation of ICE1, while SIZ1 stabilises ICE1. Cold temperature activates ICE1, which activates CBF and subsequent activation of *COR* genes, which are particularly important for dehydration stress (Tao et al. 2022; Shi and Yang 2014).

8.3.5.5 Brassinosteroids (BRs)

Brassinosteroids provide resistance against heat stress. During high-temperature stress, Bri1-EMS Supressor1(BES1) undergoes dephosphorylation, binds to the promotor HSE element of *HSP*, and produces HSP for proper protein folding. BES1 also activates the HSFA1 factor that enhances resistance against stress. BR interacts with Brassinazole-Resistant 1(BZR1) and inhibits ERF transcription factors to improve heat tolerance. Brassinosteroids-Insensitive (BIN2) causes dephosphorylation of BZR1 and activates the Abscisic Acid Insensitive 5 (ABI5) protein, which has a negative role in stress. The transcription factor BZR1 is a positive regulator of heat tolerance. It interacts with the *FER* gene, which is involved in the plant's response to heat stress. BIN2 increases the activity of ABI5, which is involved in heat response (Miao et al. 2024).

BZR1 enhances the tolerance of plants against cold stress and activates ABA. Dwarf4 (DWF4) causes upregulation of *COR15A*, which protects the cell during freezing damage. SIBRI1 plays a significant role during cold stress by regulating the level of ROS, biosynthesis of phytochrome, and expressions of *CBF* and *COR* genes (Miao et al. 2024).

During elevated ambient temperature, BR signalling is involved in modulating root growth. The brassinosteroid receptor BRI1 is crucial for integrating temperature changes with the BR signalling pathway. The loss of function mutant *bri1* and *bes1* show shorter roots and smaller meristem. This indicates that BRI1 and BES1 are essential for normal root development. The *det2* and *dwf4* mutants show shorter roots than wild-type plants at 21°C (Martins et al. 2017).

8.3.5.6 Jasmonic Acid (JA)

During heat stress conditions, the concentration of oxo-phytodienoic acid and JA-isoleucine is increased. The expression of the *Arabidopsis thaliana Oligopeptide Transporter3*(*AtOP3*) gene is higher, which activates JA biosynthesis. Due to this activation, other genes like *dehydration-responsive element binding protein* (*DREB2A*) associated with enhanced heat tolerance are activated. Prolonged exposure to heat stress reduces JA levels; in these conditions, Jasmonic acid Oxidase (JOX) and sulfotransforase2 (ST2A) are involved in modifying and degrading JA. High temperature decreases the expression of *Gossypium hirsutum Allene Oxide Cyclase2* (*GhAOC2*), which synthesises JA in anthers. Application of exogenous JA produces heat shock proteins, which prevent oxidative damage and cellular damage to cells during heat stress. SGT1 protein acts as a cofactor of HSP90. SGT1 is involved in JA signalling during stress conditions and maintaining the steady state of JA receptor Coronatine-insensitive 1(COI1). WRKY transcription factor is important in plant growth and development and stress conditions. *CaWRKY40* is induced in response to heat stress and when exogenous JA is present. This indicates that JA triggers the expressions of *CaWRKY40* during heat stress (Wang et al. 2023).

JA plays a positive role in mediating a plant's response to cold stress. JA level rises in response to cold stress. Cold stress enhances the activity of JA biosynthetic genes *Oryza sativa* Lipoxygenase 2 (*OsLOX2*) and *Oryza sativa* Allene Oxide Cyclase (*OsAOC*), which maintain endogenous levels of JA. Two important proteins, AOS and LOX2, play important roles in cold stress. If a mutation arises in these two proteins, JA synthesis is hindered in responses to cold stress. The main transcription factor of JA is MYC2. MYC2 activates the *aldehyde dehydrogenase* gene, which produces glycine and betaine, which protect plants against stress. Glycine and betaine help plants withstand low temperatures. MYC2 also activates Arginine decarboxylase1(ADC1), which helps activate polyamines to cope with low temperatures. Dehydration-Responsive Element Binding protein1/C-Repeat Binding Factor (DREB1/CBF) transcription factors bind to the (Dehydration responsive element) DRE element and produce *COR* genes, which help to withstand cold stress. ICE1 plays an important role because it activates *DREB1/CBF*s (Wang et al. 2023).

ABA and JA play an important role in cold stress. JA activates ABA during stress conditions. MYC2 activates the *9-cis-epoxycarotenoid dioxygenase* (*NCED*) ABA biosynthetic gene. Together, ABA and JA activate the *Cold Regulated* (*COR*) gene and enhance cold response to plants.

GA and JA also crosstalk during cold stress. Without GA, DELLA binds to MYC2 transcription factors and inhibits JA signalling. In the presence of GA, DELLA protein activates, and MYC2 becomes free and provides resistance against stress. JA also activates the Ethylene biosynthetic gene *ERF* (Wang et al. 2023).

8.3.5.7 Salicylic Acid (SA)

High-temperature stress reduces the reproductive and developmental phases of flowers. The male reproductive stage of anther is more vulnerable than the female reproductive structure. During heat stress, anther development, tapetum development, pollen release, and fertilisation process are inhibited. Sometimes heat stress causes programmed cell death (PCD) which leads to male infertility and causes the downregulation of many genes concerned with male fertility. SA effectively reduces high-temperature stress. SA causes spikelet differentiation in rice during heat stress which means SA maintains reproductive development during stress conditions. SA enhances photosynthetic efficiency during heat stress. During heat stress in *Arabidopsis*, SA influences activities in apical cells, which leads to flower and pollen development. SA scavenging ROS production prevents tapetum degradation and inhibits PCD. SA maintains pollen viability by reducing malondialdehyde (MDA) levels (Sangwan et al. 2022).

During high-temperature stress, proteins are degraded or misfolding occurs. To properly fold proteins, heat-shock proteins are produced in heat stress, and SA contributes to this process. When exogenous SA is applied to plants, HSF4, HSF7 and DREB proteins are produced, which helps plants cope with heat stress. SA activates HSP70, which helps proper folding and prevents protein aggregation. During Heat stress, the SA level rises and promotes the binding of heat shock factors in DNA to produce heat shock proteins. SA activates stress-responsive genes HSFA, HSFA2, and HSFB1 (Sangwan et al. 2022).

During high temperatures, photosynthetic activity decreases because the D2 protein (a part of PSII) is sensitive to heat. The disbalance of Oxygen evolving complex (OEC) and electron transport during heat stress conditions is also a contributing factor. SA increases photosynthetic activity and induces physiological changes like increased uptake of nitrogen and water potential increase during stress conditions.

ROS causes cellular damage and oxidative damage to cells. SA helps the production of antioxidants to counteract ROS. SA increases the level of ascorbate peroxidase (APX) and lowers catalase levels. SA also has roles in reducing lipid peroxidation and maintaining Ca^{2+} homeostasis during heat stress (Sangwan et al. 2022).

SA enhances the plant's survival by reducing cold injury. SA causes seed germination in low-temperature stress. The combination of SA and JA reduces chilling injury and decay during low temperatures. SA reduces chilling injury and activates the alternative oxidases (AOX) enzyme, which is involved in electron transport channel ETS in mitochondria and provides energy during stress conditions (Sangwan et al. 2022).

8.3.5.8 Strigolactones (SLs)

Strigolactone signalling and biosynthesis are critical to withstanding high-temperature stress. SL acts as a signalling molecule that regulates growth and development in response to stress. Physiological and metabolic pathways are important steps in plant growth. SL enhances photosynthetic activity by scavenging oxidative stress-mediating molecules. Quantum yield and PSII are sensitive to temperatures, and accumulation of SL recruits HSP to scavenge oxidative stress. SL enhances a symbiotic relationship with mycorrhizal fungi and enhances nutrient uptake during stress. SL enhances the production of antioxidant enzymes that scavenge ROS. SL enhances the activity of the enzyme SOD which promotes the detoxification of harmful compounds like methyl-glyoxal. SL helps to mitigate the harmful effects of heat stress (Khalid et al. 2024).

More axillary growth (MAX1) and carotenoid cleavage enzymes CCD7 and CCD8 are the main biosynthetic enzymes of SL. SL regulates gene expression to prevent cell cycle damage in response to heat stress. SL enhances the expressions of proliferating cell nuclear antigen (PCNA), CDK, and CYCLIN in heat stress. SL regulates root architecture functions by activating *AtMAX2* and *OsD14*, which are important genes for root growth and development. SL facilitates lateral root branching by activating CCD7 and CCD8 enzymes (Khalid et al. 2024).

SL enhances chlorophyll content and maintains photosynthetic activity during heat stress. SL also checks transpiration by closing stomata during heat stress. The main functions of strigolactones are hyphal branching, lateral root elongation, nodulation, stem thickening, etc. (Khalid et al. 2024).

8.3.5.9 Gibberellins

Prolonged cold treatment is required for certain plants for both stem elongation and flowering, and gibberellins can be a substitute for this cold treatment. Ent-kaurenoic acid accumulates in the shoot apex. At high temperatures, the ent-kaurenoic acid is converted to the most active form of gibberellin, GA9 (gibberellic acid 9), which stimulates the flowering process. gibberellin activates AGAMOUS-LIKE20, which further activates the LEAFY gene and promotes flowering (Srivastava and Pandey 2023).

GA promotes growth and development by degrading DELLA protein. When the temperature rises to 50°C, the seed germination process is inhibited, but when exogenous GA is applied, high-temperature stress is inhibited, and seed germination is seen. GA is inhibited when high-temperature mediated hypocotyl elongation occurs (Srivastava and Pandey 2023).

During cold temperatures below 4 °C, the plant does not attain so much growth, and the C-repeat Binding Factor (CBF) protein is activated. GA binds to a receptor DELLA and promotes proteasomal degradation. During cold stress, CBF binds to the DELLA receptor, and GA is reduced (Srivastava and Pandey 2023).

8.3.6 Lipids

Heat stress has been observed to cause changes in the membrane fluidity. The lipid metabolism during heat stress is significantly altered. ER membrane is the location of synthesis of lipid molecules like phosphatidylcholine (PC), phosphatidylethanolamine (PE), and phosphatidylinositol (PI) (Ohlrogge and Browse 1995), while triacylglycerol (TAG) is synthesised in the chloroplast envelope membranes along with ER membrane. Heat stress causes plant leaves to specifically increase levels of phospholipids in the endoplasmic reticulum and plasma membrane, galactolipids containing linoleate (18:2) in chloroplasts, and triacylglycerol containing hexadecatrienoic acid (16:3) and α-linolenic acid (18:3) as lipid droplets in *A. thaliana* leaves (Higashi and Saito 2019). Wheat (*Triticum aestivum* L) genotypes' lipid profiles significantly changed during heat stress, and plants responded to increased temperatures by rearranging their lipids and decreasing their unsaturated lipids (Narayanan et al. 2016). Variations in lipid metabolising enzymes were also noted during heat stress; Digalactosyldiacylglycerol synthase 1, which catalyses the conversion of digalactosyldiacylglycerol (DGDG) from monogalactosyldiacylglycerol (MGDG), is less active in plants that are vulnerable to high temperatures (Welti et al. 2007). *Arabidopsis fad2* mutant plants which lack the FAD2 enzyme, which converts PC-bound 18:1 to 18:2 in the ER, reported disabled homeostasis by disproportionate plasma membrane fluidity, which indicates the importance of FAD2 enzyme in heat stress tolerance (Martinière et al. 2011). Furthermore, heat stress may upregulate the expression of genes related to phospholipid production, including glycerol-3-phosphate (G3P) acyltransferase 9 (GPAT9), according to transcriptome research (Shockey et al. 2016).

Along with other metabolites, lipids are also well-known signalling molecules. Lipid molecules such as inositol-1,4,5-trisphosphate (IP3) and diacylglycerol (DAG) are crucial secondary messengers. IP3 and DAG are produced when membrane-bound phosphatidylinositol-4,5-bisphosphate (PIP2) is hydrolysed. The levels of PIP2 showed a two-fold increase in pea seedlings when exposed to 37 °C at 40 min. The expression of PIP2 was observed within a few minutes of heat stress induction, and the levels of PIP2 increased remarkably after 15 min, with a maximum amount observed at 40 min (Liu et al. 2006). In *Arabidopsis*, the activation of enzymes phospholyase-D and PIP2 kinase is observed during heat stress, which leads to the accumulation of phosphatidic acid and PIP2. It is also observed that PIP2 accumulates in the nuclear envelope, nucleolus and punctate cytoplasmic structures (Mishkind et al. 2009). In tobacco BY2 cells, a heat-associated MAP kinase (HAMK) is triggered during heat stress at 37 °C, which in turn induces the expression of several heat shock proteins such as HSP70 and initiates heat resistance responses (Suri and Dhindsa 2008). NF-YC10 is an important transcription factor which regulates the activity of many heat-responsive genes; the levels of glyceraldehyde-3-phosphate dehydrogenase (GAPDH), which is a glycolytic enzyme converting 3-phosphoglyceraldehyde to 1,3-bisphosphoglycerate interacts with NF-YC10 and promote its activity (Huang et al. 2023a, b). Overall, the synthesis of various phospholipids and glycerolipids is enhanced during heat stress for improved heat tolerance. The levels of lipid metabolising enzymes also showed variations under heat stress. The expression of signalling lipids showed remarkable increases during heat stress and helped produce HSPs. This indicates the importance that lipids hold for improved stress tolerance.

8.3.7 Heat Shock Proteins

Heat stresses resulting from abrupt, unusually high temperatures can trigger various defence mechanisms based on changing genotypic expression. Heat stress disrupts proteins which make integral structures in cell membranes, cellular compartments, tissue, cytoskeleton and various ion channels, causing a loss in functional characteristics that they maintain in the cell (Kaur et al. 2018). Higher plants resolved these issues using molecular chaperons called heat shock proteins (HSPs) as a defence mechanism.

HSPs are a set of proteins which directly make contact with other proteins and help in protein folding, preventing misfolding and aggregation of misfolded protein and stabilising protein integral structures (Sharp et al. 1999). However, under stressful heat conditions, plants undergo modification at epigenetic and transcriptomic levels, leading to the synthesis of HSPs and acquiring thermotolerance. Cell synthesising HSPs show higher levels of thermal tolerance under heat stress conditions. HSPs are induced in various environmental conditions like water deficit, salinity, flooding, and low-temperature conditions. The cell previously going through any stress condition provides protection against other stress conditions (Ahuja et al. 2010).

When plants are exposed to heat stress, the heat shock response is mediated by various signal transduction pathways. Among these, a set of heat shock transcription factors (HSF) is an important factor. When cells are returned to standard growth conditions, HSPs produced through Heat-Shock transcription factors (HSFs) shield them from injury and aid their recovery and survival (Biamonti and Caceres 2009; Morimoto 1998). These TFs modulate various epigenetic modifications and help synthesise HSP mRNA and chaperone and ROS scavengers (Ren et al. 2019; Ohama et al. 2017). In HSF family members, HSFA1s act as "master regulators" in *Arabidopsis*, stimulating *heat stress-responsive* genes (*HSR*) and increasing heat tolerance. The expression of *HSFA1* is strictly controlled by high temperature (Ohama et al. 2017).

8.3.7.1 Signalling Perception and Activation of HSP Genes

As previously explained, HSFA1 synthesis is strictly controlled by high temperatures, various signalling pathways, or a cascade of signalling events that activate HSFA1 synthesis. Below are listed pathways that lead to the synthesis of HSFA1.

Calcium Ion Influx
Under heat stress, the plasma membrane fluidity is altered, leading to the influx of Ca^{2+} ions in CNGC (ion channel). This increase in cytoplasmic Ca^{2+} is crucial in activating various signalling pathways. This Ca^{2+} serves as a secondary messenger molecule and interacts with the calmodulin (CaM) protein, and it modifies CBK3 (CaM-binding protein kinase 3) and Protein Phosphatase 7 (PP7). CBK3 then appears to phosphorylate HSFA1 and activate it. This HSFA1 then triggers HSF and other stress-responsive factors. However, it was observed that in experiments involving cam3 mutants, HSR is still retained- despite the absence of functional CAM3. Moreover, the addition of Ca^{2+} chelators showed a drastic repression of HSR. This implies the existence of other unknown pathway(s) in Ca^{2+} signal transduction (Ohama et al. 2017).

Inhibition of HSFA1 Suppressors
Under normal conditions, unnecessary activation of HSFA1 activity is negatively regulated by heat shock proteins (HSPs) such as HSP70 and HSP90. However, during heat stress, the levels of these HSPs are altered, diminishing their effect on HSFA1, which allows for its activation. After activation. HSFA1s influence the production of various HSR genes and microRNAs, including DREB2A, miR398 and HSFA2. A feedback mechanism plays a regulatory role here. These genes induce the responses against heat stress. MicroR398 suppresses the activity of the ROS scavenger gene *CSD1*(*Copper/Zinc Superoxide Dismutase 1*), *CSD2*, and *CCS1* (*Copper Chaperone for Superoxide Dismutase 1*), which increases ROS generation. Accumulation of elevated levels of ROS leads to activation of HSFA1s. The relationship between miR398 and HSFA1 is part of a complex feedback mechanism. While miR398 can enhance HSFA1 activity through ROS accumulation, HSFA1 itself can also influence the expression of miR398 and its target genes. Initially, the factors that positively modulate plant stabilisation (CSD1, CSD2 and CCS1) diminish to activate HSFA1. This interplay helps plants to deal with heat stress more efficiently. miR398's interaction is not only involved in the HSFA1 pathway activation but also interacts with other signalling pathways related to stress responses. Studies show that PAP (30-PHOSPHOADENOSINE 50-PHOSPHATE)-XRN module regulates *miR398* transcription (Chinnusamy et al. 2007; Ding et al. 2020; Suzuki et al. 2011).

HSFA1s with other factors like JUB1 (JUNGBRUNNEN 1) and MBF1c (MULTIPROTEIN BRIDGING FACTOR 1C) stimulate *DREB2A* (*Dehydration-Responsive Element Binding Protein 2A*) transcription factor, which bind DRE responsive gene which has main role in water deficit and heat stress like osmotic adjustment, protective protein and antioxidant function (Suzuki et al. 2011; Yoshida et al. 2011).

HSFA2 is another HSF that plays an important role in gene modification. It modulates the H3K4 methylation site of the heat-responsive gene. Under non-stress conditions, upregulation of *DREB2C*, similar to the *DREB2A* gene, promotes the *HSFA3* gene. HSFA3 is an important heat shock-responsive transcription factor because deletion or suppression mutants of *HSFA3* led to a decreased regulation of potential target *HSP* genes under heat stress (Chen et al. 2010; Schramm et al. 2008; Yoshida et al. 1998). *HSFA2* is a primary target of HSFA1. A splice variant of HSFA2 helps in heat stress by a positive feedback mechanism. A spliced variant of HSFA2 produced during heat stress is converted into tiny condensed HSFA2 isoforms (S-HSFA2), which then get attached to the HSFA2 locus and activate the *HSFA2* gene (Liu et al. 2013). In *Arabidopsis*, HSFBs (HSFB1 and HSFB2b) are also the downstream target of HSFA1s. In a negative feedback system, HSFBs are considered to precisely modulate the Heat Stress-Responsive gene by inhibiting the function of HSFA1s (Ikeda et al. 2011).

Other transcriptional regulators, including WRKY, bZIP, and MYB, play a modest but considerable role in evolving HS tolerance in plants. Most WRKYs TFs play a role in stress reactions and developmental and physiological events. WRKY18, WRKY25, WRKY26, WRKY33, WRKY39, WRKY40, WRKY46, and WRKY68 all work together to help plants acquire thermotolerance by favourably modulating HSP-related signalling mechanisms (e.g., HSFs, HSPs, and MBF1c) (Li et al. 2011; Li et al. 2010).

8.3.7.2 Functions of HSPs and HSFs

Heat shock proteins are classified according to their molecular weight and different functions. Each family has a specific function in different cellular stress and even in non-stress conditions maintaining protein homeostasis in the cell. in plants there is five groups of HSPs: HSP100, HSP90, HSP70, HSP60, and small heat-shock proteins (sHSPs). Plants have about 20 sHSPs in general (Qureshi et al. 2022; Al-Whaibi 2011).

HSFs (Heat Shock Factors) As discussed in the signalling part, HSFs are strictly controlled by higher temperatures. These transcription factors regulate the expression of HSPs in response to stress. They bind to heat shock elements (HSEs) in the promoters of *HSP* genes, activating their transcription when cells are exposed to elevated temperatures or other stressors. In *Arabidopsis*, HSFA1a and HSFA1b tend to govern the initial response of several genes to high temperature.

After a heat stress encounter, heat shock genes like HSP18, HSP22.0, APX2, and HSP70 will acquire histone H3 lysine 9 acetylation (H3K9Ac) and histone H3 lysine 4 trimethylation (H3K4me3). Histone modifications and DNA methylation through the RdDM (RNA-dependent DNA methylation) process also contribute to plant heat tolerance (Lämke et al. 2016). HSP24 from *Trichoderma harzianum* in *Saccharomyces cerevisiae* conferred much greater heat stress tolerance to plants.

HSP100 group of proteins play a crucial role in protein quality control in cells, especially under stress conditions. Under stress conditions, HSP100 helps prevent the aggregation of misfolded protein, helps in the folding of partially folded protein and can also assist in the degradation of misfolded proteins by facilitating their entry into the proteasomal degradation pathway (Mondal et al. 2023; Al-Whaibi 2011).

HSP90 protein has a role in protein folding and as a chaperone for the stabilisation and activation of a wide range of client proteins, including many signalling proteins and transcription factors. It is also involved in the maturation of steroid hormone receptors and protein kinase in the cell nucleus and keeps binding steroid receptors until the signalling molecule arrives. HSP90 also plays a role in regulating cell cycle progression and apoptosis (Mondal et al. 2023; Al-Whaibi 2011).

HSP70 is one of the most abundant HSPs and acts as a molecular chaperone that binds to nascent polypeptides and prevents aggregation. It assists in protein folding, translocation across membranes, and the refolding of denatured proteins. HSP70 also plays a role in targeting misfolded proteins for degradation (Mondal et al. 2023; Al-Whaibi 2011).

HSP60, also known as chaperonin, assists in the proper folding of proteins within a protected environment. It forms large complexes (e.g., the GroEL/GroES complex in bacteria) that encapsulate unfolded proteins, allowing them to fold correctly without the risk of aggregation. HSP60 is essential for mitochondrial protein folding (Mondal et al. 2023; Al-Whaibi 2011).

HSP40 proteins act as co-chaperones that assist HSP70 in recognising and binding to unfolded proteins (Mondal et al. 2023; Al-Whaibi 2011).

Small Heat Shock Proteins (sHSPs) are found to be over 20 different types like other HSPs, they do not have ATPase activity. Their main role is to prevent aggregation and protein stabilisation (Tian et al. 2021; Al-Whaibi 2011).

Overall, heat shock proteins have a huge impact on maintaining cell homeostasis by shielding the protein molecule at all levels. Together, they form a critical network that helps cells survive and adapt to stress conditions, allowing organisms to acquire thermotolerance activity.

8.3.8 Osmolytes

Plants under different abiotic stresses like drought, salinity and extreme temperature undergo cellular damages like losing cell membrane integrity and dysfunction in major enzymatic and cell signalling pathways, disrupting cellular homeostasis. To counter this, a plant's cell's first response is to influx, synthesise and sequester some solutes or metabolite molecules to maintain cellular homeostasis. These solutes are called osmolytes or cytoprotectants as they maintain the cellular integrity of plant cells in stressful conditions (Ghosh et al. 2021; Burg and Ferraris 2008).

Osmolytes are typically organic, non-charged molecules that are basically inert in nature. The characteristics of osmolytes make them compatible with cells. Their concentration does not disrupt cellular membrane integrity nor interrupt any enzymatic activity or cellular signalling pathway like any other charged ion or inorganic compound, thus maintaining cellular homeostasis by combating osmotic and oxidative stress (Ghosh et al. 2021). Plant cells can tolerate high concentrations of this molecule without any detrimental effect on their metabolic pathway. That is why it is called compatible solutes or osmolytes (Den NZUb et al. 2024).

Mainly, four different compounds, amino acids, sugar alcohol/polyols, quaternary amino compounds and sugars, serve as osmolytes in plant cells (Den NZUb et al. 2024). When plants sense stress, they activate many signalling pathways like calcium signalling, ABA signalling, MAP kinase signalling and ROS signalling to accumulate and synthesise the osmolyte compound as a fast and first response to abiotic stress (Roychoudhury and Banerjee 2017). Among these, osmolytes like proline (amino acid), glycine betaine (quaternary amino compound), sorbitol, mannitol (sugar alcohol) and methyl sulfonium (tertiary sulfonium compound) are extensively studied in plants like their signalling, synthesis and gene involvement in this (Roychoudhury and Banerjee 2016). Normally, they are synthesised in the cytosol and chloroplast and get accumulated and stored in the vacuole of plant cells without interfering with any cellular part (Ghosh et al. 2021; Andreev 2001).

Long-period exposure to high temperatures in plants leads to many other stresses, such as osmotic stress, drought stress, and the accumulation of ROS in plant cells (Ahanger et al. 2017). A continuum of water between soil, plant and atmosphere only occurs if water potential decreases along the path, which means water will only be taken up by the roots of plants if the water potential of the root is lower than the water potential of soil. Water potential is the sum of the osmotic potential and the turgor pressure (Taiz and Zeiger 2006). So, the osmotic potential is like the movement of water from a lower solute concentration (higher water potential) to a higher solute concentration (lower water potential). So, in plants, the accumulation of solutes in roots and leaves helps in adjusting water potential to lower so that soil having high water potential moves water to the roots of plants. So, during stress like high temperatures or drought stress, plants' osmolytes play a role in osmotic adjustment by accumulating more solutes in plants (Ghosh et al. 2021; Martinoia et al. 2012). In plants, as soon as they sense heat or drought stress, they activate ABA signalling, calcium signalling, and the MAP kinase network to respond rapidly and synergistically to establishing osmotic homeostasis by regulating the production and accumulation of osmolytes. ABA signalling stimulates the accumulation of proline (Roychoudhury et al. 2013). MAP kinases like MKK2 help accumulate proline in plants (Teige et al. 2004).

A recent report suggested that exogenous application of osmoprotectant modulates the antioxidant and glyoxalase systems (Hasanuzzaman et al. 2014). Using cold or heat to prime plants to confer cross-stress tolerance by inducing osmoprotectant accumulation has been well documented. Accumulation due to heat priming increased drought and salt tolerances in cotton (Kuznetsov et al. 1999). Similarly, glycine betaine and proline accumulation due to cold priming- resulted in tolerance to freezing in barley and *Arundo donax* L. (Pompeiano et al. 2015; Kishitani et al. 1994). Cold priming-induced Proline accumulation resulted in freezing tolerance in *Arabidopsis* (Xin and Browse 1998). Osmoprotectants are important for inducing stress tolerance in plants (Hossain et al. 2018; Burg and Ferraris 2008).

8.3.9 Cold Acclimation Proteins

Plants tolerate cold stress through an array of physiological and biochemical changes. Cold stress includes chilling (0–15 °C) and freezing stress (≤0 °C), the former being less destructive than the latter. When plants are exposed to freezing conditions, ice formation occurs on the apoplast of tissues, and it is evident that the plasma membrane is the prominent area of injury (Thomashow 1999; Steponkus 1984). The membrane senses cold stress, and the stimulus is transferred to the nucleus, which leads to cold acclimation responses by activating cold response genes. Cold response genes have always been a topic of great interest. Research in this field witnessed a revolutionary change when Guy and his colleagues identified various genes that encoded enzymes involved in the biosynthetic pathway of certain osmolytes, which increases freezing tolerance by depositing certain cryoprotective proteins and soluble sugars; these genes are collectively known as *Cold regulated* (*COR*) genes (Vaultier et al. 2006; Guy et al. 1985). In *Arabidopsis thaliana*, *COR* genes include the *COR*, *low-temperature induced* (*LTI*), *responsive to desiccation* (*RD*), and *early dehydration-inducible* (*ERD*) genes (Guy et al. 1985).

8.3.9.1 Cold Regulated (COR) Genes

COR genes encode several proteins, such as enzymes involved in the biosynthesis of osmoprotectants, late-embryogenesis-abundant (LEA) proteins, antifreeze proteins, chaperones, detoxification enzymes, transcription factors (Puhakainen et al. 2004; Fowler and Thomashow 2002). LEA proteins are expressed during the late embryonic stage, and there are several classes of LEA proteins, including dehydrins. All dehydrins contain three highly conserved domains and generally contribute to freeze tolerance. A wheat dehydrin WCOR410 accumulation leads to freeze tolerance during low temperatures (Danyluk et al. 1998). The transfer of WCOR410 and its overexpression was shown to increase freezing tolerance in transgenic strawberry leaves (Houde et al. 2004). Along with WCOR410, proteins such as WCS120 and WCOR14b contribute to the freeze tolerance in wheat (Denesik 2007). In *Arabidopsis*, the combined effect of RAB18 and COR47, LTI29 and LTI30 enhances the freeze tolerance (Puhakainen et al. 2004). The localisation of dehydration occurs in the vicinity of the plasma membrane in wheat; in the case of *Arabidopsis*, dehydrins were found in the cytosol, plastid envelope, tonoplast and in the plasma membrane (Puhakainen et al. 2004; Danyluk et al. 1998). Even though the exact activity of dehydrins is unknown, dehydrins are proposed to have antifreeze and cryoprotective functions (Bravo et al. 2003). Some critical transcription factors regulate the activity of COR, collectively called CBFs.

8.3.9.2 CBFs

CBF transcription factors, otherwise known as dehydration response element binding factors (DREBs), are important transcription factors for cold acclimation. CBFs recognise the TACCGACAT sequence known as the Dehydration response element (DRE), which is found in most of the *COR* genes (Jaglo-Ottosen et al. 1998). A single copy of DRE is enough for the gene expression of *CORs* (Yamaguchi-Shinozaki and Shinozaki. 2005). In *Arabidopsis*, the CBF family has three genes: *CBF1*, *CBF2* and *CBF3*. These genes are arranged as a direct repeat in chromosome 4 (Gilmour et al. 1998). Overexpression of CBF transcription factors in many plant species such as *Arabidopsis thaliana* (Gilmour et al. 2000), *Solanum lycopersicum* (Zhang et al. 2004), *Zea mays* (Qin et al. 2004), *Brassica napus* (Savitch et al. 2005) and *Oryza sativa* (Ito et al. 2006) has resulted in increased freezing tolerance compared to wild type. *CBF* expression is regulated by an array of transcription regulators that recognise cis-elements in *CBF* promoters. Transcriptional activator ICE1 binds to MYC binding sites (CANNTG) of *CBF1-3* and enhances expression during cold stress. The mutant *ice1* in *Arabidopsis* showed defective chilling tolerance and cold acclimation. However, overexpression of ICE1 induced higher expression of CBFs and CORs. Although ICE1 is constitutively produced and localised in the nucleus, cold therapy is necessary to activate CBF expression. This finding implies that ICE1 alteration brought on by cold is required for the target genes it activates (Chinnusamy et al. 2003). Apart from ICE1, a calmodulin-binding site containing transcriptional activators is named Calmodulin-binding transcriptional activators (CAMTA). Initially, it was found that the CAMTA3 protein bound to the *CBF2* promoter's CM2 motif (CGCG-box) was activating the gene expression (Doherty et al. 2009). This finding corresponds with the increased production of cytosolic Ca^{2+} ions during cold stress. Plant hormones also play a significant role in CBF regulation. Treatment of ABA in *Arabidopsis* has resulted in the increased expression of *CBF1-3*. The 125-bp region between −189 and −65 of the *CBF2* promoter might include a potential ABRE/ZIP binding site. However, this expression is considerably lower than the cold-induced expression (Knight et al. 2004). Circadian rhythm also influences the expression of *CBF*. Clock regulatory elements have been found in *CBF* promoters. In *Arabidopsis* mutant plants containing mutations on day clock regulatory elements CCA1 and LHY, expression of *CBF1*, *CBF2* and *CBF3* was immensely reduced, and the plants were susceptible to freezing stress. The expression of CBF-regulated genes *COR15A*, *COR47* and *COR78* were greatly reduced (Dong et al. 2011). CBF proteins are also self-regulated. The *Arabidopsis* null mutant *cbf2* showed increased expression of *CBF1* and *CBF3* during cold stress; this indicates that CBF2 is involved in the feedback regulation of CBF1 and CBF3 (Novillo et al. 2004). Overall, CBFs are critical transcription factors that regulate *COR* gene expression; many cold-induced transcription regulators regulate the expression of *CBFs*. Besides cold response, *CBF* expression is regulated by phytohormones and circadian rhythm, and they are also self-regulated. Studies implicate other proteins, such as TTL1

(TETRATRICOPEPTIDE-REPEAT THIOREDOXIN-LIKE1), to be involved in the cold response. Further studies are required to validate their mechanisms (Manasa et al. 2023).

8.3.10 Anti-freeze Proteins

In the biosphere, plants are nearly ubiquitous, and plants have evolved to withstand the stress they are exposed to in their habitat. Overwintering plants, which are predominantly seen in the Arctic, Antarctic, and Alpine regions, are exposed to temperatures dropping below −50 °C. To withstand these harsh conditions, overwintering plants have developed highly efficient freeze tolerance mechanisms. Cereal crops such as winter wheat and winter rye have intricate mechanisms to survive freezing temperatures below −20 °C (Antikainen and Griffith 1997; Yoshida et al. 1997). Intercellular ice crystals exert physical stress on the plasma membrane and rupture it, which leads to the cell's death (Steponkus 1984). One of the major factors that aid overwintering plants in surviving cold conditions is the presence of anti-freeze proteins (AFP). AFPs are diverse proteins that can interact with intracellular ice formations and prevent their growth (Yang et al. 1998). AFPs prevent ice formation through a phenomenon known as thermal hysteresis, where the AFPs depress the freezing temperature, thereby preventing ice formation. A unique property of AFP is their ability to lower freezing temperature; their efficiency is about 500 times more than other solute molecules. During thermal hysteresis, ice is maintained as a supercooled solution (Gupta and Deswal 2014; DeVries 1971). In an alternate scenario, if ice crystals are formed, AFPs perform another function known as ice recrystallisation inhibition. AFPs prevent the recrystallisation of smaller ice crystals to form large ice crystals (Balcerzak et al. 2014; Knight et al. 1988).

Accumulation of most of the AFPs is on apoplast (Marentes et al. 1993; Griffith et al. 1992). However, in *Prunus persica* and *Forysthia suspensa,* the accumulation of AFP is intracellular (Simpson et al. 2005; Wisniewski and Fuller 199;). AFPs are homologous to PR-proteins, including β-1,3-glucanase, class I and II chitinases, and thaumatin-like proteins (Hon et al. 1995). Additionally, a resemblance between AFPs from *F. suspensa* and *P. persica* with dehydrins has been found (Simpson et al. 2005; Wisniewski and Fuller 1999). A peculiar property of plant AFP is their dual nature; they bear homology with PR proteins and have both antifreeze and hydrolytic activity (Gupta and Deswal 2014). It was discovered that the purified chitinases from seabuckthorn and winter rye that have been cold-acclimated show both hydrolytic and antifreeze properties. Similarly, at subzero temperatures, a β-1,3-glucacold adapted enzyme also maintained its partial hydrolytic activities in addition to its antifreeze activity, suggesting that some sort of post-translational process is involved in controlling the activities. This indicates that winter rye glucanases provide a dual function of prevention of ice formation at low temperatures and degradation of glucans at high temperatures (Yaish et al. 2006).

Overall, antifreeze proteins seen in overwintered plants interact with inter- and intracellular ice crystals, preventing the formation of ice crystals by thermal hysteresis and the recrystallisation of ice.

8.3.11 Calcium Signalling

Calcium is an important metabolite, acting as a secondary metabolite in numerous plant processes. Cytoplasmic divalent calcium levels are regulated strictly, with calcium levels inside the cells much lower than outside. Thus, small changes in cellular calcium levels can act as potent signals for several downstream processes.

In ambient temperatures, Calcium signalling is important for plants to respond to temperature elevation. The Cyclic Nucleotide Gated Calcium Channel (CNGC), such as CNGC2, is an important part of the thermosensory and thermotolerance mechanism in *Arabidopsis*, which is present in non-vascular plants like *Physcomitrium* as well. It is suggested that they trigger an optimal heat shock response to ensure acquired thermotolerance (Finka et al. 2012).

When the plant experiences heat stress, calcium levels go up. Plants treated with exogenous calcium are more tolerant to heat stress. Increased calcium levels act with CaM to form GABA from glutamate via the activation of L-glutamate decarboxylase (GAD). GABA improves leaf turgor and concentration of protective osmolytes and reduces oxidative damage, thus imparting thermotolerance. In the case of cold stress, the plant experiences an increase in intracellular calcium levels as well. Experiments in *Arabidopsis* roots (Plieth et al. 1999) show that plants are more sensitive to the rate of temperature change, which goes hand in hand with calcium response, than the absolute temperature. The response becomes more attenuated if the temperature change is more gradual compared to a sudden temperature change. The increased calcium levels lead to a cascade in which several genes are activated, leading up to the activation of several *COR* genes. The several components involved in these signalling pathways and their roles are still being elucidated (Kang et al. 2023; Patra et al. 2021; Yuan et al. 2018).

8.3.12 Alterations in Nucleic Structure

Nucleic acids- DNA and RNA- are essential molecules that control and coordinate various biological processes. The ecological and metabolic processes of the plant can be fine-tuned with respect to the environment by modifying the structures of the nucleic acids.

In the nucleus, the DNA is bound to positively charged proteins called histones, which form a highly condensed complex called chromosomes. The part of the chromosomes that are highly condensed and hence transcriptionally silent is called heterochromatin, while the less condensed, transcriptionally active chromatin is called euchromatin. Chromatin undergoes several changes under stressful conditions. This includes histone modifications by methylation, acetylation, and phosphorylation, as well as DNA modifications by methylation and acetylation. Such chromatin modifications under different conditions give rise to selective gene expression.

During cold stress, the histone H3K27me3 is observed to accumulate at FLOWERING LOCUS C (FLC), the flower repressor. This is facilitated by the polycomb repressive complex 2. The accumulation of H3K27me3 represses FLC by inhibition of transcription. In *Arabidopsis* leaf tissue, prolonged heat stress causes centromeric repeats and decondensation of 5S rDNA. Extended periods of heat stress in plants result in the short-term transcriptional activation of both transgenic and specific endogenous repeats. These repeats are regulated by transcriptional gene silencing (TGS). Remarkably, demethylation of DNA is not necessary for heat-induced transcriptional activation (Bhadouriya et al. 2021).

Temperature also affects the activity of transposable elements (TE). For instance, the TAM element (Transposable Element of *Antirrhinum majus*) in Snapdragon undergoes hypermethylation in higher temperatures and demethylation in lower temperatures, which significantly affects the TE activity (Pecinka et al. 2010). It is important to note that the methylation states are reversible within one generation.

RNA is a desirable candidate for sensing temperature because of its inherent thermosensitivity. A temperature gradient can cause significant rearrangements in certain RNA structures, and cells can use these structures to sense their environment's temperature. However, many conformational changes brought on by temperature changes are probably minor and have few functional ramifications. A recent genome-wide analysis of rice seedlings (*Oryza sativa* L.) reveals that unfolding of the mRNA 5′ and 3′ untranslated regions (UTRs) causes global structural changes in the transcriptome in response to high temperature. This results in decreased mRNA stability and downregulation of global translation in rice seedlings (Thomas et al. 2022).

Experiments were done to understand mRNA stability under heat stress. It was observed that while the mRNAs of certain genes were less stable, the mRNAs of certain other genes became relatively highly stabilised under high temperatures. 5′-UTRs were noticeably longer in the comparatively stabilised transcripts. The 5′-UTR lengths of comparatively destabilised transcripts and the 3′-UTR lengths of relatively stabilised transcripts, on the other hand, showed minor but significant variations (Chiba et al. 2013).

mRNA structural regulation can also be done at the splicing level. Changes in pre-mRNA secondary structures at intron-exon junctions can mediate temperature-dependent splicing by regulating the activity of splicing regulators or the spliceosome complex. One such newly identified example is AS at intron 2 of *Heat Shock Factor* A2 (*HSFA2*) pre-mRNA from *Solanum lycopersicum.* Thermosensitive RNA structures directly influence protein synthesis in *Arabidopsis* (Thomas et al. 2022).

It has been demonstrated that elevated RNA G-quadruplex (RG4) formation in plant transcript 3′-UTRs in response to low temperatures improves mRNA stability and cold environment adaptation. Analysis of plant transcriptomes showed that plants growing in cold climates have guanine (G)-enriched transcripts. These transcripts, in turn, are highly likely to form RG4 structures. These RG4 complexes were observed to form in the 3′-UTR regions of the mRNA of *Arabidopsis* under cold stress and are now understood to aid in mRNA stability and tolerance to cold stress (Yang et al. 2022a, b).

PIF7 is facilitated by forming an mRNA hairpin in the 5′-UTR, approximately 30 nucleotides upstream of the start codon. At lower temperatures- below 22°- the hairpin is stable. As the temperature reaches 27–32°, it transitions to a partially melted open conformation. These structural changes aid in increased PIF7 synthesis and the subsequent synthesis of the downstream genes that regulate normal plant growth and metabolism (Chung et al. 2020).

8.3.13 Alterations in Protein Structure

Proteins are long chains of amino acids that constitute essential metabolomes of an organism. Alterations in protein structure, concentration and production are necessary to fine-tune an organism's developmental and metabolic pathways. Plants, under thermal stress, need to upregulate certain proteins while downregulating others. This is achieved mostly by post-translational modifications (Kim et al. 2015).

Post-translational modifications (PTMs) are modifications added to the polypeptide after it is translated. It is the final step that determines the protein function. PTM also functions as a rapid regulatory mechanism when the plant is exposed to sudden rapid stress. acetyl groups, phosphoric acids, and other small peptides are some of the different molecular groups involved in PTM. Ubiquitination is a PTM involved in thermal stress tolerance in plants. Ubiquitin (Ub) protein targets a substrate for proteasomal degradation by covalently binding to the lysine residues of its target protein. Ubiquitination is facilitated by E3 ligase, another protein that helps transfer ubiquitin to the substrate. It is observed that E3 ligase genes are induced by heat shock, and hence ubiquitination is noticed to increase under high temperatures (Kim et al. 2015).

SUMO (small ubiquitin-related modifier) is a reversible post-translational modifier significantly influencing protein interactions in eukaryotic cells. SUMOylation of histones during thermal stress helps gene regulation at the transcriptional level (Kim et al. 2015).

Widespread transcriptional changes in the organism are induced by exposure to low temperatures. It is estimated that cold stress affects 3–20% of the genes in *Arabidopsis*. Selective unsilencing of heterochromatic tandem repeats increases H3K9ac accumulation while lowering DNA methylation and H3K9me2 levels in these regions (Kim et al. 2015). The protein HOS15 in *Arabidopsis* functions as a repressor, modifying chromatin to control the expression of genes linked to cold tolerance.

Protein kinases also play a major role in thermotolerance by phosphorylating certain proteins. For instance, under warm temperatures, BR-regulated transcription factor brassinazole-resistant 1 (BZR1) is translocated to the nucleus, where it binds to the promoter of PIF4 to promote expression and cell elongation (Han et al. 2022).

HSPs represent a significant class of proteins implicated in heat stress. Their expression may start from a few seconds after heat shock and continue to increase until they reach their maximum at 1–2 h from the start of heat exposure, after which they begin to decline. HSPs function as molecular chaperones to other proteins. Higher temperatures can cause protein misfolding, which interferes with the normal functioning of the plant. HSP functions to ensure the proper folding of proteins under heat stress. HSP chaperones help the misfolded proteins to overcome the activation barrier necessary for their proper conformation (Nievola et al. 2017).

8.3.14 ROS and Antioxidants

Reactive oxygen species are constantly produced as a byproduct of several metabolic processes in the plant. Also known as ROS, the plant maintains these species in equilibrium via endogenous antioxidative mechanisms that scavenge and remove said compounds (Halliwell 1982) from chloroplasts (Dietz et al. 2016), mitochondria (Huang et al. 2016), and peroxisomes (Sandalio and Romero-Puertas 2015). Under stress conditions, however, the loss of cellular homeostasis and the closing of stomata are initiated, which decreases CO_2 assimilation in the plant whilst light reaction and electron transport pathways continue unhindered. When the depletion of NADP available to accept electrons becomes pronounced, a new electron acceptor is in demand. Oxygen rescues the plant by performing a key electron acceptor role. This rescue, nonetheless, is only temporarily beneficial. Prolonged stress conditions can lead to an accumulation of oxygen and its reactive forms. The byproducts, superoxide and hydrogen peroxide are toxic molecules that can undergo an oxidative burst and produce far more reactive oxygen species, such as hydroxyl radicals, that inactivate enzymes, bleach pigments, degrade proteins (such as Rubisco) and even cause membrane destabilising lipid-peroxidation (Apel and Hirt 2004).

In ambient temperatures, antioxidant enzymes regulate ROS at equilibrium with its scavenging, where ROS can participate productively, primarily in signalling and defence pathways (Rensburg and Krüger 1994). The plant upregulates the protective antioxidant mechanisms that tackle the oxidative stress caused by temperature stress. The production of enzymes such as catalase (CAT), superoxide dismutase (SOD), guaiacol peroxidase (GPX), ascorbate peroxidase (APX) and glutathione reductase (GR) and non-enzymatic low-molecular antioxidants for example, ascorbate (ASA), glutathione (GSH), tocopherols (vitamin E) and carotenoids, as a result, are instantly upregulated to provide the plant resistance to temperature stresses (Nievola et al. 2017; Sairam et al. 1997). Glutathione, for example, plays a key role in the signalling of cold acclimation in plants (Kocsy et al. 2001).

GSH and GR are found in the chloroplast stroma to stabilise low molecular weight thiol enzymes used in the Calvin cycle (Kraus et al. 1995; Foyer and Halliwell 1976). GR facilitates the scavenging of ROS produced from heat stress, such as hydrogen peroxide, by participating in the Glutathione-Ascorbate Cycle which prevents oxidative injury (Gamble and Burke 1984; Rensburg and Krüger 1994; Nievola et al. 2017). In cold stress, GR also prevents enzyme and membrane oxidation (Mahan et al. 1990). SOD instigates ROS detoxification of superoxide radicals into hydrogen peroxide. CAT and GPX are found in chloroplasts, endoplasmic reticulum, mitochondria and cytosol (Halliwell 1982) and can further reduce hydrogen peroxide to water along with APX. APX and GR regenerate ASA and GSSG (reduced GSH). *APX* gene expression was 2.5 times more in rubber plants grown at 10 °C. An increase was observed even with only an hour of cold stress treatment (Nievola et al. 2017). APX is reported to scavenge hydrogen peroxide in chloroplasts (Nakano and Asada 1981).

Short but severe temperature stress has been associated with decreased levels of ASA, GSH and other low molecular weight antioxidants, correlating to an added demand for scavenging ROS molecules. However, under prolonged exposure to temperature stress, acclimation resulted in an overall increase in antioxidant levels (Szymańska et al. 2017). Heat-tolerant cultivars/lines of susceptible plant species have been identified with an increased standard level of antioxidant enzymes and non-enzymes in *Triticum aestivum* and *Zea mays* (Almeselmani et al. 2006; Sairam, et al. 1997; Kraus et al. 1995; Pastori and Trippi 1995); *Brassica juncea* (Wilson et al. 2014) and *Solanum lycopersicum* (Camejo et al. 2006).

The pathways they follow to remove ROS from the plant system cannot be restricted as a temperature mechanism, for they mostly follow the same pathways for abiotic stresses, regardless of the character of the stress. The antioxidants do not partake in pathways that provide them with a distinctive identity; instead, they primarily function by coordinating the processing of ROS and regeneration of ASA and GSH redox metabolites (Almeselmani et al. 2006).

Further research continues to elucidate the ways in which ROS and antioxidants function to enable plants to respond to temperature.

8.3.15 Other Factors

MecPP

High temperatures can activate stress response gene transcriptions through a signal from the chloroplast, called 2-C-methyl-D-erythritol-2,4-cyclopyrophosphate or MEcPP. MEcPP is an isoprenoid biosynthesis intermediate metabolite that behaves as a retrograde signal, along with hydrogen peroxide, in regulating those heat stress response genes. These chloroplast retrograde signals are essential for heat stress tolerance and acclimation. Recent studies suggest the involvement of chloroplast-MEcPP in regulating PIFs and phyB levels; however, more such cross-talks are yet to be uncovered. One such cross-talk is how MEcPP activates CAMTA3 to regulate phyB abundance (Hayes et al. 2021; Jiang et al. 2019).

Sugar Signalling
Stress stimuli affect the source and sink metabolism of sugars in plants. However, not a lot has been studied about the regulation of enzyme activity and gene transcription in response to these stimuli (Roitsch 1999). Root-to-shoot signalling is critical for the coordination of growth and development of the whole plant. Studies have yet to deduce the pathways driven by sugars and hormones together. There have been few attempts at linearly defining the network of sugars with hormones that have been met with incredible difficulty. PIFs and DELLAs have been identified as playing key roles in these cross-regulations. PIFs integrate sugar signalling with auxin, GA, and BR pathways (Ljung et al. 2015).

Cytoskeletal Dynamics
Plants can sense cold stress based on organisational changes in the cytoskeleton. This makes the cytoskeleton a low-temperature sensor. Microtubules make up the cell's cytoskeleton along with actin filaments and other intermediate filaments. In freeze-resistant plants, microtubules partially disassemble as the plant initially acclimatises to the cold. However, this does not happen in freeze-sensitive plants. This shows that there is an increase in microtubule dynamics in all plants in response to chilling, but it does not suffice. Efficient cold resistance is characterised by microtubules that can disassemble easily on chilling and reassemble into a cold stress tolerant organisation (Abdrakhamanova et al. 2003).

Abiotic stimuli induce ROS production, followed by an increase in the endogenous calcium ion (Ca^{2+}) in the cell. Microtubule remodelling occurs due to the acztivation of different MAPK signalling pathways that induce several transcription enzymes regulating PTMs (Livanos et al. 2014). SAMK and HAMK, which are activated by temperature, are also activated by cytoskeletal destabilisers. The cytoskeleton translates the temperature stress signal from the plasma membrane (sensed by structural changes in the membrane) to activate distinct MAPK cascades (Sangwan et al. 2002).

Epigenetic Mechanisms
Plants finally also have epigenetic regulation to withstand temperature stresses. DNA methylation is one such mechanism where a methyl group is attached to the cytosine bases of the DNA to form 5-methylcytosine to regulate gene expression. Global methylation is increased in *A. thaliana* (Thale cress; Boyko et al. 2010), *Quercus suber* (cork oak; Correia et al. 2013), *B. napus* (rapeseed; Gao et al. 2014). However, hypomethylation is observed in *G. hirsutum* (cotton; Min et al. 2014), and both upregulation and downregulation of different genes in *Z. mays* (maize; Qian et al. 2019) and *O. sativa* (rice; Folsom et al. 2014). Different species respond differently to heat stress, making no methylation patterns discernible as of now (Liu et al. 2015).

Histone modifications have been found to have the same case, their roles are yet to be identified under heat stress. Transcription factors such as HSF1 regulate histone acetylation, methylation and transcription initiation/elongation with heat stress (Strenkert et al. 2011). Conclusion, plants respond to heat stress by inducing covalent modifications of histones and DNA, histone variants, histone chaperones, sRNAs, lncRNAs and ATP-regulated chromatin remodelling and other possible mechanisms, but a lot is yet to be researched into identifying any global patterns across species (Liu et al. 2015).

Emerging research continues to reveal the role of several other components in plant thermal sensing, response and acclimation.

8.4 Conclusion

This chapter has provided an overview of the mechanisms of temperature perception, signalling, and acclimation in plants, highlighting the intricate web of molecular, cellular, and physiological processes that enable plants to respond and adapt to thermal fluctuations. Light-sensing molecules like phytochromes and phototropins are implicated in temperature sensation. The functionality of other such molecules may have some influence of temperature that is yet to be uncovered. The circadian clock components integrate temperature cues as input signals to accommodate plant processes according to the environment. Proteins such as PIFs act to mediate several temperature-dependent processes. Several metabolites and proteins are involved in downstream signalling and response to temperature, which is still under study.

While significant progress has been made in elucidating these mechanisms, it is evident that many factors have only been scratched on the surface, with much remaining to be uncovered. The complex interplay between temperature-sensing pathways, hormone regulation, and gene expression, for instance, warrants further investigation to fully grasp the nuances of plant thermal responses.

As research in plant functional genomics continues to evolve, it is crucial that we acknowledge the existing knowledge gaps and pursue a deeper understanding of these mechanisms. Future investigations should prioritise the elucidation of temperature-specific signalling cascades, the characterisation of thermosensing molecules, and the exploration of epigenetic regulation in plant thermal acclimation. Moreover, the functional implications of recently discovered factors, such as non-coding RNAs and chromatin modifications, remain to be fully elucidated. By acknowledging the complexity and depth of these mechanisms, we can refine our approaches to improving crop resilience to abiotic stress, ultimately enhancing global food security and sustainability in the face of climate uncertainty.

Acknowledgements We are grateful to the Department of Atomic Energy, Government of India and National Institute of Science Education and Research (NISER), Bhubaneswar, Odisha, India for providing the infrastructure and core grant.

References

Abdrakhamanova A, Wang QY, Khokhlova L, Nick P (2003) Is microtubule disassembly a trigger for cold acclimation? Plant Cell Physiol 44(7):676–686

Ahanger MA, Akram NA, Ashraf M, Alyemeni MN, Wijaya L, Ahmad P (2017) Plant responses to environmental stresses—from gene to biotechnology. AoB Plants 9(4):plx025

Ahuja I, de Vos RC, Bones AM, Hall RD (2010) Plant molecular stress responses face climate change. Trends Plant Sci 15(12):664–674

Aiqing S, Somayanda I, Sebastian SV, Singh K, Gill K, Prasad PVV, Jagadish SK (2018) Heat stress during flowering affects time of day of flowering, seed set, and grain quality in spring wheat. Crop Sci 58(1):380–392

Al-Hawija BN, Partzsch M, Hensen I (2012) Effects of temperature, salinity and cold stratification on seed germination in halophytes. Nord J Bot 30(5):627–634

Almeselmani M, Deshmukh PS, Sairam RK, Kushwaha SR, Singh TP (2006) Protective role of antioxidant enzymes under high temperature stress. Plant Sci 171(3):382–388

Almeselmani M, Deshmukh P, Sairam R (2009) High temperature stress tolerance in wheat genotypes: role of antioxidant defence enzymes. Acta Agron Hungarica 57(1):1–14

Al-Whaibi MH (2011) Plant heat-shock proteins: a mini review. J King Saud Univ Sci 23(2): 139–150

Andreev IM (2001) Functions of the vacuole in higher plant cells. Russ J Plant Physiol 48:672–680

Antikainen M, Griffith M (1997) Antifreeze protein accumulation in freezing-tolerant cereals. Physiol Plant 99(3):423–432

Apel K, Hirt H (2004) Reactive oxygen species: metabolism, oxidative stress, and signal transduction. Annu Rev Plant Biol 55(1):373–399

Ashraf M, Hafeez M (2004) Thermotolerance of pearl millet and maize at early growth stages: growth and nutrient relations. Biol Plant 48:81–86

Balasubramanian S, Sureshkumar S, Lempe J, Weigel D (2006) Potent induction of Arabidopsis thaliana flowering by elevated growth temperature. PLoS Genet 2(7):e106

Balcerzak AK, Capicciotti CJ, Briard JG, Ben RN (2014) Designing ice recrystallization inhibitors: from antifreeze (glyco) proteins to small molecules. RSC Adv 4(80):42682–42696

Bastow R, Mylne JS, Lister C, Lippman Z, Martienssen RA, Dean C (2004) Vernalization requires epigenetic silencing of FLC by histone methylation. Nature 427(6970):164–167

Bellstaedt J, Trenner J, Lippmann R, Poeschl Y, Zhang X, Friml J, Quint M, Delker C (2019) A mobile auxin signal connects temperature sensing in cotyledons with growth responses in hypocotyls. Plant Physiol 180(2):757–766

Bhadouriya SL, Mehrotra S, Basantani MK, Loake GJ, Mehrotra R (2021) Role of chromatin architecture in plant stress responses: an update. Front Plant Sci 11:603380

Biamonti G, Caceres JF (2009) Cellular stress and RNA splicing. Trends Biochem Sci 34(3): 146–153

Bianchimano L, De Luca MB, Borniego MB, Iglesias MJ, Casal JJ (2023) Temperature regulation of auxin-related gene expression and its implications for plant growth. J Exp Bot 74(22): 7015–7033

Blázquez MA, Ahn JH, Weigel D (2003) A thermosensory pathway controlling flowering time in Arabidopsis thaliana. Nat Genet 33(2):168–171

Bochenek A, Golaszewski J, Piotrowicz-Cieslak AI, Górecki RJ (2009) The effects of temperature on the dormancy and germination of Cirsium arvense [L.] Scop. seeds. Acta Soc Bot Pol 78(2): 105–114

Box MS, Huang BE, Domijan M, Jaeger KE, Khattak AK, Yoo SJ, Sedivy EL, Jones DM, Hearn TJ, Webb AA, Grant A (2015) ELF3 controls thermoresponsive growth in Arabidopsis. Curr Biol 25(2):194–199

Boyko A, Blevins T, Yao Y, Golubov A, Bilichak A, Ilnytskyy Y, Hollander J, Meins F Jr, Kovalchuk I (2010) Transgenerational adaptation of Arabidopsis to stress requires DNA methylation and the function of Dicer-like proteins. PLoS One 5(3):e9514

Bravo LA, Gallardo J, Navarrete A, Olave N, Martínez J, Alberdi M, Close TJ, Corcuera LJ (2003) Cryoprotective activity of a cold-induced dehydrin purified from barley. Physiol Plant 118(2): 262–269
Brewster JL, de Valoir T, Dwyer ND, Winter E, Gustin MC (1993) An osmosensing signal transduction pathway in yeast. Science 259(5102):1760–1763
Burg MB, Ferraris JD (2008) Intracellular organic osmolytes: function and regulation. J Biol Chem 283(12):7309–7313
Camejo D, Jiménez A, Alarcón JJ, Torres W, Gómez JM, Sevilla F (2006) Changes in photosynthetic parameters and antioxidant activities following heat-shock treatment in tomato plants. Funct Plant Biol 33(2):177–187
Catalá R, Medina J, Salinas J (2011) Integration of low temperature and light signaling during cold acclimation response in Arabidopsis. Proc Natl Acad Sci 108(39):16475–16480
Chen H, Hwang JE, Lim CJ, Kim DY, Lee SY, Lim CO (2010) Arabidopsis DREB2C functions as a transcriptional activator of HsfA3 during the heat stress response. Biochem Biophys Res Commun 401(2):238–244
Cheng J, Huang H, Liu W, Zhou Y, Han W, Wang X, Zhang Y (2022) Unraveling the effects of cold stratification and temperature on the seed germination of invasive Spartina alterniflora across latitude. Front Plant Sci 13:911804
Chiba Y, Mineta K, Hirai MY, Suzuki Y, Kanaya S, Takahashi H, Onouchi H, Yamaguchi J, Naito S (2013) Changes in mRNA stability associated with cold stress in Arabidopsis cells. Plant Cell Physiol 54(2):180–194
Chinnusamy V, Ohta M, Kanrar S, Lee BH, Hong X, Agarwal M, Zhu JK (2003) ICE1: a regulator of cold-induced transcriptome and freezing tolerance in Arabidopsis. Genes Dev 17(8): 1043–1054
Chinnusamy V, Zhu J, Zhou T, Zhu JK (2007) Small RNAs: big role in abiotic stress tolerance of plants. In: Advances in molecular breeding toward drought and salt tolerant crops, pp 223–260
Chung BY, Balcerowicz M, Di Antonio M, Jaeger KE, Geng F, Franaszek K, Marriott P, Brierley I, Firth AE, Wigge PA (2020) An RNA thermoswitch regulates daytime growth in Arabidopsis. Nature Plants 6(5):522–532
Correia B, Valledor L, Meijon M, Rodriguez JL, Dias MC, Santos C, Canal MJ, Rodriguez R, Pinto G (2013) Is the interplay between epigenetic markers related to the acclimation of cork oak plants to high temperatures? PLoS One 8(1):e53543
Cosgrove DJ (1997) Assembly and enlargement of the primary cell wall in plants. Annu Rev Cell Dev Biol 13(1):171–201
Danyluk J, Perron A, Houde M, Limin A, Fowler B, Benhamou N, Sarhan F (1998) Accumulation of an acidic dehydrin in the vicinity of the plasma membrane during cold acclimation of wheat. Plant Cell 10(4):623–638
Den NZUbAM, Sa Z, Ac U, Ma J, Sd G, Me K, Af Z, Ng A, Kg A, Mh I, Zi A (2024) Multifarious role of osmolytes in plants: signaling and defense
Denesik TJ (2007) Quantitative expression analysis of four low-temperature-tolerance-associated genes during cold acclimation in wheat (Triticum aestivum L.). Doctoral dissertation, University of Saskatchewan
DeVries Arthur L (1971) Freezing resistance in fishes. Fish Physiol (Academic Press 1971) 6:157–190
Dietz KJ, Turkan I, Krieger-Liszkay A (2016) Redox-and reactive oxygen species-dependent signaling into and out of the photosynthesizing chloroplast. Plant Physiol 171(3):1541–1550
Ding Y, Shi Y, Yang S (2020) Molecular regulation of plant responses to environmental temperatures. Mol Plant 13(4):544–564
Doherty CJ, Van Buskirk HA, Myers SJ, Thomashow MF (2009) Roles for Arabidopsis CAMTA transcription factors in cold-regulated gene expression and freezing tolerance. Plant Cell 21(3): 972–984

Dong MA, Farré EM, Thomashow MF (2011) Circadian clock-associated 1 and late elongated hypocotyl regulate expression of the C-repeat binding factor (CBF) pathway in Arabidopsis. Proc Natl Acad Sci 108(17):7241–7246

Dove N, Mangan ME (2010) The effect of increasing temperature on germination of native plant species in the North Woods region. University of Vermont, Burlington

Driesen E, Van den Ende W, De Proft M, Saeys W (2020) Influence of environmental factors light, CO2, temperature, and relative humidity on stomatal opening and development: a review. Agronomy 10(12):1975

Dürr C, Dickie JB, Yang XY, Pritchard HW (2015) Ranges of critical temperature and water potential values for the germination of species worldwide: contribution to a seed trait database. Agric For Meteorol 200:222–232

Edwards KD, Lynn JR, Gyula P, Nagy F, Millar AJ (2005) Natural allelic variation in the temperature-compensation mechanisms of the Arabidopsis thaliana circadian clock. Genetics 170(1):387–400

Edwards KD, Anderson PE, Hall A, Salathia NS, Locke JC, Lynn JR, Straume M, Smith JQ, Millar AJ (2006) FLOWERING LOCUS C mediates natural variation in the high-temperature response of the Arabidopsis circadian clock. Plant Cell 18(3):639–650

Erwin JE, Heins RD, Karlsson MG (1989) Thermomorphogenesis in Lilium longiflorum. Am J Bot 76(1):47–52

Ezer D, Jung JH, Lan H, Biswas S, Gregoire L, Box MS, Charoensawan V, Cortijo S, Lai X, Stöckle D, Zubieta C (2017) The evening complex coordinates environmental and endogenous signals in Arabidopsis. Nature Plants 3(7):1–12

Fahad S, Hussain S, Saud S, Hassan S, Tanveer M, Ihsan MZ, Shah AN, Ullah A, Khan F, Ullah S, Alharby H (2016) A combined application of biochar and phosphorus alleviates heat-induced adversities on physiological, agronomical and quality attributes of rice. Plant Physiol Biochem 103:191–198

Fahad S, Bajwa AA, Nazir U, Anjum SA, Farooq A, Zohaib A, Sadia S, Nasim W, Adkins S, Saud S, Ihsan MZ (2017) Crop production under drought and heat stress: plant responses and management options. Front Plant Sci 8:1147

Feller U (2006) Stomatal opening at elevated temperature: an underestimated regulatory mechanism. General Appl Plant Physiol 32:19–31

Feurtado JA, Ambrose SJ, Cutler AJ, Ross AR, Abrams SR, Kermode AR (2004) Dormancy termination of western white pine (Pinus monticola Dougl. Ex D. Don) seeds is associated with changes in abscisic acid metabolism. Planta 218:630–639

Finch-Savage WE, Leubner-Metzger G (2006) Seed dormancy and the control of germination. New Phytol 171(3):501–523

Finka A, Cuendet AFH, Maathuis FJ, Saidi Y, Goloubinoff P (2012) Plasma membrane cyclic nucleotide gated calcium channels control land plant thermal sensing and acquired thermotolerance. Plant Cell 24(8):3333–3348

Finnegan EJ, Sheldon CC, Jardinaud F, Peacock WJ, Dennis ES (2004) A cluster of Arabidopsis genes with a coordinate response to an environmental stimulus. Curr Biol 14(10):911–916

Finnegan EJ, Kovac KA, Jaligot E, Sheldon CC, Peacock WJ, Dennis ES (2005) The downregulation of FLOWERING LOCUS C (FLC) expression in plants with low levels of DNA methylation and by vernalization occurs by distinct mechanisms. Plant J 44(3):420–432

Fleishon S, Shani E, Ori N, Weiss D (2011) Negative reciprocal interactions between gibberellin and cytokinin in tomato. New Phytol 190(3):609–617

Folsom JJ, Begcy K, Hao X, Wang D, Walia H (2014) Rice fertilization-independent endosperm1 regulates seed size under heat stress by controlling early endosperm development. Plant Physiol 165(1):238–248

Fowler S, Thomashow MF (2002) Arabidopsis transcriptome profiling indicates that multiple regulatory pathways are activated during cold acclimation in addition to the CBF cold response pathway. Plant Cell 14(8):1675–1690

Foyer CH, Halliwell B (1976) The presence of glutathione and glutathione reductase in chloroplasts: a proposed role in ascorbic acid metabolism. Planta 133:21–25

Fujii Y, Tanaka H, Konno N, Ogasawara Y, Hamashima N, Tamura S, Hasegawa S, Hayasaki Y, Okajima K, Kodama Y (2017) Phototropin perceives temperature based on the lifetime of its photoactivated state. Proc Natl Acad Sci 114(34):9206–9211

Gamble PE, Burke JJ (1984) Effect of water stress on the chloroplast antioxidant system: I. Alterations in glutathione reductase activity. Plant Physiol 76(3):615–621

Gao G, Li J, Li H, Li F, Xu K, Yan G, Chen B, Qiao J, Wu X (2014) Comparison of the heat stress induced variations in DNA methylation between heat-tolerant and heat-sensitive rapeseed seedlings. Breed Sci 64(2):125–133

Gendall AR, Levy YY, Wilson A, Dean C (2001) The VERNALIZATION 2 gene mediates the epigenetic regulation of vernalization in Arabidopsis. Cell 107(4):525–535

Ghosh UK, Islam MN, Siddiqui MN, Khan MAR (2021) Understanding the roles of osmolytes for acclimatizing plants to changing environment: a review of potential mechanism. Plant Signal Behav 16(8):1913306

Gilmour SJ, Zarka DG, Stockinger EJ, Salazar MP, Houghton JM, Thomashow MF (1998) Low temperature regulation of the Arabidopsis CBF family of AP2 transcriptional activators as an early step in cold-induced COR gene expression. Plant J 16(4):433–442

Gilmour SJ, Sebolt AM, Salazar MP, Everard JD, Thomashow MF (2000) Overexpression of the Arabidopsis CBF3 transcriptional activator mimics multiple biochemical changes associated with cold acclimation. Plant Physiol 124(4):1854–1865

Greenboim-Wainberg Y, Maymon I, Borochov R, Alvarez J, Olszewski N, Ori N, Eshed Y, Weiss D (2005) Cross talk between gibberellin and cytokinin: the Arabidopsis GA response inhibitor SPINDLY plays a positive role in cytokinin signaling. Plant Cell 17(1):92–102

Griffith M, Ala P, Yang DS, Hon WC, Moffatt BA (1992) Antifreeze protein produced endogenously in winter rye leaves. Plant Physiol 100(2):593–596

Grimstad SO, Frimanslund E (1993) Effect of different day and night temperature regimes on greenhouse cucumber young plant production, flower bud formation and early yield. Sci Hortic 53(3):191–204

Gulen H, Eris A (2004) Effect of heat stress on peroxidase activity and total protein content in strawberry plants. Plant Sci 166(3):739–744

Gupta R, Deswal R (2014) Antifreeze proteins enable plants to survive in freezing conditions. J Biosci 39:931–944

Guy CL, Niemi KJ, Brambl R (1985) Altered gene expression during cold acclimation of spinach. Proc Natl Acad Sci 82(11):3673–3677

Halliwell B (1982) Superoxide-dependent formation of hydroxyl radicals in the presence of iron salts is a feasible source of hydroxy radicals in vivo. Biochem J 205(2):461

Han D, Yu Z, Lai J, Yang C (2022) Post-translational modification: a strategic response to high temperature in plants. aBIOTECH 3(1):49–64

Hasanuzzaman M, Nahar K, Alam MM, Roychowdhury R, Fujita M (2013) Physiological, biochemical, and molecular mechanisms of heat stress tolerance in plants. Int J Mol Sci 14(5): 9643–9684

Hasanuzzaman M, Alam MM, Rahman A, Hasanuzzaman M, Nahar K, Fujita M (2014) Exogenous proline and glycine betaine mediated upregulation of antioxidant defense and glyoxalase systems provides better protection against salt-induced oxidative stress in two rice (Oryza sativa L.) varieties. Biomed Res Int 2014(1):757219

Hayes S, Schachtschabel J, Mishkind M, Munnik T, Arisz SA (2021) Hot topic: thermosensing in plants. Plant Cell Environ 44(7):2018–2033

Higashi Y, Saito K (2019) Lipidomic studies of membrane glycerolipids in plant leaves under heat stress. Prog Lipid Res 75:100990

Hon WC, Griffith M, Mlynarz A, Kwok YC, Yang DS (1995) Antifreeze proteins in winter rye are similar to pathogenesis-related proteins. Plant Physiol 109(3):879–889

Hossain MA, Li ZG, Hoque TS, Burritt DJ, Fujita M, Munné-Bosch S (2018) Heat or cold priming-induced cross-tolerance to abiotic stresses in plants: key regulators and possible mechanisms. Protoplasma 255:399–412

Houde M, Dallaire S, N'Dong D, Sarhan F (2004) Overexpression of the acidic dehydrin WCOR410 improves freezing tolerance in transgenic strawberry leaves. Plant Biotechnol J 2(5):381–387

Huang S, Van Aken O, Schwarzländer M, Belt K, Millar AH (2016) The roles of mitochondrial reactive oxygen species in cellular signaling and stress response in plants. Plant Physiol 171(3): 1551–1559

Huang J, Huang J, Feng Q, Shi Y, Wang F, Zheng K, Huang Q, Jiang J, Luo S, Xie Y, Han D (2023a) SUMOylation facilitates the assembly of a Nuclear Factor-Y complex to enhance thermotolerance in Arabidopsis. J Integr Plant Biol 65(3):692–702

Huang J, Zhao X, Bürger M, Chory J, Wang X (2023b) The role of ethylene in plant temperature stress response. Trends Plant Sci 28(7):808–824

Ikeda M, Mitsuda N, Ohme-Takagi M (2011) Arabidopsis HsfB1 and HsfB2b act as repressors of the expression of heat-inducible Hsfs but positively regulate the acquired thermotolerance. Plant Physiol 157(3):1243–1254

Ito Y, Katsura K, Maruyama K, Taji T, Kobayashi M, Seki M, Shinozaki K, Yamaguchi-Shinozaki K (2006) Functional analysis of rice DREB1/CBF-type transcription factors involved in cold-responsive gene expression in transgenic rice. Plant Cell Physiol 47(1):141–153

Jaglo-Ottosen KR, Gilmour SJ, Zarka DG, Schabenberger O, Thomashow MF (1998) Arabidopsis CBF1 overexpression induces COR genes and enhances freezing tolerance. Science 280(5360): 104–106

Jiang J, Zeng L, Ke H, De La Cruz B, Dehesh K (2019) Orthogonal regulation of phytochrome B abundance by stress-specific plastidial retrograde signaling metabolite. Nat Commun 10(1): 2904

Jung JH, Domijan M, Klose C, Biswas S, Ezer D, Gao M, Khattak AK, Box MS, Charoensawan V, Cortijo S, Kumar M (2016) Phytochromes function as thermosensors in Arabidopsis. Science 354(6314):886–889

Jung JH, Barbosa AD, Hutin S, Kumita JR, Gao M, Derwort D, Silva CS, Lai X, Pierre E, Geng F, Kim SB (2020) A prion-like domain in ELF3 functions as a thermosensor in Arabidopsis. Nature 585(7824):256–260

Kanaoka MM, Pillitteri LJ, Fujii H, Yoshida Y, Bogenschutz NL, Takabayashi J, Zhu JK, Torii KU (2008) SCREAM/ICE1 and SCREAM2 specify three cell-state transitional steps leading to Arabidopsis stomatal differentiation. Plant Cell 20(7):1775–1785

Kang X, Zhao L, Liu X (2023) Calcium signaling and the response to heat shock in crop plants. Int J Mol Sci 25(1):324

Kardailsky I, Shukla VK, Ahn JH, Dagenais N, Christensen SK, Nguyen JT, Chory J, Harrison MJ, Weigel D (1999) Activation tagging of the floral inducer FT. Science 286(5446):1962–1965

Karlsson LM, Tamado T, Milberg P (2008) Inter-species comparison of seed dormancy and germination of six annual Asteraceae weeds in an ecological context. Seed Sci Res 18(1):35–45

Kaur A, Vishwakarma H, Maibam A, Padaria JC (2018) Cloning, characterization and in silico studies on abiotic stress responsive Hsp17. 9 from Prosopis cineraria. Indian J Plant Physiol 23: 731–740

Khalid MF, Shafqat W, Khan RI, Jawaid MZ, Hussain S, Saqib M, Rizwan M, Ahmed T (2024) Unveiling the resilience mechanism: strigolactones as master regulators of plant responses to abioti stresses. Plant Stress 100490

Kim JM, Sasaki T, Ueda M, Sako K, Seki M (2015) Chromatin changes in response to drought, salinity, heat, and cold stresses in plants. Front Plant Sci 6:114

Kim S, Hwang G, Kim S, Thi TN, Kim H, Jeong J, Kim J, Kim J, Choi G, Oh E (2020) The epidermis coordinates thermoresponsive growth through the phyB-PIF4-auxin pathway. Nat Commun 11(1):1053

Kishitani S, Watanabe K, Yasuda S, Arakawa K, Takabe TJPC (1994) Accumulation of glycinebetaine during cold acclimation and freezing tolerance in leaves of winter and spring barley plants. Plant Cell Environ 17(1):89–95

Knight CA, Hallett J, DeVries AL (1988) Solute effects on ice recrystallization: an assessment technique. Cryobiology 25(1):55–60

Knight H, Zarka DG, Okamoto H, Thomashow MF, Knight MR (2004) Abscisic acid induces CBF gene transcription and subsequent induction of cold-regulated genes via the CRT promoter element. Plant Physiol 135(3):1710–1717

Kobayashi Y, Kaya H, Goto K, Iwabuchi M, Araki T (1999) A pair of related genes with antagonistic roles in mediating flowering signals. Science 286(5446):1960–1962

Kocsy G, Galiba G, Brunold C (2001) Role of glutathione in adaptation and signalling during chilling and cold acclimation in plants. Physiol Plant 113(2):158–164

Kostaki KI, Coupel-Ledru A, Bonnell VC, Gustavsson M, Sun P, McLaughlin FJ, Fraser DP, McLachlan DH, Hetherington AM, Dodd AN, Franklin KA (2020) Guard cells integrate light and temperature signals to control stomatal aperture. Plant Physiol 182(3):1404–1419

Kraus TE, McKersie BD, Fletcher RA (1995) Paclobutrazol-induced tolerance of wheat leaves to paraquat may involve increased antioxidant enzyme activity. J Plant Physiol 145(4):570–576

Kumar S, Kaur R, Kaur N, Bhandhari K, Kaushal N, Gupta K, Bains TS, Nayyar H (2011) Heat-stress induced inhibition in growth and chlorosis in mungbean (Phaseolus aureus Roxb.) is partly mitigated by ascorbic acid application and is related to reduction in oxidative stress. Acta Physiol Plant 33:2091–2101

Kumar SV, Lucyshyn D, Jaeger KE, Alós E, Alvey E, Harberd NP, Wigge PA (2012) Transcription factor PIF4 controls the thermosensory activation of flowering. Nature 484(7393):242–245

Kuznetsov VV, Rakitin VY, Zholkevich VN (1999) Effects of preliminary heat-shock treatment on accumulation of osmolytes and drought resistance in cotton plants during water deficiency. Physiol Plant 107(4):399–406

Lämke J, Brzezinka K, Altmann S, Bäurle I (2016) A hit-and-run heat shock factor governs sustained histone methylation and transcriptional stress memory. EMBO J 35(2):162–175

Lau OS, Song Z, Zhou Z, Davies KA, Chang J, Yang X, Wang S, Lucyshyn D, Tay IHZ, Wigge PA, Bergmann DC (2018) Direct control of SPEECHLESS by PIF4 in the high-temperature response of stomatal development. Curr Biol 28(8):1273–1280

Lee H, Suh SS, Park E, Cho E, Ahn JH, Kim SG, Lee JS, Kwon YM, Lee I (2000) The AGAMOUS-LIKE 20 MADS domain protein integrates floral inductive pathways in Arabidopsis. Genes Dev 14(18):2366–2376

Lee HJ, Jung JH, Cortés Llorca L, Kim SG, Lee S, Baldwin IT, Park CM (2014) FCA mediates thermal adaptation of stem growth by attenuating auxin action in Arabidopsis. Nat Commun 5(1):5473

Legris M, Klose C, Burgie ES, Rojas CCR, Neme M, Hiltbrunner A, Wigge PA, Schäfer E, Vierstra RD, Casal JJ (2016) Phytochrome B integrates light and temperature signals in Arabidopsis. Science 354(6314):897–900

Li Y, Han D, Hu G, Sommerfeld M, Hu Q (2010) Inhibition of starch synthesis results in overproduction of lipids in Chlamydomonas reinhardtii. Biotechnol Bioeng 107(2):258–268

Li S, Fu Q, Chen L, Huang W, Yu D (2011) Arabidopsis thaliana WRKY25, WRKY26, and WRKY33 coordinate induction of plant thermotolerance. Planta 233:1237–1252

Liu X, Huang B (2000) Heat stress injury in relation to membrane lipid peroxidation in creeping bentgrass. Crop Sci 40(2):503–510

Liu HT, Huang WD, Pan QH, Weng FH, Zhan JC, Liu Y, Wan SB, Liu YY (2006) Contributions of PIP2-specific-phospholipase C and free salicylic acid to heat acclimation-induced thermotolerance in pea leaves. J Plant Physiol 163(4):405–416

Liu J, Sun N, Liu M, Liu J, Du B, Wang X, Qi X (2013) An autoregulatory loop controlling Arabidopsis HsfA2 expression: role of heat shock-induced alternative splicing. Plant Physiol 162(1):512–521

Liu J, Feng L, Li J, He Z (2015) Genetic and epigenetic control of plant heat responses. Front Plant Sci 6:267
Liu Y, Zhang M, Meng Z, Wang B, Chen M (2020) Research progress on the roles of cytokinin in plant response to stress. Int J Mol Sci 21(18):6574
Livanos P, Galatis B, Apostolakos P (2014) The interplay between ROS and tubulin cytoskeleton in plants. Plant Signal Behav 9(3):e28069
Ljung K, Nemhauser JL, Perata P (2015) New mechanistic links between sugar and hormone signalling networks. Curr Opin Plant Biol 25:130–137
Ma D, Li X, Guo Y, Chu J, Fang S, Yan C, Noel JP, Liu H (2016) Cryptochrome 1 interacts with PIF4 to regulate high temperature-mediated hypocotyl elongation in response to blue light. Proc Natl Acad Sci 113(1):224–229
Mahan JR, Burke JJ, Orzech KA (1990) Thermal dependence of the apparent K m of glutathione reductases from three plant species. Plant Physiol 93(2):822–824
Manasa LS, Panigrahy M, Panigrahi KC, Mishra G, Panda SK, Rout GR (2023) Cold tolerance mechanisms in mungbean (Vigna radiata L.) genotypes during germination. Agriculture 13(2): 315
Marentes E, Griffith M, Mlynarz A, Brush RA (1993) Proteins accumulate in the apoplast of winter rye leaves during cold acclimation. Physiol Plant 87(4):499–507
Martinière A, Shvedunova M, Thomson AJ, Evans NH, Penfield S, Runions J, McWatters HG (2011) Homeostasis of plasma membrane viscosity in fluctuating temperatures. New Phytol 192(2):328–337
Martinoia E, Meyer S, De Angeli A, Nagy R (2012) Vacuolar transporters in their physiological context. Annu Rev Plant Biol 63(1):183–213
Martins S, Montiel-Jorda A, Cayrel A, Huguet S, Roux CPL, Ljung K, Vert G (2017) Brassinosteroid signaling-dependent root responses to prolonged elevated ambient temperature. Nat Commun 8(1):309
Miao R, Li C, Liu Z, Zhou X, Chen S, Zhang D, Luo J, Tang W, Wang C, Wu J, Chen Z (2024) The role of endogenous brassinosteroids in the mechanisms regulating plant reactions to various abiotic stresses. Agronomy 14(2):356
Michaels SD, Amasino RM (1999) FLOWERING LOCUS C encodes a novel MADS domain protein that acts as a repressor of flowering. Plant Cell 11(5):949–956
Michaels SD, Amasino RM (2000) Memories of winter: vernalization and the competence to flower. Plant Cell Environ 23(11):1145–1153
Milbau A, Graae BJ, Shevtsova A, Nijs I (2009) Effects of a warmer climate on seed germination in the subarctic. Ann Bot 104(2):287–296
Min L, Li Y, Hu Q, Zhu L, Gao W, Wu Y, Ding Y, Liu S, Yang X, Zhang X (2014) Sugar and auxin signaling pathways respond to high-temperature stress during anther development as revealed by transcript profiling analysis in cotton. Plant Physiol 164(3):1293–1308
Mishkind M, Vermeer JE, Darwish E, Munnik T (2009) Heat stress activates phospholipase D and triggers PIP2 accumulation at the plasma membrane and nucleus. Plant J 60(1):10–21
Mizuno T, Kitayama M, Takayama C, Yamashino T (2015) Insight into a physiological role for the EC night-time repressor in the Arabidopsis circadian clock. Plant Cell Physiol 56(9):1738–1747
Mondal S, Karmakar S, Panda D, Pramanik K, Bose B, Singhal RK (2023) Crucial plant processes under heat stress and tolerance through heat shock proteins. Plant Stress 100227
Morimoto RI (1998) Regulation of the heat shock transcriptional response: cross talk between a family of heat shock factors, molecular chaperones, and negative regulators. Genes Dev 12(24): 3788–3796
Myster J, Moe R (1995) Effect of diurnal temperature alternations on plant morphology in some greenhouse crops—a mini review. Sci Hortic 62(4):205–215
Nagel DH, Pruneda-Paz JL, Kay SA (2014) FBH1 affects warm temperature responses in the Arabidopsis circadian clock. Proc Natl Acad Sci 111(40):14595–14600
Nakano Y, Asada K (1981) Hydrogen peroxide is scavenged by ascorbate-specific peroxidase in spinach chloroplasts. Plant Cell Physiol 22(5):867–880

Narayanan S, Prasad PV, Welti R (2016) Wheat leaf lipids during heat stress: II. Lipids experiencing coordinated metabolism are detected by analysis of lipid co-occurrence. Plant Cell Environ 39(3):608–617

Nieto C, López-Salmerón V, Davière JM, Prat S (2015) ELF3-PIF4 interaction regulates plant growth independently of the evening complex. Curr Biol 25(2):187–193

Nievola CC, Carvalho CP, Carvalho V, Rodrigues E (2017) Rapid responses of plants to temperature changes. Temperature 4(4):371–405

Novillo F, Alonso JM, Ecker JR, Salinas J (2004) CBF2/DREB1C is a negative regulator of CBF1/DREB1B and CBF3/DREB1A expression and plays a central role in stress tolerance in Arabidopsis. Proc Natl Acad Sci 101(11):3985–3990

Nusinow DA, Helfer A, Hamilton EE, King JJ, Imaizumi T, Schultz TF, Farré EM, Kay SA (2011) The ELF4–ELF3–LUX complex links the circadian clock to diurnal control of hypocotyl growth. Nature 475(7356):398–402

Ohama N, Sato H, Shinozaki K, Yamaguchi-Shinozaki K (2017) Transcriptional regulatory network of plant heat stress response. Trends Plant Sci 22(1):53–65

Ohlrogge J, Browse J (1995) Lipid biosynthesis. Plant Cell 7(7):957

Omae H, Kumar A, Shono M (2012) Adaptation to high temperature and water deficit in the common bean (Phaseolus vulgaris L.) during the reproductive period. J Bot 2012(1):803413

Onouchi H, Igeño MI, Périlleux C, Graves K, Coupland G (2000) Mutagenesis of plants overexpressing CONSTANS demonstrates novel interactions among Arabidopsis flowering-time genes. Plant Cell 12(6):885–900

Ortiz C, Cardemil L (2001) Heat-shock responses in two leguminous plants: a comparative study. J Exp Bot 52(361):1711–1719

Pastori GM, Trippi VS (1995) Fatty acid composition in water-and oxygen-stressed leaves of maize and wheat strains. Phytochemistry 40(1):45–48

Patel D, Franklin KA (2009) Temperature-regulation of plant architecture. Plant Signal Behav 4(7): 577–579

Patra N, Hariharan S, Gain H, Maiti MK, Das A, Banerjee J (2021) TypiCal but DeliCate Ca++ re: dissecting the essence of calcium signaling network as a robust response coordinator of versatile abiotic and biotic stimuli in plants. Front Plant Sci 12:752246

Pecinka A, Dinh HQ, Baubec T, Rosa M, Lettner N, Scheid OM (2010) Epigenetic regulation of repetitive elements is attenuated by prolonged heat stress in Arabidopsis. Plant Cell 22(9): 3118–3129

Penfield S (2008) Temperature perception and signal transduction in plants. New Phytol 179(3): 615–628

Plieth C, Hansen UP, Knight H, Knight MR (1999) Temperature sensing by plants: the primary characteristics of signal perception and calcium response. Plant J 18(5):491–497

Pompeiano A, Vita F, Miele S, Guglielminetti L (2015) Freeze tolerance and physiological changes during cold acclimation of giant reed [A rundo donax (L.)]. Grass Forage Sci 70(1):168–175

Prasad PV, Boote KJ, Allen LH Jr (2006) Adverse high temperature effects on pollen viability, seed-set, seed yield and harvest index of grain-sorghum [Sorghum bicolor (L.) Moench] are more severe at elevated carbon dioxide due to higher tissue temperatures. Agric For Meteorol 139(3–4):237–251

Pritchard HW (2000) Review of seeds. Ecology, biogeography, and evolution of dormancy and germination, by C Baskin & J Baskin. Ann Bot 86(3):705–707

Puhakainen T, Hess MW, Mäkelä P, Svensson J, Heino P, Palva ET (2004) Overexpression of multiple dehydrin genes enhances tolerance to freezing stress in Arabidopsis. Plant Mol Biol 54: 743–753

Qian Y, Ren Q, Zhang J, Chen L (2019) Transcriptomic analysis of the maize (Zea mays L.) inbred line B73 response to heat stress at the seedling stage. Gene 692:68–78

Qin F, Sakuma Y, Li J, Liu Q, Li YQ, Shinozaki K, Yamaguchi-Shinozaki K (2004) Cloning and functional analysis of a novel DREB1/CBF transcription factor involved in cold-responsive gene expression in Zea mays L. Plant Cell Physiol 45(8):1042–1052

Qiu Y, Li M, Kim RJA, Moore CM, Chen M (2019) Daytime temperature is sensed by phytochrome B in Arabidopsis through a transcriptional activator HEMERA. Nat Commun 10(1):140

Qiu Y, Pasoreck EK, Yoo CY, He J, Wang H, Bajracharya A, Li M, Larsen HD, Cheung S, Chen M (2021) RCB initiates Arabidopsis thermomorphogenesis by stabilizing the thermoregulator PIF4 in the daytime. Nat Commun 12(1):2042

Qureshi H, Abbas MH, Jan T, Mumtaz K, Mukhtar H, Khan U (2022) Plant responses to heat stress. J Adv Nutr Sci Technol (ANST) 2(2)

Ren R, Jiang X, Di W, Li Z, Li B, Xu J, Liu Y (2019) HSP70 improves the viability of cryopreserved Paeonia lactiflora pollen by regulating oxidative stress and apoptosis-like programmed cell death events. Plant Cell Tissue Organ Cult (PCTOC) 139:53–64

Rensburg LV, Krüger GHJ (1994) Evaluation of components of oxidative stress metabolism for use in selection of drought tolerant cultivars of Nicotiana tabacum L. J Plant Physiol 143(6):730–737

Reynolds-Henne CE, Langenegger A, Mani J, Schenk N, Zumsteg A, Feller U (2010) Interactions between temperature, drought and stomatal opening in legumes. Environ Exp Bot 68(1):37–43

Roberts EH (1988) Temperature and seed germination. Symp Soc Exp Biol 42:109–132

Rodríguez M, Canales E, Borrás-Hidalgo O (2005) Molecular aspects of abiotic stress in plants. Biotecnol Apl 22(1):1–10

Roitsch T (1999) Source-sink regulation by sugar and stress. Curr Opin Plant Biol 2(3):198–206

Roychoudhury A, Banerjee A (2016) Endogenous glycine betaine accumulation mediates abiotic stress tolerance in plants. Trop Plant Res 3(1):105–111

Roychoudhury A, Banerjee A (2017) Abscisic acid signaling and involvement of mitogen activated protein kinases and calcium-dependent protein kinases during plant abiotic stress. Mechanism Plant Hormone Signal Under Stress 1:197–241

Roychoudhury A, Paul S, Basu S (2013) Cross-talk between abscisic acid-dependent and abscisic acid-independent pathways during abiotic stress. Plant Cell Rep 32:985–1006

Sairam RK, Deshmukh PS, Shukla DS (1997) Tolerance of drought and temperature stress in relation to increased antioxidant enzyme activity in wheat. J Agron Crop Sci 178(3):171–178

Salomé PA, Weigel D, McClung CR (2010) The role of the Arabidopsis morning loop components CCA1, LHY, PRR7, and PRR9 in temperature compensation. Plant Cell 22(11):3650–3661

Sánchez B, Rasmussen A, Porter JR (2014) Temperatures and the growth and development of maize and rice: a review. Glob Chang Biol 20(2):408–417

Sandalio LM, Romero-Puertas MC (2015) Peroxisomes sense and respond to environmental cues by regulating ROS and RNS signalling networks. Ann Bot 116(4):475–485

Sangwan V, Örvar BL, Beyerly J, Hirt H, Dhindsa RS (2002) Opposite changes in membrane fluidity mimic cold and heat stress activation of distinct plant MAP kinase pathways. Plant J 31(5):629–638

Sangwan S, Shameem N, Yashveer S, Tanwar H, Parray JA, Jatav HS, Sharma S, Punia H, Sayyed RZ, Almalki WH, Poczai P (2022) Role of salicylic acid in combating heat stress in plants: insights into modulation of vital processes. Front Biosci Landmark 27(11):310

Savitch LV, Allard G, Seki M, Robert LS, Tinker NA, Huner NP, Shinozaki K, Singh J (2005) The effect of overexpression of two Brassica CBF/DREB1-like transcription factors on photosynthetic capacity and freezing tolerance in Brassica napus. Plant Cell Physiol 46(9):1525–1539

Schramm F, Larkindale J, Kiehlmann E, Ganguli A, Englich G, Vierling E, Von Koskull-Döring P (2008) A cascade of transcription factor DREB2A and heat stress transcription factor HsfA3 regulates the heat stress response of Arabidopsis. Plant J 53(2):264–274

Sharp FR, Massa SM, Swanson RA (1999) Heat-shock protein protection. Trends Neurosci 22(3):97–99

Sheldon CC, Burn JE, Perez PP, Metzger J, Edwards JA, Peacock WJ, Dennis ES (1999) The FLF MADS box gene: a repressor of flowering in Arabidopsis regulated by vernalization and methylation. Plant Cell 11(3):445–458

Sheldon CC, Rouse DT, Finnegan EJ, Peacock WJ, Dennis ES (2000) The molecular basis of vernalization: the central role of FLOWERING LOCUS C (FLC). Proc Natl Acad Sci 97(7): 3753–3758

Sheldon CC, Finnegan EJ, Dennis ES, Peacock WJ (2006) Quantitative effects of vernalization on FLC and SOC1 expression. Plant J 45(6):871–883

Shi Y, Yang S (2014) ABA regulation of the cold stress response in plants. In: Abscisic acid: metabolism, transport and signaling, pp 337–363

Shockey J, Regmi A, Cotton K, Adhikari N, Browse J, Bates PD (2016) Identification of Arabidopsis GPAT9 (At5g60620) as an essential gene involved in triacylglycerol biosynthesis. Plant Physiol 170(1):163–179

Silva CS, Nayak A, Lai X, Hutin S, Hugouvieux V, Jung JH, López-Vidriero I, Franco-Zorrilla JM, Panigrahi KC, Nanao MH, Wigge PA (2020) Molecular mechanisms of evening complex activity in Arabidopsis. Proc Natl Acad Sci 117(12):6901–6909

Simpson DJ, Smallwood M, Twigg S, Doucet CJ, Ross J, Bowles DJ (2005) Purification and characterisation of an antifreeze protein from Forsythia suspensa (L.). Cryobiology 51(2): 230–234

Smertenko A, Dráber P, Viklický V, Opatrný Z (1997) Heat stress affects the organization of microtubules and cell division in Nicotiana tabacum cells. Plant Cell Environ 20(12):1534–1542

Sobol S, Chayut N, Nave N, Kafle D, Hegele M, Kaminetsky R, Wuensche JN, Samach A (2014) Genetic variation in yield under hot ambient temperatures spotlights a role for cytokinin in protection of developing floral primordia. Plant Cell Environ 37(3):643–657

Sohn SO, Back K (2007) Transgenic rice tolerant to high temperature with elevated contents of dienoic fatty acids. Biol Plant 51:340–342

Song J, Yang J, Jeong BR (2023) Difference between day and night temperature (DIF) and light intensity affect growth and photosynthetic characteristics of Panax ginseng meyer sprouts. Plants 12(15):2820

Srivastava A, Pandey GC (2023) Role of cytokinins and gibberellins in crops response to heat and drought stress. J Cereal Res 15(2)

Steponkus PL (1984) Role of the plasma membrane in freezing injury and cold acclimation. Annu Rev Plant Physiol 35:543–584

Strenkert D, Schmollinger S, Sommer F, Schulz-Raffelt M, Schroda M (2011) Transcription factor–dependent chromatin remodeling at heat shock and copper-responsive promoters in Chlamydomonas reinhardtii. Plant Cell 23(6):2285–2301

Sung S, Amasino RM (2004) Vernalization in Arabidopsis thaliana is mediated by the PHD finger protein VIN3. Nature 427(6970):159–164

Suri SS, Dhindsa RS (2008) A heat-activated MAP kinase (HAMK) as a mediator of heat shock response in tobacco cells. Plant Cell Environ 31(2):218–226

Suzuki N, Sejima H, Tam R, Schlauch K, Mittler R (2011) Identification of the MBF1 heat-response regulon of Arabidopsis thaliana. Plant J 66(5):844–851

Szymańska R, Ślesak I, Orzechowska A, Kruk J (2017) Physiological and biochemical responses to high light and temperature stress in plants. Environ Exp Bot 139:165–177

Taiz L, Zeiger E (2006) Plant physiology. Sinauer Associates Inc, Sunderland

Tan W, Chen J, Yue X, Chai S, Liu W, Li C, Yang F, Gao Y, Gutiérrez Rodríguez L, Resco de Dios V, Zhang D (2023) The heat response regulators HSFA1s promote Arabidopsis thermomorphogenesis via stabilizing PIF4 during the day. Sci Adv 9(44):eadh1738

Tao Z, Yan P, Zhang X, Wang D, Wang Y, Ma X, Yang Y, Liu X, Chang X, Sui P, Chen Y (2022) Physiological mechanism of abscisic acid-induced heat-tolerance responses to cultivation techniques in wheat and maize. Agronomy 12(7):1579

Teige M, Scheikl E, Eulgem T, Doczi R, Ichimura K, Shinozaki K, Dangl JL, Hirt H (2004) The MKK2 pathway mediates cold and salt stress signaling in Arabidopsis. Mol Cell 15(1):141–152

Thingnaes E, Torre S, Ernstsen A, Moe R (2003) Day and night temperature responses in Arabidopsis: effects on gibberellin and auxin content, cell size, morphology and flowering time. Ann Bot 92(4):601–612

Thomas SE, Balcerowicz M, Chung BYW (2022) RNA structure mediated thermoregulation: what can we learn from plants? Front Plant Sci 13:938570

Thomashow MF (1999) Plant cold acclimation: freezing tolerance genes and regulatory mechanisms. Annu Rev Plant Biol 50(1):571–599

Thomson J (1936) Vernalisation. Sci Progr (1933) 30(120):644–651

Tian F, Hu XL, Yao T, Yang X, Chen JG, Lu MZ, Zhang J (2021) Recent advances in the roles of HSFs and HSPs in heat stress response in woody plants. Front Plant Sci 12:704905

Toh S, Imamura A, Watanabe A, Nakabayashi K, Okamoto M, Jikumaru Y, Hanada A, Aso Y, Ishiyama K, Tamura N, Iuchi S (2008) High temperature-induced abscisic acid biosynthesis and its role in the inhibition of gibberellin action in Arabidopsis seeds. Plant Physiol 146(3): 1368–1385

Toledo-Ortiz G, Johansson H, Lee KP, Bou-Torrent J, Stewart K, Steel G, Rodríguez-Concepción M, Halliday KJ (2014) The HY5-PIF regulatory module coordinates light and temperature control of photosynthetic gene transcription. PLoS Genet 10(6):e1004416

Ullah A, Nadeem F, Nawaz A, Siddique KH, Farooq M (2022) Heat stress effects on the reproductive physiology and yield of wheat. J Agron Crop Sci 208(1):1–17

Vaultier MN, Cantrel C, Vergnolle C, Justin AM, Demandre C, Benhassaine-Kesri G, Çiçek D, Zachowski A, Ruelland E (2006) Desaturase mutants reveal that membrane rigidification acts as a cold perception mechanism upstream of the diacylglycerol kinase pathway in Arabidopsis cells. FEBS Lett 580(17):4218–4223

Vertucci CW (1989) The kinetics of seed imbibition: controlling factors and relevance to seedling vigor. Seed Moist 14:93–115

Wahid A, Gelani S, Ashraf M, Foolad MR (2007) Heat tolerance in plants: an overview. Environ Exp Bot 61(3):199–223

Wang M, Fan X, Ding F (2023) Jasmonate: a hormone of primary importance for temperature stress response in plants. Plants 12(24):4080

Welti R, Shah J, Li WQ, Li MY, Chen JP, Burke JJ, Fauconnier ML, Chapman K, Chye ML, Wang XM (2007) Plant lipidomics: discerning biological function by profiling plant complex lipids using mass spectrometry. Front Biosci 12

Went FW (1944) Plant growth under controlled conditions. II. Thermoperiodicity in growth and fruiting of the tomato. Am J Bot:135–150

Wilson RA, Sangha MK, Banga SS, Atwal AK, Gupta S (2014) Heat stress tolerance in relation to oxidative stress and antioxidants in Brassica juncea. J Environ Biol 35(2):383

Wisniewski M, Fuller M (1999) Ice nucleation and deep supercooling in plants: new insights using infrared thermography. In: Wisniewski M, Fuller M (eds) Cold-adapted organisms: ecology, physiology, enzymology and molecular biology. Springer, Berlin/Heidelberg, pp 105–118

Xin Z, Browse J (1998) Eskimo1 mutants of Arabidopsis are constitutively freezing-tolerant. Proc Natl Acad Sci 95(13):7799–7804

Yaish MW, Doxey AC, McConkey BJ, Moffatt BA, Griffith M (2006) Cold-active winter rye glucanases with ice-binding capacity. Plant Physiol 141(4):1459–1472

Yamaguchi-Shinozaki K, Shinozaki K (2005) Organization of cis-acting regulatory elements in osmotic-and cold-stress-responsive promoters. Trends Plant Sci 10(2):88–94

Yamashino T, Nomoto Y, Lorrain S, Miyachi M, Ito S, Nakamichi N, Fankhauser C, Mizuno T (2013) Verification at the protein level of the PIF4-mediated external coincidence model for the temperature-adaptive photoperiodic control of plant growth in Arabidopsis thaliana. Plant Signal Behav 8(3):e23390

Yang DS, Hon WC, Bubanko S, Xue Y, Seetharaman J, Hew CL, Sicheri F (1998) Identification of the ice-binding surface on a type III antifreeze protein with a "flatness function" algorithm. Biophys J 74(5):2142–2151

Yang LE, Peng DL, Li ZM, Huang L, Yang J, Sun H (2020) Cold stratification, temperature, light, GA3, and KNO3 effects on seed germination of Primula beesiana from Yunnan, China. Plant Divers 42(3):168–173

Yang J, Qu X, Ji L, Li G, Wang C, Wang C, Zhang Y, Zheng L, Li W, Zheng X (2022a) PIF4 promotes expression of HSFA2 to enhance basal thermotolerance in Arabidopsis. Int J Mol Sci 23(11):6017

Yang X, Yu H, Duncan S, Zhang Y, Cheema J, Liu H, Benjamin Miller J, Zhang J, Kwok CK, Zhang H, Ding Y (2022b) RNA G-quadruplex structure contributes to cold adaptation in plants. Nat Commun 13(1):6224

Yoshida M, Abe J, Moriyama M, Shimokawa S, Nakamura Y (1997) Seasonal changes in the physical state of crown water associated with freezing tolerance in winter wheat. Physiol Plant 99(3):363–370

Yoshida M, Abe J, Moriyama M, Kuwabara T (1998) Carbohydrate levels among winter wheat cultivars varying in freezing tolerance and snow mold resistance during autumn and winter. Physiol Plant 103(1):8–16

Yoshida T, Ohama N, Nakajima J, Kidokoro S, Mizoi J, Nakashima K, Maruyama K, Kim JM, Seki M, Todaka D, Osakabe Y (2011) Arabidopsis HsfA1 transcription factors function as the main positive regulators in heat shock-responsive gene expression. Mol Gen Genomics 286: 321–332

Yuan P, Yang T, Poovaiah BW (2018) Calcium signaling-mediated plant response to cold stress. Int J Mol Sci 19(12):3896

Zhang X, Fowler SG, Cheng H, Lou Y, Rhee SY, Stockinger EJ, Thomashow MF (2004) Freezing-sensitive tomato has a functional CBF cold response pathway, but a CBF regulon that differs from that of freezing-tolerant Arabidopsis. Plant J 39(6):905–919

Zhu J, Zhang KX, Wang WS, Gong W, Liu WC, Chen HG, Xu HH, Lu YT (2015) Low temperature inhibits root growth by reducing auxin accumulation via ARR1/12. Plant Cell Physiol 56(4): 727–736

Zhu JY, Oh E, Wang T, Wang ZY (2016) TOC1–PIF4 interaction mediates the circadian gating of thermoresponsive growth in Arabidopsis. Nat Commun 7(1):13692

Zoulias N, Harrison EL, Casson SA, Gray JE (2018) Molecular control of stomatal development. Biochem J 475(2):441–454

Chapter 9
Conservation and Diversification of the Cold Stress Response in Higher Plants

Arabindu Debbarma, Rakhi Chaturvedi, Kyonoshin Maruyama, and Yoshiharu Y. Yamamoto

Abstract Cold stress poses a significant challenge to plant development, leading to significant reductions in crop yield. While a few plants that are temperate, exemplified by *Arabidopsis thaliana*, exhibit internal adjustments to maintain metabolic machinery during cold temperatures through a procedure called "cold acclimation", plants that are tropical or subtropical, like rice, show sensitivity to chilling stress. Temperate plants must first undergo cold acclimation, which involves to be exposed to low, freezing temperatures in order to prevent damage during subsequent freezing stress and to ensure the plant's general health under low temperatures. Intricate physiological, biochemical, and molecular systems have been developed by plants to detect and react to low temperatures. Notably, two essential elements of this response are cytoskeletal rearrangement and membrane modifications. Calcium-binding proteins and calcium-dependent protein kinases are activated by the little increase in cytosolic calcium levels, which adds to the complex network of regulation during low-temperature stress. Furthermore, the plant response to cold stress is greatly affected by epigenetic regulation. This review compiles the most recent findings and advancements in our understanding of the tactics used by plants to

A. Debbarma
The United Graduate School of Agricultural Science, Gifu University, Gifu, Japan

Department of Bioscience and Bioengineering, Indian Institute of Technology Guwahati, Guwahati, India

R. Chaturvedi
The United Graduate School of Agricultural Science, Gifu University, Gifu, Japan

K. Maruyama
Japan International Research Center for Agricultural Sciences (JIRCAS), Tsukuba, Ibaraki, Japan

Y. Y. Yamamoto (✉)
The United Graduate School of Agricultural Science, Gifu University, Gifu, Japan

Faculty of Applied Biological Sciences, Gifu University, Gifu, Japan

RIKEN CSRS, Yokohama, Japan
e-mail: yamamoto.yoshiharu.h5@f.gifu-u.ac.jp

S. K. Panda, M. A. Khan (eds.), *Plant Functional Genomics for Abiotic Stress Resilience*, Plant in Challenging Environments 7,
https://doi.org/10.1007/978-3-032-01704-8_9

withstand low temperatures. By examining the conservation and diversification of the cold stress response in higher plants, the main aim of the review is to provide valuable insights into the intricate molecular networks governing plant adaptation to cold environments. The integration of physiological, biochemical, and molecular perspectives enhances our understanding of the adaptive mechanisms that contribute to plant resilience under challenging low-temperature conditions.

Keywords Cold acclimation · Conservation · Diversification · Epigenetics · Freezing stress

9.1 Introduction

Abiotic stressors that negatively impact growth and productivity includes heat and cold stress (CS), drought, heavy metal toxicity, nutrient shortage, acidity and salinity of the soil, and structure, characteristics, biochemical, and genetics changes in plants (Imran et al. 2021). Temperature effects on the growth of plants differs between species or populations, since their reactions to different temperature ranges vary widely (Shi and Yang 2014). Certain plants native to tropical or subtropical regions, accustomed to warmer temperatures, might experience cold injuries or stress when exposed to cooler temperatures below 15 °C. Conversely, Antarctic algae, adapted to extremely cold environments, might exhibit signs of heat stress or physiological disruption when temperatures rise above 5 °C (Guy et al. 1985).

CS is the substantial abiotic stresses that significantly affect crop productivity as well as the geographic distribution of many plant species. It can inflict fatal harm, significantly impacting their productivity and survival rates. Low-temperature stress impacts various facets of plant life, encompassing germination, growth, development, and reproductive processes. In general, plants face chilling (0–15 °C) and freezing (>0 °C) as the most common CS (Albertos et al. 2019; Pan et al. 2020). In temperate zones, rice is especially susceptible to chilling stress, especially in high-elevation areas and at high latitudes. Rice seedlings that are exposed for more than four days to temperatures below freezing experience reduced germination, slowed growth, and occasionally even death. Furthermore, during times of chilling stress, researchers have observed a widespread outbreak of rice blast disease, resulting in significant losses in rice production. Sensitive plants exposed to chilling injury often exhibit observable signs such as chlorosis (yellowing of leaves), rigidity or stiffness in seedlings, wilting, and even the occasional seedling death. These observable phenotypes are helpful markers for differentiating between plant varieties that are sensitive and tolerance to chilling (Xie et al. 2023).

To acclimate amidst profoundly challenging environmental circumstances plants have evolved efficient survival mechanisms. Numerous temperate plants, including *Arabidopsis thaliana*, rapeseed, wheat, and rye, have evolved mechanisms to sense and react to freezing stress through a mechanism called cold acclimation. Cold acclimation necessitates low exposure, but not subject to freezing temperatures, which triggers the plant capability to cope with subsequent CS. This adaptive mechanism helps the plant maintain its overall health and functionality despite the unfavourable low-temperature conditions (Huang et al. 2017). Perennial plants, such as trees, perceive seasonal changes by monitoring alterations in both photoperiod (day length) and temperature through the year. These plants go from active growth to dormant phases and develop frost resistance for cold hardiness in response to signals from variations in day length and temperature. In temperate climates, this transformation is essential for woody plants to survive winter conditions. Perennial plants have morphological and physiological characteristics that enable them to synchronize with climatic rhythms, which helps them survive in harsh cold environments, unlike annual herbaceous plants (Sharma et al. 2022).

Since the initial discovery of cold tolerance genes by Guy et al. (1985) and their naming as *cold-regulated* (COR) genes (Hassan et al. 2021), substantial advancements have occurred in comprehending the molecular mechanisms underlying plants' reactions to CS. Recent studies have isolated and characterized, numerous genes that respond to freezing stress, highlighting their importance in both chill tolerance and cold acclimation processes. *The CBF/DREB1* pathway is among the most well-studied and important cold signalling pathways out of several others (Lv et al. 2020). Plant responses to stress are largely dependent on regulatory genes, particularly transcription factors (TFs). By inducing the expression of useful genes, their activation either directly or indirectly regulates the signal transduction network and genes that respond to stress. Osmo-regulators and anti-freeze proteins are direct regulators while proteins like chaperones, functional proteins, TFs, and kinases are indirect regulators (Miura and Furumoto 2013).

In order to select CS resistant crop cultivars, the availability of molecular markers assist quick screening, bypassing the need to wait for observable phenotypic changes. Compared to other stresses, the reactions of plants to CS are complex, what makes difficult to increase crop tolerance to cold. Nonetheless, over the previous 20 years, efforts have led to the creation of transgenic crop lines, exhibiting enhanced CS tolerance (Jiang et al. 2017). To achieve improved outcomes, it is crucial to utilize appropriate tools and genotypes that surpass existing varieties and are accessible to farmers (Aslam et al. 2022). This review covers emerging research areas, technological advancements, and tools for studying plant cold tolerance additionally to reviewing various genes and mechanisms participating in higher plants' adaptive responses to CS.

9.2 Cold Stress Resilience in Plants: Signalling Pathways, Small Molecules, Hormonal Dynamics, and Gene Regulation

9.2.1 *Cellular Adaptation and Membrane Dynamics in Cold Stress Resilience*

Biological membranes are essential for cell protection, homeostasis, and metabolism. The cell membranes composed of lipids, sterols, and proteins, collaborate and actively participate in cellular metabolism (Cacas et al. 2012). The cell membranes adaptability is frequently hinged due to cold temperature resilience. Metabolite homeostasis re-establishing occurs due to compositional changes in both lipids and membrane protein and represent a process whereby cells detect cold temperatures. The alterations induced by low temperatures in membrane physical conformation primarily increased unsaturated fatty acids enhancing membrane liquidity and solidity. This adaptation enables cells to mechanically adjust to the cold environment (Khodakovskaya et al. 2006). Proteins and membrane lipids are stabilized by the presence of higher amount of sugars, nitrogen compounds, and proteins to preserving membrane integrity in cold acclimation (Jiang et al. 2021).

When the plasma membrane stiffens during cold acclimation, it influences ion concentration and metabolite transport from apoplasts, endo-membranes, and organelles. This starts the cold signaling cascade downstream. Plants sense their surroundings and convert the signal into subsequent biological reactions by decreasing the fluidity of their membranes. This affects the cellular activities associated with the membranes. Calcium channels, a transmembrane protein complex, is primary pathway for calcium ions to enter cells, that are vital sensors for abiotic stress. Plants sense CS and then a complex cascade of signal transduction processes in the nucleus and cytosol begins. For intracellular signaling secondary messengers such as nitric oxide (NO), reactive oxygen species (ROS), and Ca^{2+} are necessary as well as for cell-to-cell communication during these events. The regulation of *CBF/DREB1* pathway, in response to CS, causes modifications in post-transcriptional and post-translational modifications and triggers subsequent downstream reactions (Liu et al. 2019).

The freezing process is initiated by accumulation of water molecules in plant cells, which causes stable ice nuclei at freezing temperatures. The process of ice formation starts at the nucleation site which is formed due to the assimilate of microscopic ice nuclei by formation of ice-nucleating proteins (INPs). These sites help water molecules align correctly, which in turn encourages the creation of extracellular ice crystals. Lack of ice-binding proteins (IBPs), big ice crystals developed in the apoplast that has potential to physically harm plasma membranes. Cell dehydration results from an osmotic gradient caused due to ice crystallization followed by isolation of intracellular water. Cell collapse or rupture may arise from this loss of cell volume (Denais et al. 2016). Through a process unique to plants called supercooling, avoidance mechanisms lower the intracellular freezing temperature by utilizing a range of cryoprotectant molecules. These molecules help prolong or prevent the formation of extracellular ice crystals, by keeping intracellular liquids cooler than zero (Bhandari and Nayyar 2013). By using IBPs as part of its survival strategy, certain plant species have created intricate systems that reduce the freeze damage. IBPs, also known as antifreeze proteins (AFPs) or ice recrystallization inhibition (IRI) proteins, are a class of low temperature-associated proteins found in a range of cold-adapted organisms, including plants like *Ammopiptanthus nanus* and *Loliurn perenne* (Wisniewski et al. 2020). Plant IBPs able to communicate with ice crystals because of their special structural characteristics and functional traits. Therefore, they cannot grow, and water molecules are unable to exit the intracellular compartment into the extracellular space. This binding ability keeps the cell from freezing and helps to prevent it from becoming dehydrated. It has been noted that these ice-binding proteins are diverse in multiple AFPs from different plant species. For example, *Fs*AFP from weeping forsythia (*Forsythia suspensa*) has sequence similarity to dehydrin, and *Pa*AFP from Norway spruce (*Picea abies*), *Ta*IRI1-2 from winter wheat (*Triticum aestivum*), *Ba*AFP from malting barley (*Hordeum vulgare* L.) have similarity to chitinase, thaumatin-like protein, and alpha-amylase inhibitor protein, respectively (Bredow and Walker 2017).

9.2.2 Role of Small Signalling Molecules in Responses to Cold Stress Adaptation in Plants

Small signalling molecules function as pivotal regulators in various stress conditions within plants, generated upon the detection of environmental stress via signal transduction pathways (Tuteja and Sopory 2008). These molecules are important since they transmit and amplify stress signals, start cascade reactions that regulate gene expression, osmoregulation, function of antioxidant enzymes, and communicate with other signal pathways. These signalling molecules under the influence of CS orchestrate diverse actions, mediating gene regulation, adjusting osmotic balance, modulating antioxidant enzyme functions, and engaging in crosstalk with other signalling pathways (Lamers et al. 2020). Identified signalling molecules such as ROS, Ca^{2+}, NO, cyclic guanosine monophosphate (cGMP), hydrogen sulfide (H_2S), and lipid-derived signals, notably diacylglycerol (DAG), phosphatidic acid (PA), sphingolipids, and inositol triphosphate (IP3), synergistically operate through crosstalk mechanisms (Barrero-Sicilia et al. 2017).

Secondary messenger Ca^{2+} is known to participate in several biological processes in eukaryotic cells, among which cold stress response is its major role plants. A strong influx of Ca^{2+} into the cell was caused by the activation of Ca^{2+} channels by the cold signal. The sensors that are activated by Ca2+ ions include calcineurin B-like proteins (CBLs), calmodulins (CaMs), CaM-like proteins (CMLs), and Ca^{2+}-dependent protein kinases (CPKs/CDPKs). These sensors further trigger the downstream process through phosphorylation, transcriptional reprogramming, and activation of MAPKs cascade (Yuan et al. 2018). Conversely, the constitutive suppression of MPK3 and MPK6 kinase activities and proteins is due to a *MEKk1-MKK2MPK4* cascade that also contributes positively to regulating temperature response (Zhao et al. 2017a, b). CS induced plasma membrane rigidification in alfalfa and *Brassica napus* causes cytoskeletal actin rearrangement, Ca^{2+} channel induction, and a rise in cytosolic Ca^{2+} levels. These circumstances cause cold acclimation and the expression of COR genes (Chinnusamy et al. 2010; Sangwan et al. 2001)

9.2.3 Effects of Phytohormones in Response to Cold Stress in Plant

Phytohormones performed a crucial task in helping CS tolerance by intricating physiological as well as biochemical mechanisms, to foster overall growth and development of the plant. Among these, abscisic acid (ABA) is pivotal in orchestrating physiological stress responses in plants during CS (De Smet et al. 2006). Normally, plants require low levels of ABA for growth, with excess ABA regulated via its endogenous biosynthesis. However, low temperatures induce osmotic stress due to decreased water availability in plant cells, prompting increased ABA production and biosynthesis. Endogenous ABA is transported from roots to leaves, leading to its accumulation in guard cells, resulting in stomatal closure, reduced transpiration rates, and suppressed cell growth (Cotelle and Leonhardt 2019).

Elevated levels of ABA within cold-tolerant plants serve as an intracellular signal, facilitating the process of creating fresh proteins during the process of cold acclimation. Furthermore, the application of exogenous ABA can enhance CS tolerance in various plants by triggering ABA signalling pathways. These pathways induce the transcription of antioxidant enzymes that aid in scavenging ROS, thus reducing CS-induced damage. ABA's role in CS signalling is regulated via pathways independent of *CBF* (Zhang et al. 2013). Plant responses to CS can be modulated through either ABA-dependent or ABA-independent pathways (Roychoudhury et al. 2013). The most commonly reported ABA-regulated genes are the *Late Embryogenesis Abundant (LEA)* or *LEA*-like genes (Banerjee and Roychoudhury 2016). Abscisic acid responsive element (*ABRE*) is an 8 bp cis-acting element with the ACGT core sequence in the middle. The *Em1a* element (GGACACGTGGC) of the *Em* (*Early-Methionine-labelled*) gene is recognized as a potent *ABRE* (Guiltinan et al. 1990).

9.2.4 Function of ABA in Cold Stress Response

According to Xue-Xuan et al. (2010), ABA is a significant regulator of growth and development, impacting a variety of processes including leaf senescence, seed dormancy and germination, cell division, and elongation. Since it causes stomatal closure and osmolyte accumulation in plants under environmental stresses like drought and chilling stress. According to reports, ABA also serves as the primary phytohormone involved in stress signalling (Daszkowska-Golec and Szarejko 2013). ABA must be essential for cold acclimatization because of the brief rise in ABA during cold stress and the application of ABA's ability to increase freezing tolerance (Xue-Xuan et al. 2010). An effective way to combat winter temperatures is through cold stress-induced leaf senescence, which is brought on by ABA and cytokinins participates in preventing the breakdown of chlorophyll.

The ratio of ABA to cytokinins (ABA/CKs) in senescent *Iris pseudocorus* leaves increased significantly under natural conditions; this was most noticeable in November and December when ABA concentration was between 72% and 75% and CKs content was lower (Caselles et al. 2021). Higher ABA concentrations, as well as higher soluble sugar and non-structural carbon (NSC) contents, were observed in tetraploid *Fragaria moupinensis*. This suggests that ABA regulates the aggregation of NSC and soluble sugar during the species' reactions to CS. In their study, Karimi (2017) discovered that ABA can enhance total soluble protein (TSP) and polypeptide biosynthesis in unfavorable environmental circumstances. TSP then functions as a protective mechanism to maintain intracellular osmotic pressure, allowing leaf cells to minimize damage when exposed to low temperatures. All the grapevine leaves treated with ABA had higher TSP concentrations than the control leaves. It is commonly known that proteins have anti-freeze properties on cell membranes and various other cell structures.

In *Arabidopsis,* cold conditions were used to induce the expression levels of the ABA responsive transcription factors *ABF1* and *ABF4*. Huang et al. (2017) observed the expression of COR genes like *ABF1*, *CBF1*, and *LEA* is induced by both native and exogenous ABA application on bermudagrass (*Cynodon dactylon* (L), Pers.). Binding of *ABA* to *PP2C* blocks its activity, causing phosphorylate of *SnRK2* and there by activate its enzymatic activity. Subsequently, *SnRK2* controls the expression of ABA-induced genes and phosphorylates *ABF* transcription factors (Fujita et al. 2013). Earlier studies, suggests that *ABF2* is crucial for the transduction of the glucose signal and its overexpression can increase the sensitivity to ABA and glucose, which strengthens their resistance to these stresses (Kim et al. 2004).

For *ABA*-dependent transcriptional regulation, a single copy of *ABRE* is insufficient. Multiple copies of *ABREs* may exist upstream of stress-inducible genes, or *ABRE(s)* might act in conjunction with another GC-rich sequence known as the coupling element (CE) (Hobo et al. 1999; Shen et al. 1996). These *ABRE* and CE elements, when located within less than a 20 bp distance, together form an abscisic acid-responsive complex (ABRC). For instance, *CE1*, *CE3*, and motif III constitute an ABA-responsive complex in the regulation of wheat *HVA22*, *HVA1*, and *Rab16B* genes (Chakraborty and Roychoudhury 2022). Transcription factors (TFs) of the bZIP and *APELATA2* (AP2) types, and *VIVIPAROUS1 (VP1)/ABI3* a key transcriptional regulator, participate in gene expression. The bZIP TFs engage in dimer interactions with ABREs, which are "G-boxes" in promoter elements that contain ACGT (Francko et al. 2011). Transcription factors belonging to that of basic leucine zipper (bZIP) family predominantly bind to *ABREs/CEs* inside its target genes, controlling their stress-inducible expression (Sanghera et al. 2011).

9.2.5 The Function of **CBF/DREB1** *Pathway in Cold Stress Response*

In CS response, the *CBF/DREB1*-dependent pathway stands out as a well-defined regulatory mechanism (Knight and Knight 2012). Previous research indicates that the CBF3/DREB1A pathway is governed by *ICE1* (*INDUCER OF CBF EXPRESSION 1*); however, recent studies came to the conclusion that ICE is not involved in *DREB1A* activation. In their research on mutant ice1-1, Kidokoro et al. (2020) have confirmed that transgene-induced DNA methylation, rather than gene activation, was the cause of DREB1A repression. Further study verifies the involvement of a circadian-clock related MYB family, *LATE ELONGATED HYPOCOTYL* (*LHY*) and *CIRCADIAN CLOCK-ASSOCIATED1* (*CCA1*) as indirect expression of *DREB1* in *Arabidopsis* during cold stress. *REVEILLE4/LHY-CCA1-Like1* (*RVE4/LCL1*) and RVE8/LCL5 belonging to MYB proteins serve as direct DREB1 expression during cold stress (Kidokoro et al. 2021). The team in their further investigation found out light-inducible and clock-regulated (LNK) proteins in cooperation with RVEs under heat and CS. Among four Arabidopsis LNK genes, LNK3 and LNK4 function in response to cold stress, which is activated through phosphorylation under the conditions (Kidokoro et al. 2023).

The DREB proteins, DREB1 and DREB2, partake in distinct signal transduction pathways that react to low temperatures and dehydration, respectively. The *DREB* family encompasses novel *DREB1* and *DREB2*-related genes, including *DREB1D/CBF4* and *DREB1F*, which respond to high-salinity treatments. Notably, *DREB1A/CBF3*, *DREB1B/CBF1*, and *DREB1C/CBF2* are specifically trigger by low-temperature stress in an ABA-independent manner (Liu et al. 1998). Research has indicated rapid accumulation of *CBF* gene transcripts within minutes of exposure to cold, with varying expression patterns among *CBF1/DREB1B*, *CBF3/DREB1A*, and *CBF2/DREB1C*. While *CBF1/DREB1B* and *CBF3/DREB1A* show swift induction within 15 min, *CBF2/DREB1C* transcripts accumulate more gradually, peaking after 2.5 h of cold exposure (Novillo et al. 2004). The promoters of these target genes often contain the *DRE/CRT* or related core motifs. The swift induction of *CBF* transcripts upon cold exposure suggests the presence of a transcription factor at normal temperatures that recognizes *CBF* promoters and triggers *CBF* expression upon exposure to cold stress (Maruyama et al. 2004).

9.2.6 Role of Circular RNAs (circRNAs) in Plant Responses to Cold Stress

The advent of high-throughput sequencing technology and advanced big data analysis has led to new findings and characterization of numerous non-protein-coding genes in plant stress responses. Noncoding RNAs, capable of interacting directly with DNA, RNA, and proteins, control the expression of genes responsive to environmental stress (Khaldun et al. 2016). Circular RNAs (circRNAs), a unique class of noncoding RNAs lacking 5′ caps or 3′ tails, have gained attention in plant research, including species like *Arabidopsis*, rice, tomato, wheat, and soybean (Chen et al. 2019).

Research studies focusing on circRNAs in plants, such as tomato and grape, have indicated their involvement in chilling response and cold resistance (Yang et al. 2020; Gao et al. 2019). Studies in soybean have highlighted the tissue-specific expression of circRNAs (Zhao et al. 2017a, b). Notably, in tomato, 163 circRNAs have been confirmed to respond to chilling injury, suggesting their potential regulatory role under low-temperature stress conditions (Zuo et al. 2016)

In a study by Wang et al. (2020a, b), 14 circRNAs were investigated in soybean, revealing that nine were upregulated and four were downregulated under stress conditions. These expression patterns align with the stress-specific expression observed in most plant circRNAs, emphasizing the potential participation of circRNAs in responses to low-temperature stress in soybean.

Figure 9.1 overviews signaling pathways activated by the cold stress. As we have seen, transcriptional regulation in plant cells is achieved by at least four distinct signaling pathways.

9.3 Conservation of Cold Stress Response Across Plant Species: Key Regulators and Transcription Factors Involved

9.3.1 CBF/DREB1*-Dependent Signalling Components Conferring Plant Cold Tolerance*

The *DRE/DREB1A*-dependent pathway is crucial for cold-inducible gene expression (Yamaguchi-Shinozaki and Shinozaki 2006). The C-repeat binding factor/dehydration-responsive element-binding protein 1 (*CBF/DREB1*) genes are swiftly activated during CS, playing vital function in plant cold acclimation, notably through the *CBF*-COR signalling pathway. These CBF proteins, acting as DNA-binding factors, specifically recognize the DNA-regulatory element *CRT/DRE* situated within the COR gene promoters, thereby regulating COR gene expression (Gilmour et al. 1998). The COR genes encompass various classes such as cold regulated, low

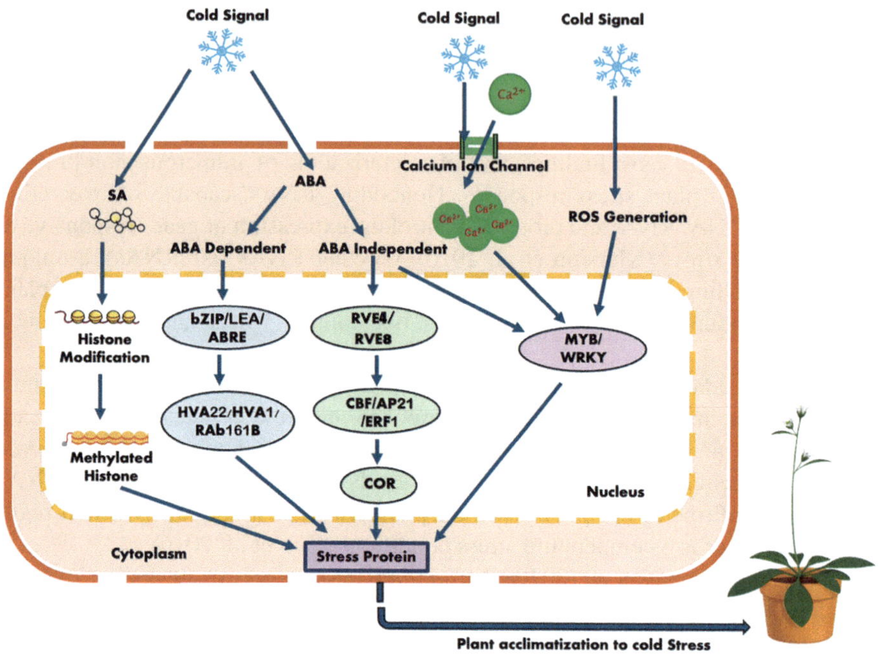

Fig. 9.1 Different mechanism of plant in response to cold stress
Salicylic acid under cold stress condition modifies through DNA methylation and downstream regulation preventing from CS. ABA also mediates the cold signal and activates ABRE-containing genes in the promoter region. There is another, ABA-independent signaling pathway activating the DREB/CBF regulon. In parallel, the ABA-independent pathway also activates MYB/WRKY genes, which are also activated by the Ca signal and also the ROS signal stimulated by the cold signal. Both ABA-dependent and ABA-independent genes get activated and took part in cold acclimatization and freezing tolerance. *SA* salicylic acid, *ABA* abscisic acid, *ROS* reactive oxygen species, *bZIP* basic region-leucine zipper type transcription factors, *LEA* late embryonic abundant proteins, *ABRE* ABA-responsive element, *RVE4 and 8* Protein REVIELLE4 and 8, *CBF* C-repeat binding factor, *AP21* ??, *ERF1* ethylene-responsive factor1, *MYB* MYB transcription factor, *WRKY* WRKY transcription factor, *COR* cold responsive

temperature induced, and cold inducible, encoding protective proteins like osmolytes and cryoprotectants to safeguard plants from freezing injuries (Yamaguchi-Shinozaki and Shinozaki 2006).

CBF1–3 (*CBF1/DREB1B*, *CBF2/DREB1C*, and *CBF3/DREB1A*), are the three *Arabidopsis CBF* genes organized in repeated sequence on chromosome IV (Medina et al. 1999). Freezing stress tolerance is enhanced in overexpressed *DREB1A,* transgenic *Arabidopsis. Arabidopsis*, rice, and soybeans exhibit cold-inducible *ABRE*-dependent gene expression (Maruyama et al. 2012). Another *CBF* gene in *Arabidopsis*, *CBF4*, is not cold-induced, yet its overexpression augments to withstand freezing and drought in plants (Haake et al. 2002). These *CBF* genes, categorized as *AP2/ERF1*-type transcription factors, specifically attached to conserved

CRT/DRE motifs in COR gene promoters (*CBF* regulons), and express under cold conditions (Maruyama et al. 2012).

Murayama et al. (2004) also identified the downstream gene that *DREB1A* regulates and divided the genes into two categories. While one group deals with drought, freezing stress, and high salt, another group has antifreezing and LEA protein functions. According to Wang et al. (2022), rice *OsDREB1C*, *OsDREB1E*, and *OsDREB1G* mutants were able to acclimate to cold, suggesting that these rice *DREB1* genes are crucial for basal chilling tolerance but not for cold acclimation. Overexpression of transgenic rice *OsDREB1A or DREB1A* accumulates different sugars and osmoprotectants like proline (Ito et al. 2006). *OsDREB1A* targets genes that may be involved in chilling stress tolerance, such as dehydrin genes. The existence of the low-temperature responsive (LTR) element within the promoter of *PnaDREBs* highlights its significance in responding to CS (Qian et al. 2023). Overexpression of *DREB1/CBF* genes from different kinds of plants, including rice, tomato, and barley, significantly enhances cold tolerance, suggesting evolutionary conservation in higher plants. *MaDREB1F* overexpression in bananas increases resistance to drought and CS by activating protective metabolites, antioxidant systems, and jasmonate and ethylene syntheses (Xu et al. 2023)

Additionally, the overexpression of transcription factors like *BpERF13* enhances cold tolerance in woody plants like birch, apple, and radish, thereby modulating osmotic adjustment and cell wall mechanical strength (Lv et al. 2020).

For a comprehensive overview of *CBF/DREB1*-dependent COR gene expression in various plant species, refer to Table 9.1.

9.3.2 Role of MYB Genes in Conferring Plant Cold Tolerance

The MYB domain consists of 1–4 repetitive sequences (R) containing 50–52 amino acids have a highly conserved region in the DNA-binding called as MYB transcription factor family. Based on the sequences MYB transcription factors are categorize into four separate subfamilies: *R1-MYB/MYB*, *R2R3-MYB*, *R1R2R3-MYB*, and *4R-MYB*. Among these, the extensively studied subfamily is the *R2R3-typ*e MYB (Yang et al. 2022). In the context of plant cold tolerance, investigations into the regulatory mechanisms of *R2R3-MYBs* have primarily focused on apple and pear. For instance, in apple overexpression of *MdMYB308L* significantly heightened cold tolerance (An et al. 2020). Similarly, in pear, expression *of PbrMYB5* was triggered under CS, and its overexpression enhancing the cold resistivity was study in transgenic tobacco (Xing et al. 2019).

Notably, a few R2R3-MYB transcription factors, such as HOS10 in *Arabidopsis* and Osmyb4 in rice, have been suggested to play pivotal roles in CS responses (Dai et al. 2007). Additionally, in chrysanthemum overexpressing *DgMYB2,* a cold-induced R1-MYB transcription factor, increased cold resistance. DgMYB2 directly act on *DgGPX1* to boosting activity of GPX enzyme, subsequently it reduces accumulation of ROS and bolstering chrysanthemum's cold resistance (Yang et al.

Table 9.1 Signalling components dependent on CBF/DREB conferring plant cold tolerance

Sl. no.	Name of the gene	Source plant	Effect on plants	References
1.	*AtCBF1*	*Arabidopsis thaliana*	Freezing, salt and drought tolerance	Jaglo-Ottosen et al. (1998)
2.	*AtCBF3*	*Arabidopsis thaliana*	Freezing, salt and drought tolerance	Gilmour et al. (1998)
3.	*ZmDREB1A*	*Zea mays*	Salt, drought and freezing tolerance	Feng et al. (2004)
4.	*OsDREB1A* *OsDREB1B* *OsDREB1C*	*Oryza sativa*	Salt, drought and chilling tolerance	Ito et al. (2006)
5.	*HvCBF4*	*Hordeum vulgare*	Salt, drought and chilling tolerance	Oh et al. (2007)
6.	*BpCBF1*	*Betula pendula*	Freezing tolerance	Welling and Palva (2008)
7.	*LpCBF3*	*Lolium perenne*	Freezing tolerance	Zhao and Bughrara (2008)
8.	*GmDREB3*	*Glycine max*	Drought, salt and freezing tolerance	Chen et al. (2009)
9.	*TaDREB2* *TaDREB3*	*Triticum aestivum*	Drought and frost tolerance	Morran et al. (2011)
10.	*MbDREB1*	*Malus baccata*	Chilling, drought and salt tolerance	Yang et al. (2011)
11.	*PbeNAC1*	*Pyrus betulifolia*	Drought and cold tolerance	Jin et al. (2017)
12.	*SlCBF1*	*Solanum lycopersicum*	Chilling tolerance	Li et al. (2018)
13.	*BpERF13*	*Betula platyphylla*	Cold tolerance	Lv et al. (2020)
14.	*OsSAPK6*	*Oryza sativa*	Chilling tolerance	Jia et al. (2022)
15.	*MbCBF2*	*Malus baccata*	Cold and salt tolerance	Li et al. (2022)
16.	*MbCBF1*	*Malus baccata*	Salt and cold tolerance	Liang et al. (2022)
17.	*LmTrxh2*	*Lobularia maritima*	Cold tolerance	Ben Saad et al. (2023)
18.	*RsERF40*	*Raphanus sativus*	Cold tolerance	Li et al. (2023a, b)
19.	*MaDREB1F*	*Musa accuminata*	Cold and drought tolerance	Xu et al. (2023)

2022). Moreover, in cold sensitive rice, cold-inducible R1R2R3 MYB TF, *OsMYB3R-2*, has been reported. This factor exhibited increased expression after cold treatment and demonstrated accumulation over 72 h. Unlike Osmyb4, which responds solely to CS, *OsMYB3R-2* was discovered to be caused by drought and salt stress as well (Dai et al. 2007). For a comprehensive overview of MYB genes associated with conferring cold tolerance among different plant species, refer to Table 9.2.

Table 9.2 MYB genes conferring plant cold tolerance

Sl no.	Name of the gene	Source plant	Effect on plants	References
1.	*Osmyb4*	*Oryza sativa*	Increased freezing and chilling tolerance	Vannini et al. (2007)
2.	*OsMYB3R-2*	*Oryza sativa*	Enhanced tolerance towards drought, freezing, and salt stress	Dai et al. (2007)
3.	*OsMYB3R-2*	*Oryza sativa*	Resistance to cold stress	Ma et al. (2009)
4.	*OsMYBS3*	*Oryza sativa*	Enhanced cold tolerance	Su et al. (2010)
5.	*TaMYB2A*	*Triticum aestivum*	Enhanced tolerance towards drought, salt, and freezing stress	Mao et al. (2011)
6.	*MdoMYB121*	*Malus domestica*	salinity, drought, and cold stresses	Cao et al. (2013)
7.	*GmMYBJ1*	*Glycine max*	Enhanced cold and drought tolerance	Su et al. (2014)
8.	*MdMYB4*	*Malus domestica*	Increased cold and salt stress	Wu et al. (2017)
9.	*MdMYB23*	*Malus domestica*	Enhanced cold tolerance	An et al. (2018)
10.	*MbMYB4*	*Malus baccata*	Drought and cold tolerance	Yao et al. (2022)
11.	*DgMYB2*	*Dendranthema grandiflora*	Cold tolerance	Yang et al. (2022)

9.3.3 Role of WRKY *Genes in Cold Tolerance*

In plants the WRKY transcription factor (TF) family, is considered to be largest, primarily occurs in higher plants, but rarely found in lower plants. It plays diverse roles in various biological processes. These genes have the potential to enhance an individual's ability to survive in CS conditions by raising antioxidant activity levels, preserving membrane stability, and improving osmotic regulation ability (Bakshi and Oelmuller 2014; Khoso et al. 2022). The WRKY or Sweet Potato Factor 1 (SPF1) gene was first studied from sweet potatoes (Ishiguro and Nakamura 1994). Subsequence studies has identified and characterized WRKY from a variety of plants, which including 72 WRKY TFs in *Arabidopsis thaliana* and 103 WRKY members in *Oryza sativa* (Eulgem and Somssich 2007; Ramamoorthy et al. 2008). These TFs share a highly conserved DNA-binding domain of approximately 60 amino acids known as the WRKY domain. This domain comprises the conserved WRKYGQK sequence at the N-terminal and a zinc-finger structure at the C-termina (Jiang et al. 2017).

Although the core DNA binding motif of WRKY proteins is mostly WRKYGQK, variations such as WKKYGQK, WRKYGMK, WSKYGQK, WKRYGQK, WVKYGQK, and WRKYGKK have been observed (Huang et al.

2012; Xie et al. 2005). These differences in the WRKY domain, along with the structural features of zinc-finger motifs, categorize WRKY proteins into three groups: I, II, and III. Group I WRKY proteins has two WRKY domains and a C2H2 zinc-finger structure, while Group II comprises WRKY proteins with one WRKY domain and a C2H2 zinc-finger structure. Group III consists of WRKY proteins having only single WRKY domain and a C2HC-type zinc finger (Wu et al. 2005). The WRKY34 (WRKY group I) TF might take part in the C-repeat binding factor (CBF) signal cascade in mature pollen and mature *Arabidopsis* pollen it negatively regulated the cold sensitivity (Zou et al. 2010). SlWRKY50, a Group II WRKY transcription factor that is cold inducible, reacts to cold stimuli, and is essential for the biosynthesis of JA, according to research by Wang et al. (2023). The tomato *allene oxide synthase* gene (*SlAOS*) promoter is directly bound by *SlWRKY50*, and overexpressing *SlWRKY50* increased tomato resistance to chilling.

WRKY genes have been found to be identified to orchestrate multiple biological processes. For example, *AtWRKY33* regulates disease resistance, NaCl tolerance (Birkenbihl et al. 2012), and thermotolerance, while *CaWRKY40* in pepper modulates resistance to *Ralstonia solanacearum* and heat stress tolerance (Dang et al. 2019). Specifically, in rice plants, *OsWRKY71* responds specifically to cold stress, targeting potential downstream genes *OsTGFR* and *WSI76* upon cold treatment. Overexpression of *OsWRKY71* enhances cold tolerance in transgenic rice, validated by assessments such as chlorophyll fluorescence (Liu et al. 2007). Expression of *GhWRKY41* was observed significantly at 4 °C. After the cotton was exposed to cold treatment, the level of *GhWRKY15* expression was also elevated (Guo et al. 2022). *V. bonariensis* demonstrated a higher induction of *VbWRKY32* overexpression (OE) in leaves compared to wild-type (WT), which increased the plant's resistance to cold. The results showed that *VbWRKY32* functioned as a positive regulator by upregulating the levels of gene transcription that responded to cold (Wang et al. 2020a, b). In white clover under cold stress, *TrWRKY* genes were sharply up-regulated; specifically, at 30 min, 43 *TrWRKY* genes were up-regulated, suggesting that these *TrWRKY* genes responded to cold stress rapidly (Li et al. 2023a, b).

Research has demonstrated the induction of WRKY gene expression under both cold and salt stresses in plants. Investigating the role of *KoWRKY40* in cold stress response, transgenic *Arabidopsis* plants overexpressing KoWRKY40 were developed. Under CS, *KoWRKY40* expression was notably heightened in the roots and leaves of *K. obovate* (Fei et al. 2022). Table 9.3 summarizes WRKY genes associated with conferring cold tolerance in various plant species.

Table 9.3 WRKY genes conferring plant cold tolerance

Sl no.	Name of the gene	Source plant	Phenotype and effect	References
1.	*GmWRKY21*	*Glycine max*	Freezing tolerance	Zhou et al. (2008)
2.	*TaWRKY19*	*Triticum aestivum*	Tolerance to freezing, salt and drought stress	
3.	*BcWRKY46*	*Verbena bonariensis*	Increased tolerance towards cold, salt and dehydration	Wang et al. (2012)
4.	*OsWRKY76*	*Oryza sativa*	Cold tolerance	Yokotani et al. (2013)
5.	*OsWRKY74*	*Oryza sativa*	Cold tolerance	Dai et al. (2016)
6.	OsWRKY71	*Oryza sativa*	Cold tolerance	Kim et al. (2016)
7.	*CsWRKY46*	*Cucumis sativus*	Highly sensitivity to ABA; Freezing tolerance	Zhang et al. (2016)
8.	*VbWRKY32*	*Verbena bonariensis*	Increased cold tolerance	Wang et al. (2020a)
9.	KoWRKY40	*Kandelia obovata*	Cold tolerance	Fei et al. (2022)

9.4 Genetic and Epigenetic Factors Contributing to the Diversification of Cold Stress Response

Since plant lack a brain to retain memories, it senses and reacts to heat and CS, which is controlled by epigenetic modifications and have an important function in gene regulation. The ability to develop suitable molecular and physiological changes results from this. Furthermore, stress memories that are epigenetically regulated are passed down to subsequent generation (Ramakrishnan et al. 2022). Salicylic acid is essential for the activation of the salicylic acid responsive gene, and downstream signalling proteins influence both the memory process and alteration of the chromatin for these target genes. High levels of H3 acetylation, H4 acetylation, and H3K4 methylation brought on by salicylic acid (Bhadouriya et al. 2021). Modifications of Flowering Locus C (FLC) gene's to H3K27me3 preserve CS memory. This shows that accessibility of chromatin and structure, which are frequently linked to epigenetic codes or epigenetic regulations like DNA methylation, histone variants, and posttranslational modifications (PTMs), are crucial for controlling the expression of genes(Xie et al. 2023).

9.4.1 Histone Modification

Plant acclimation to cold conditions gradually diminishes after vernalization, representing a prominent instance of plant epigenetic regulation in reaction to environmental cues (Kang et al. 2022). Genetic and molecular investigations have demonstrated the pivotal part of epigenetic markers in vernalization, where an epigenetic memory governs the plant's response (Sharma et al. 2022). Research has demonstrated that changes to histones, including methylation and acetylation, undergo alterations during CS, impacting various molecular mechanisms, especially genes that respond to cold and can be genetically altered (Bhadouriya et al. 2021).

Post-translational modifications (PTMs) of histones entail the chemical modification of specific amino acid residues—lysine, arginine serine and threonine—in the histone tail. These PTMs have capacity to modulate chromatin dynamics and structure, consequently impacting the way that stress-responsive genes are expressed by recruiting histone modifiers (Ramakrishnan et al. 2022). Chromatin conformation plays a crucial part in determining the accessibility of DNA sequences, including regulatory and coding parts, to the transcriptional apparatus, thereby governing gene expression. Nucleosome positioning mediates DNA accessibility for transcription, with each core nucleosome comprising H2A, H2B, H3, and H4 histone proteins and roughly 146 DNA base pairs (Mariño-Ramírez et al. 2005).

In plants during CS histone modifications and DNA methylation are prominent epigenetic markers. During acetylation an acetyl group is transfer to lysine residues of the histone tail, performed by histone acetyltransferases (HATs). This modification leads to an "open" chromatin configuration, facilitating transcriptional activation. On exposure to abiotic stress in plants, stress-related genes are activated by the interaction between HATs and transcription factors (Pandey et al. 2002). Through chromatin configuration, genes expression is silence through histone deacetylation which regulated by histone deacetylases (HDACs). Histone methylation, conducted by histone methyltransferases (HMTs), also plays a significant role. In potato, genomic regions housing active regulatory elements, recognized as DNase I hypersensitive sites (DHSs) via chromatin analysis, exercise significant regulatory authority over cold stress response (Zeng et al. 2019).

Specific histone modifications including H3K4me3 and H3K27me3 act as memory marks for stress-responsive genes of *Arabidopsis* (Kwon et al. 2009). CS promotes these modifications and improve chromatin accessibility, making it easier for regulatory proteins needed for up or down-regulation of genes (Zeng et al. 2019). Histone acetylation typically facilitates gene expression. Histone H3 acetylation stimulates COR gene activation during cold stress, with the HAC1 histone acetyltransferase being essential for stress memory and augmenting cold memory (Roy et al. 2014). Conversely, cold exposure triggers histone deacetylation through the increased control of histone deacetylases (HDACs), resulting in deacetylation of histone subunits H3 and H4, which represses genes sensitive to cold expression such as *ZmDREB1* and *ZmCOR413* in Zea mays (Hu et al. 2011). The gene product of

HOS15, operating as a histone deacetylation factor, specifically interacts with histone H4 during cold acclimatization in *Arabidopsis* (Park et al. 2018).

9.4.2 DNA Methylation

DNA methylation, involving in addition of methyl group to cytosine at CpG sites, orchestrates DNA stability, chromatin architecture, and gene expression without altering in its DNA sequence. In plant genomes, it spans heterochromatic and euchromatic regions, encompassing gene promoters, intergenic spaces, transposable elements (TEs), and gene bodies (Zilberman and Henikoff 2007). Studies highlight 5-methylcytosine (5mC) and N6-methyladenine (6 mA) as regulators of transcription during cold stress responses (Zhang et al. 2018), moreover, cold-induced DNA methylation changes can be passed across generations or as epigenetic memories for subsequent cold stress responses in plants (Verma et al. 2022). DNA methylation as well as demethylation, induces during cold acclimation, impacting freezing tolerance in *Arabidopsis* and RNA-directed DNA methylation (RdDM) in cucumbers. Vernalization-triggered demethylation varies among plants, influencing non-CG methylation via mitosis (Liu and He 2020). DNA methylation contributes to transposable element (TE) silencing and environmentally driven alterations in the DNA methylome may promote or suppress TE activity. While DNA methylation in *Arabidopsis* is minimally influenced by temperature and other stresses, it exhibits weak environmental sensitivity, potentially not inducing extensive gene expression changes within a single generation (Ganguly 2018).

9.4.3 Non-coding RNAs

Non-coding RNAs (ncRNAs) are essential for many characteristics of plant biology, such as growth, development, and response to diverse environmental stressors (Bhogireddy et al. 2021) These ncRNAs are categorized based on its nucleotides length. ncRNA with 20–200 nucleotides are small and above 200 nucleotides are considered as long. Trans-acting small interfering RNAs (ta-siRNAs), micro-RNAs (miRNAs), heterochromatic (hc-RNAs), and small natural-antisense siRNAs (na-tsiRNAs) are few of the small ncRNAs in plants. The significance of long non-coding RNAs (lncRNAs) has been highlighted bi-genome-scale surveys in plants in regulating responses to CS. For example, in cassava, biosynthesis of hormone signaling, sucrose metabolism, and secondary metabolite biosynthesis were associated 318 lncRNAs, indicating their part in CS tolerance (Li et al. 2017). Similarly, studies in grapevine identified 284 lncRNAs lncRNAs that respond to cold, alongside genes coding for 326 proteins (Wang et al. 2019).

Furthermore, Under CS conditions certain lncRNAs act as goal imitators for miRNAs. For instance, LincRNA159 functions as a goal imitator for miR164 in cassava, which lowers the expression of NAC genes (Li et al. 2017). In *Brassica napus*, a recent study by Waseem et al. (2022) investigated the expression alterations of lncRNAs in cold-sensitive and cold-tolerant varieties under CS. They identified 11,073 lncRNAs, of which 4819 were expressed differently in response to CS. Interestingly, in silico investigation identified 16 lncRNAs acting as predecessors of seven B. napus particular miRNA families and 15 lncRNAs expected to be targets of 23 miRNAs. Numerous miRNAs have been recognized as important participants in the way plants react to CS. In rice, downregulation of Osa-miR319b due to CS leads to the downregulation of *OsPCF6* and *OsTCP21*, which in turn upregulate cold stress-responsive genes like *DREB1/CBF*, *DREB2A*, and *TPP1/2*, ultimately enhancing CS tolerance (Wang et al. 2014) Additionally, in *Arabidopsis*, several hundred lncRNAs were differentially expressed under various abiotic stresses, while 813 lncRNAs in grapevine, under CS (Wang et al. 2019).

9.4.4 Circular RNAs

Circular RNAs (circRNAs) describe a distinct class of endogenous non-coding RNAs define by its unique closed-loop arrangement. These circular transcripts are predominantly derived from genes coding proteins and can comprise of either only exons or multiple exons (Santer et al. 2019). circRNAs lack conventional 5′ caps and poly(A) tails, yet it is generated inside the nucleus, typically transported to cytoplasm (Salzman et al. 2012). Plants under CS response has a distinct expression of circRNAs. For instance, analyses in chilled bell pepper and chilled tomato fruit identified 36 and 163 differentially expressed circRNAs, respectively (Zuo et al. 2016). Similarly, 475 differently expressed circRNAs were found in grape leaves after it was exposed to CS (Gao et al. 2019). Zhang et al. (2020) elucidated the mechanism by which Vv-circATS1 operates. It was found that this circRNA regulates the expression of various stress-responsive genes, including *CSD2*, *PRXCA*, *PME41*, *LOX3*, and *WRKY48*, thereby improving cold tolerance in *Arabidopsis*. The identified grapevine circRNA, Vv-circATS1, originating from glycerol-3-P acyltransferase, exhibits a potential role in enhancing CS resilience in plants by regulating the genes expression which is crucial in responding to environmental stimuli, as evidenced in *Arabidopsis* (Gao et al. 2019)

9.5 Comparative Analysis Between Arabidopsis and Rice During Cold Stress Response

The DREB1-type ERF/AP2 proteins in monocots share a close phylogenetic relationship, in compare to their dicot counterparts. Noteworthy is the binding affinity difference exhibited by the *Arabidopsis* AtDREB1A and rice OsDREB1A proteins.

AtDREB1A binds ACCGAC and GCCGAC equally, while OsDREB1A prefers GCCGAC over ACCGAC (Dubouzet et al. 2003). In rice It is clear the *DREB1/CBF* cold-responsive pathway is conserved. In transgenic *Arabidopsis* excessive expression of *DREB1* or *OsDREB1* induces robust expression of stress-responsive genes, enhancing high-salt and freezing stresses tolerance. Transgenic rice with upregulated *OsDREB1* or *DREB1* levels exhibit lower growth under natural conditions but enhanced resilience to drought, low-temperature and high-salt (Ito et al. 2006).

The CBF/DREB1 family, comprising six genes in *Arabidopsis* and ten in rice (*OsDREB1A* to *OsDREB1J*), shows distinct responses to CS. *OsDREB1A*, *OsDREB1B*, and *OsDREB1F* are induced, while *OsDREB1D* remains unaffected. Overexpression of *OsDREB1A*, *OsDREB1D*, and *OsDREB1F* enhances cold resistance in *Arabidopsis* through elevated COR expression (Mao and Chen 2012). Rice, under cold stress, sequentially upregulates *OsDREB1B*, *OsHsfA3*, and trehalose-6-phosphate phosphatase, contributing to cold acclimation *via* trehalose synthesis. In *Arabidopsis*, *ATGOLS3* and *BBX14* are upregulated during CS (Nakamura et al. 2011).

Distinct cold-inducible promoter sequences are identified in *Arabidopsis* and rice (Maruyama et al. 2012). ACACGT and AATATC are the most conserved sequences in *Arabidopsis*, while CGTACG and GTAGTA are most conserved in rice. Promoter analyses reveal the absence of highly conserved motifs akin to DRE or ABRE in rice promoters. 22 genes *OsDREB1c* regulon was identified by Zhang et al. (2012) with 18 differentially expressed in cold-stressed *japonica* rice (cv. Jumli Marshi), highlighting the crucial part of the *OsDREB1c* regulon in cold tolerance. Comparative transcription profiles under CS highlight similarities between rice and *Arabidopsis*. Both species exhibit upregulation of dehydrin and Late Embryogenesis Abundant Protein genes (*LEA*) and genes relating downregulation of photosynthesis activity. In the promoters of cold-inducible *Arabidopsis* and the DREB1A regulated downstream genes have highly conserved DRE motif. This comparison underscores the conservation and divergence within plants in cold-responsive pathways between monocots and dicots, providing insights into the intricate molecular process controlling stress responses (Mikołajczak et al. 2023).

Gene expressional examination profiles in *Arabidopsis* indicated a noteworthy representation of the transcription factors from the AP2/ERF family. In action against low temperatures rice deploys three principal signal cascades: the DREB1/CBF pathway, a MAPK cascade, and an ABA-dependent pathway. This intricate network involves the activation of various genes, such as calcium-dependent protein kinases (*CDPKs*), including *OsCDPK4*, *OsCDPK5*, and OsCDPK7, as well as a cytochrome P450 monooxygenase gene, *CYP76M8*, and a peroxidase gene, *POX-1*, during chilling exposure. Concomitantly, OsMYB3R-2 and OsMYB2 have been identified as contributors to enhanced tolerance to low temperatures. Notably, the upregulation of a heat shock protein analogous to HSP70B (Os01g0688900) is observed early in the chilling stress response (Buti et al. 2019).

The role of lipid transfer proteins (LTPs) in freezing tolerance is highlighted by the overexpression of LTP3 in *Arabidopsis*. In rice, specific LTPs, including LTP7, LTP8, LTP12, and LTP25, demonstrate an early and maximal activation following

3 h of cold exposure, particularly in the cold-tolerant cultivar *Nipponbare* (Moraes de Freitas et al. 2019). Cyclic nucleotide-gated channels (CNGCs), functioning as general cation channels, has a critical role in different type of stress responses, such as low temperatures, in *Arabidopsis*. However, in rice, the chilling tolerance divergence is connected to *CHILLING-TOLERANCE DIVERGENCE 1 (COLD1)*. In rice G-protein a subunit 1 (RGA1) and G-protein signalling (RGS), regulatory effects of COLD1 by increasing GTPase activity, is similar to *AtRGS1*. Although COLD1's subcellular localization pattern partially overlaps with its *Arabidopsis* orthologs GTG1/2, distinctions exist, especially in intrinsic GTPase activity (Wei et al. 2021).

In absence of a low-temperature stimulus it was found transgenic *Arabidopsis* plants overexpressed *SCOF-1* and exhibit constitutive expression of *Cold Response (COR)* genes, such as *COR15a*, *Rd29B*, and *COR47*. This overexpression enhances cold tolerance in non-acclimated transgenic plants. In the context of dehydration and ABA response pathways, the bHLH protein rd22BP1 is identified in *Arabidopsis*. A comparative analysis with its rice counterpart, *OsbHLH1*, reveals divergence in identity (25%) and amino acid conservation, suggesting different target elements. *OsbHLH1* likely to be ABA-independent cold-induce, highly expressed in roots while it's expression patterns is also different. In contrast, *rd22BP1* is highly expressed in siliques and stems induce mainly by dehydration, in ABA-dependent manner, but not in leaves and roots. These indicates that *OsbHLH1* and *rd22BP1* participate distinctly in abiotic stress responses, by means of ABA-independent and ABA-dependent signal-transduction pathways (Yang et al. 2023).

9.6 Conclusions

The distribution of plants across latitudes and altitudes is influenced by low temperatures, however, less than half of the plant families belonging to the angiosperm are able to withstand seasonal low temperatures. Forecasts predict an increased global temperature rise around 1.1–6.4 °C in the next 80 years, potentially increasing the constancy and rigidity of warm winter spells. Traits induced by cold conditions, like cold acclimation, endodormancy, and vernalization responsiveness, have developed autonomously several times in angiosperms, modifying shared genetic pathways. Many key regulators of these traits belong to sizable gene families involved in stress responses and developmental transitions.

The understanding of CS response in plants has significantly progressed over the years, delving into intricate molecular mechanisms, signalling pathways, and genetic regulation. Abiotic stressors, especially those related to cold, pose substantial challenges to plant growth, development, and productivity. Plants exhibit diverse adaptive strategies, mechanisms, and genetic pathways to deal with low-temperature stress. Knowing the molecular processes that underlie plant reactions to cold stress is essential for the advancement of resilient crop varieties with improved cold tolerance. Advances in molecular tools, genome editing techniques, and biotechnological approaches have provided avenues for the manipulation and enhancement of cold stress tolerance in crops.

Recent research on circRNA identification in plants highlights a potential area for knowing their respective functions in growth and development of the plant, biotic and abiotic stress responses, akin to their roles in animals' gene expression regulation. Pathways like the *CBF/DREB1*-dependent pathway perform a central function in controlling cold signalling, involving factors such as Ca^{2+} signalling, chloroplast status, clock genes, and phytohormones. Phytohormone induced metabolic pathway manipulation via genetic engineering mechanism growth patterns can be modified to withstand cold stress.

Given the threat of biotic and abiotic stresses to agriculture-based economies and food security, understanding CS adaptation at the genomics and proteomics stage can identify genes and proteins associated with cold tolerance. Environmental changes necessitate plants to adapt growth physiology against various stresses. Extensive research on epigenetic marks (DNA methylation, histone modifications, diverse RNA forms, ATP-mediated regulations) aims to modulate the accessibility of stress-responsive genes. Sophisticated sequencing technologies and chromatin profiling have propelled epigenetic studies, especially in crop plants, addressing food security challenges.

The exploration of transcription factors like CBF/DREB1, and MYB and WRKY genes has shed light on their pivotal roles in conferring cold tolerance in various plant species. The *CBF/DREB1* pathway, in particular, emerges as a key regulator, inducing the genes sensitive to cold expression and enhancing freezing tolerance. Additionally, MYB and WRKY transcription factors play critical function in modulating plant responses to CS, contributing to the activation of protective methods.

Moreover, the study of non-coding RNAs, specifically circular RNAs (circRNAs), has drawn notice for their involvement in plant stress responses, including cold stress. Differentially expressed circRNAs, such as Vv-circATS1 originating from glycerol-3-P acyltransferase in grapevines, have shown the ability to withstand cold in Arabidopsis by regulating the stress-responsive genes expression. DNA methylation and histone modification are examples of epigenetic modifications showing crucial for supporting plants to adapt in CS. These changes affect the expression of genes and the structure of chromatin, which helps the plant tolerate low temperatures.

This comprehensive review highlights the diverse molecular pathways, signalling molecules, hormonal dynamics, epigenetic modifications, and genetic elements that is responsible in plant CS. Future research in this field should continue to explore novel genetic targets, regulatory networks, and signalling pathways to further elucidate and enhance our knowledge on how plants respond and adapt to CS, ultimately opening the door for the creation of crop varieties resistant to stress, which are essential for maintaining global food security in the face of shifting environmental conditions.

References

Albertos P, Wagner K, Poppenberger B (2019) Cold stress signalling in female reproductive tissues. Plant Cell Environ 42(3):846–853

An JP, Li R, Qu FJ et al (2018) R2R3-MYB transcription factor MdMYB23 is involved in the cold tolerance and proanthocyanidin accumulation in apple. Plant J 96(3):562–577

An JP, Wang XF, Zhang XW, Xu HF, Bi SQ, You CX, Hao YJ (2020) An apple MYB transcription factor regulates cold tolerance and anthocyanin accumulation and undergoes MIEL1-mediated degradation. Plant Biotechnol J 18(2):337–353

Aslam M, Fakher B, Ashraf MA, Cheng Y, Wang B, Qin Y (2022) Plant low-temperature stress: signaling and response. Agronomy 12(3):702

Bakshi M, Oelmüller R (2014) WRKY transcription factors: jack of many trades in plants. Plant Signal Behav 9(2)

Banerjee A, Roychoudhury A (2016) Group II late embryogenesis abundant (LEA) proteins: structural and functional aspects in plant abiotic stress. Plant Growth Regul 79:1–17

Barrero-Sicilia C, Silvestre S, Haslam RP, Michaelson LV (2017) Lipid remodelling: unravelling the response to cold stress in *Arabidopsis* and its extremophile relative Eutrema salsugineum. Plant Sci 263:194–200

Ben Saad R, Ben Romdhane W, Baazaoui N, Bouteraa MT, Chouaibi Y, Mnif W, Ben Hsouna A, Kačániová M (2023) Functional characterization of *Lobularia maritima LmTrxh2* gene involved in cold tolerance in tobacco through alleviation of ROS damage to the plasma membrane. Int J Mol Sci 24(3):3030

Bhadouriya SL, Mehrotra S, Basantani MK, Loake GJ, Mehrotra R (2021) Role of chromatin architecture in plant stress responses: an update. Front Plant Sci 11:603380

Bhandari K, Nayyar H (2013) Low temperature stress in plants: an overview of roles of cryoprotectants in defense. In: Physiological mechanisms and adaptation strategies in plants under changing environment: volume 1, pp 193–265

Bhogireddy S, Mangrauthia SK, Kumar R, Pandey AK, Singh S, Jain A, Budak H, Varshney RK, Kudapa H (2021) Regulatory non-coding RNAs: a new frontier in regulation of plant biology. Funct Integr Genomics 21:313–330

Birkenbihl RP, Diezel C, Somssich IE (2012) *Arabidopsis WRKY33* is a key transcriptional regulator of hormonal and metabolic responses toward *Botrytis cinerea* infection. Plant Physiol 159(1):266–285

Bredow M, Walker VK (2017) Ice-binding proteins in plants. Front Plant Sci 8:2153

Buti M, Baldoni E, Formentin E, Milc J, Frugis G, Lo Schiavo F, Genga A, Francia E (2019) A meta-analysis of comparative transcriptomic data reveals a set of key genes involved in the tolerance to abiotic stresses in rice. Int J Mol Sci 20(22):5662

Cacas JL, Furt F, Le Guédard M, Schmitter JM, Buré C, Gerbeau-Pissot P, Moreau P, Bessoule JJ, Simon-Plas F, Mongrand S (2012) Lipids of plant membrane rafts. Prog Lipid Res 51(3): 272–299

Cao ZH, Zhang SZ, Wang RK et al (2013) Genome wide analysis of the apple MYB transcription factor family allows the identification of *MdoMYB121* gene conferring abiotic stress tolerance in plants. PLoS One 8(7):e69955

Caselles V, Casadesús A, Munné-Bosch S (2021) A dual role for abscisic acid integrating the cold stress response at the whole-plant level in Iris *pseudacorus* L. growing in a natural wetland. Front Plant Sci 12:722525

Chakraborty S, Roychoudhury A (2022) Functional regulation of responsive to abscisic acid (*Rab*) genes from representative plant species and their stress response. Plant Physiol Rep 27(4): 653–664

Chen M, Xu Z, Xia L, Li L, Cheng X, Dong J, Wang Q, Ma Y (2009) Cold-induced modulation and functional analyses of the DRE-binding transcription factor gene, *GmDREB3*, in soybean (Glycine max L.). J Exp Bot 60(1):121–135

Chen X, Yang T, Wang W, Xi W, Zhang T, Li Q, Yang A, Wang T (2019) Circular RNAs in immune responses and immune diseases. Theranostics 9(2):588

Chinnusamy V, Zhu JK, Sunkar R (2010) Gene regulation during cold stress acclimation in plants. Methods Mol Biol (Clifton, NJ) 639:39–55

Cotelle V, Leonhardt N (2019) ABA signalling in guard cells. In: Advances in botanical research, vol 92. Academic, London, pp 115–170

Dai X, Xu Y, Ma Q, Xu W, Wang T, Xue Y, Chong K (2007) Overexpression of an *R1R2R3 MYB* gene, *OsMYB3R-2*, increases tolerance to freezing, drought, and salt stress in transgenic Arabidopsis. Plant Physiol 143(4):1739–1751

Dai X, Wang Y, Zhang WH et al (2016) OsWRKY74, a WRKY transcription factor, modulates tolerance to phosphate starvation in rice. J Exp Bot 67(3):947–960

Dang F, Lin J, Chen Y, Li GX, Guan D, Zheng SJ, He S (2019) A feedback loop between *CaWRKY41* and H2O2 coordinates the response to *Ralstonia solanacearum* and excess cadmium in pepper. J Exp Bot 70(5):1581–1595

Daszkowska-Golec A, Szarejko I (2013) The molecular basis of ABA-mediated plant response to drought. In: Abiotic stress-plant responses and applications in agriculture, pp 103–134

De Smet I, Zhang H, Inzé D, Beeckman T (2006) A novel role for abscisic acid emerges from underground. Trends Plant Sci 11(9):434–439

Denais CM, Gilbert RM, Isermann P, McGregor AL, Te Lindert M, Weigelin B, Davidson PM, Friedl P, Wolf K, Lammerding J (2016) Nuclear envelope rupture and repair during cancer cell migration. Science 352(6283):353–358

Dubouzet JG, Sakuma Y, Ito Y, Kasuga M, Dubouzet EG, Miura S, Seki M, Shinozaki K, Yamaguchi-Shinozaki K (2003) *OsDREB* genes in rice, Oryza sativa L., encode transcription activators that function in drought-, high-salt-and cold-responsive gene expression. Plant J 33(4):751–763

Eulgem T, Somssich IE (2007) Networks of WRKY transcription factors in defense signalling. Curr Opin Plant Biol 10(4):366–371

Fei J, Wang YS, Cheng H, Su YB, Zhong YJ, Zheng L (2022) The Kandelia obovata transcription factor KoWRKY40 enhances cold tolerance in transgenic *Arabidopsis*. BMC Plant Biol 22(1): 1–15

Feng Q, Yoh S, Jie L, Qiang L, Yi-Qin L, Kazuo S, Kazuko YS (2004) Cloning and functional analysis of a novel DREB1/CBF transcription factor involved in cold-responsive gene expression in *Zea mays* L. Plant Cell Physiol 45(8):1042–1052

Francko DA, Wilson KG, Li QQ, Equiza MA (2011) A topical spray to enhance plant resistance to cold injury and mortality. HortTechnology 21(1):109–118

Fujita Y, Yoshida T, Yamaguchi-Shinozaki K (2013) Pivotal role of the AREB/ABF-*SnRK2* pathway in ABRE-mediated transcription in response to osmotic stress in plants. Physiol Plant 147(1):15–27

Ganguly DR (2018) Training memory: exploring the intersection of plant stress signalling and DNA methylation. Doctoral dissertation, The Australian National University (Australia)

Gao Z, Li J, Luo M, Li H, Chen Q, Wang L, Song S, Zhao L, Xu W, Zhang C, Wang S (2019) Characterization and cloning of grape circular RNAs identified the cold resistance-related Vv-circATS1. Plant Physiol 180(2):966–985

Gilmour SJ, Zarka DG, Stockinger EJ, Salazar MP, Houghton JM, Thomashow MF (1998) Low temperature regulation of the *Arabidopsis* CBF family of AP2 transcriptional activators as an early step in cold-induced COR gene expression. Plant J 16(4):433–442

Guiltinan MJ, Marcotte WR Jr, Quatrano RS (1990) A plant leucine zipper protein that recognizes an abscisic acid response element. Science 250(4978):267–271

Guo X, Ullah A, Siuta D, Kukfisz B, Iqbal S (2022) Role of WRKY transcription factors in regulation of abiotic stress responses in cotton. Life 12(9):1410

Guy CL, Niemi KJ, Brambl R (1985) Altered gene expression during cold acclimation of spinach. Proc Natl Acad Sci 82(11):3673–3677

Haake V, Cook D, Riechmann J, Pineda O, Thomashow MF, Zhang JZ (2002) Transcription factor CBF4 is a regulator of drought adaptation in *Arabidopsis*. Plant Physiol 130(2):639–648
Hassan MA, Xiang C, Farooq M, Muhammad N, Yan Z, Hui X, Yuanyuan K, Bruno AK, Lele Z, Jincai L (2021) Cold stress in wheat: plant acclimation responses and management strategies. Front Plant Sci 12:676884
Hobo T, Asada M, Kowyama Y, Hattori T (1999) ACGT-containing abscisic acid response element (ABRE) and coupling element 3 (CE3) are functionally equivalent. Plant J 19(6):679–689
Hu Y, Zhang L, Zhao L, Li J, He S, Zhou K, Yang F, Huang M, Jiang L, Li L (2011) Trichostatin A selectively suppresses the cold-induced transcription of the *ZmDREB1* gene in maize. PLoS One 6(7):e22132
Huang S, Gao Y, Liu J, Peng X, Niu X, Fei Z, Cao S, Liu Y (2012) Genome-wide analysis of WRKY transcription factors in *Solanum lycopersicum*. Mol Gen Genomics 287:495–513
Huang X, Shi H, Hu Z, Liu A, Amombo E, Chen L, Fu J (2017) ABA is involved in regulation of cold stress response in bermudagrass. Front Plant Sci 8:1613
Imran QM, Falak N, Hussain A, Mun BG, Yun BW (2021) Abiotic stress in plants; stress perception to molecular response and role of biotechnological tools in stress resistance. Agronomy 11(8):1579
Ishiguro S, Nakamura K (1994) Characterization of a cDNA encoding a novel DNA-binding protein, SPF1, that recognizes SP8 sequences in the 5′ upstream regions of genes coding for sporamin and β-amylase from sweet potato. Mol Gen Genet MGG 244:563–571
Ito Y, Katsura K, Maruyama K, Taji T, Kobayashi M, Seki M, Shinozaki K, Yamaguchi-Shinozaki K (2006) Functional analysis of rice DREB1/CBF-type transcription factors involved in cold-responsive gene expression in transgenic rice. Plant Cell Physiol 47(1):141–153
Jaglo-Ottosen KR, Gilmour SJ, Zarka DG, Schabenberger O, Thomashow MF (1998) *Arabidopsis* CBF1 overexpression induces COR genes and enhances freezing tolerance. Science 280(5360): 104–106
Jia M, Meng X, Song X, Zhang D, Kou L, Zhang J, Jing Y, Liu G, Liu H, Huang X, Wang Y (2022) Chilling-induced phosphorylation of IPA1 by OsSAPK6 activates chilling tolerance responses in rice. Cell Discov 8(1):71
Jiang J, Ma S, Ye N, Jiang M, Cao J, Zhang J (2017) WRKY transcription factors in plant responses to stresses. J Integr Plant Biol 59(2):86–101
Jiang J, Hou R, Yang N, Li L, Deng J, Qin G, Ding D (2021) Physiological and TMT-labeled proteomic analyses reveal important roles of sugar and secondary metabolism in Citrus junos under cold stress. J Proteome 237:104145
Jin C, Li KQ, Xu XY, Zhang HP, Chen HX, Chen YH, Hao J, Wang Y, Huang XS, Zhang SL (2017) A novel NAC transcription factor, PbeNAC1, of *Pyrus betulifolia* confers cold and drought tolerance via interacting with *PbeDREBs* and activating the expression of stress-responsive genes. Front Plant Sci 8:1049
Kang H, Fan T, Wu J, Zhu Y, Shen WH (2022) Histone modification and chromatin remodeling in plant response to pathogens. Front Plant Sci 13:986940
Karimi R (2017) Potassium-induced freezing tolerance is associated with endogenous abscisic acid, polyamines and soluble sugars changes in grapevine. Sci Hortic 215:184–194
Khaldun ABM, Huang W, Lv H, Liao S, Zeng S, Wang Y (2016) Comparative profiling of miRNAs and target gene identification in distant-grafting between tomato and lycium (Goji Berry). Front Plant Sci 7:1475
Khodakovskaya M, McAvoy R, Peters J, Wu H, Li Y (2006) Enhanced cold tolerance in transgenic tobacco expressing a chloroplast ω-3 fatty acid desaturase gene under the control of a cold-inducible promoter. Planta 223:1090–1100
Khoso MA, Hussain A, Ritonga FN, Ali Q, Channa MM, Alshegaihi RM, Meng Q, Ali M, Zaman W, Brohi RD, Liu F (2022) WRKY transcription factors (TFs): molecular switches to regulate drought, temperature, and salinity stresses in plants. Front Plant Sci 13:1039329

Kidokoro S, Kim JS, Ishikawa T, Suzuki T, Shinozaki K, Yamaguchi-Shinozaki K (2020) DREB1A/CBF3 is repressed by transgene-induced DNA methylation in the *Arabidopsis ice1-1* mutant. Plant Cell 32(4):1035–1048

Kidokoro S, Hayashi K, Haraguchi H, Ishikawa T, Soma F, Konoura I, Toda S, Mizoi J, Suzuki T, Shinozaki K, Yamaguchi-Shinozaki K (2021) Posttranslational regulation of multiple clock-related transcription factors triggers cold-inducible gene expression in *Arabidopsis*. Proc Natl Acad Sci 118(10):e2021048118

Kidokoro S, Konoura I, Soma F, Suzuki T, Miyakawa T, Tanokura M, Shinozaki K, Yamaguchi-Shinozaki K (2023) Clock-regulated coactivators selectively control gene expression in response to different temperature stress conditions in *Arabidopsis*. Proc Natl Acad Sci 120(16):e2216183120

Kim S, Kang JY, Cho DI, Park JH, Kim SY (2004) ABF2, an ABRE-binding bZIP factor, is an essential component of glucose signaling and its overexpression affects multiple stress tolerance. Plant J 40(1):75–87

Kim CY, Vo KTX, Nguyen CD, Jeong DH, Lee SK, Kumar M, Kim SR, Park SH, Kim JK, Jeon JS (2016) Functional analysis of a cold-responsive rice WRKY gene, OsWRKY71. Plant Biotechnol Rep 10:13–23

Knight MR, Knight H (2012) Low-temperature perception leading to gene expression and cold tolerance in higher plants. New Phytol 195(4):737–751

Kwon CS, Lee D, Choi G, Chung WI (2009) Histone occupancy-dependent and-independent removal of H3K27 trimethylation at cold-responsive genes in *Arabidopsis*. Plant J 60(1): 112–121

Lamers J, Van Der Meer T, Testerink C (2020) How plants sense and respond to stressful environments. Plant Physiol 182(4):1624–1635

Li S, Yu X, Lei N, Cheng Z, Zhao P, He Y, Wang W, Peng M (2017) Genome-wide identification and functional prediction of cold and/or drought-responsive lncRNAs in cassava. Sci Rep 7(1): 45981

Li R, Zhang L, Wang L, Chen L, Zhao R, Sheng J, Shen L (2018) Reduction of tomato-plant chilling tolerance by CRISPR–Cas9-mediated SlCBF1 mutagenesis. J Agric Food Chem 66(34):9042–9051

Li X, Liang X, Li W, Yao A, Liu W, Wang Y, Yang G, Han D (2022) Isolation and functional analysis of MbCBF2, a *Malus baccata* (L.) Borkh CBF transcription factor gene, with functions in tolerance to cold and salt stress in transgenic *Arabidopsis thaliana*. Int J Mol Sci 23(17):9827

Li C, Mao B, Wang K, Xu L, Fan L, Wang Y, Li Y, Ma Y, Wang L, Liu L (2023a) RsERF40 contributes to cold stress tolerance and cell expansion of taproot in radish (*Raphanus sativus* L.). Horticult Res 10(3):uhad013

Li M, Zhang X, Zhang T, Bai Y, Chen C, Guo D, Guo C, Shu Y (2023b) Genome-wide analysis of the WRKY genes and their important roles during cold stress in white clover. PeerJ 11:e15610

Liang X, Luo G, Li W, Yao A, Liu W, Xie L, Han M, Li X, Han D (2022) Overexpression of a *Malus baccata* CBF transcription factor gene, MbCBF1, Increases cold and salinity tolerance in *Arabidopsis thaliana*. Plant Physiol Biochem 192:230–242

Liu J, He Z (2020) Small DNA methylation, big player in plant abiotic stress responses and memory. Front Plant Sci 11:595603

Liu Q, Kasuga M, Sakuma Y, Abe H, Miura S, Yamaguchi-Shinozaki K, Shinozaki K (1998) Two transcription factors, DREB1 and DREB2, with an EREBP/AP2 DNA binding domain separate two cellular signal transduction pathways in drought-and low-temperature-responsive gene expression, respectively. Arabidopsis Plant Cell 10(8):1391–1406

Liu X, Bai X, Wang X, Chu C (2007) OsWRKY71, a rice transcription factor, is involved in rice defense response. J Plant Physiol 164(8):969–979

Liu Y, Dang P, Liu L, He C (2019) Cold acclimation by the CBF–COR pathway in a changing climate: lessons from *Arabidopsis thaliana*. Plant Cell Rep 38:511–519

Lv K, Li J, Zhao K, Chen S, Nie J, Zhang W, Liu G, Wei H (2020) Overexpression of an *AP2/ERF* family gene, *BpERF13*, in birch enhances cold tolerance through upregulating CBF genes and mitigating reactive oxygen species. Plant Sci 292:110375

Ma Q, Dai X, Xu Y, Guo J, Liu Y, Chen N, Xiao J, Zhang D, Xu Z, Zhang X, Chong K (2009) Enhanced tolerance to chilling stress in OsMYB3R-2 transgenic rice is mediated by alteration in cell cycle and ectopic expression of stress genes. Plant Physiol 150(1):244–256

Mao D, Chen C (2012) Colinearity and similar expression pattern of rice *DREB1s* reveal their functional conservation in the cold-responsive pathway. e47275

Mao X, Jia D, Li A, Zhang H, Tian S, Zhang X, Jia J, Jing R (2011) Transgenic expression of TaMYB2A confers enhanced tolerance to multiple abiotic stresses in Arabidopsis. Funct Integr Genomics 11:445–465

Mariño-Ramírez L, Kann MG, Shoemaker BA, Landsman D (2005) Histone structure and nucleosome stability. Expert Rev Proteomics 2(5):719–729

Maruyama K, Sakuma Y, Kasuga M, Ito Y, Seki M, Goda H, Shimada Y, Yoshida S, Shinozaki K, Yamaguchi-Shinozaki K (2004) Identification of cold-inducible downstream genes of the *Arabidopsis* DREB1A/CBF3 transcriptional factor using two microarray systems. Plant J 38(6):982–993

Maruyama K, Todaka D, Mizoi J, Yoshida T, Kidokoro S, Matsukura S, Takasaki H, Sakurai T, Yamamoto YY, Yoshiwara K, Kojima M (2012) Identification of cis-acting promoter elements in cold-and dehydration-induced transcriptional pathways in *Arabidopsis*, rice, and soybean. DNA Res 19(1):37–49

Medina J, Bargues M, Terol J, Pérez-Alonso M, Salinas J (1999) The Arabidopsis CBF gene family is composed of three genes encoding AP2 domain-containing proteins whose expression is regulated by low temperature but not by abscisic acid or dehydration. Plant Physiol 119(2): 463–470

Mikołajczak K, Kuczyńska A, Krajewski P, Kempa M, Nuc M (2023) Transcriptome profiling disclosed the effect of single and combined drought and heat stress on reprogramming of genes expression in barley flag leaf. Front Plant Sci 13:1096685

Miura K, Furumoto T (2013) Cold signalling and cold response in plants. Int J Mol Sci 14(3): 5312–5337

Moraes de Freitas GP, Basu S, Ramegowda V, Thomas J, Benitez LC, Braga EB, Pereira A (2019) Physiological and transcriptional responses to low-temperature stress in rice genotypes at the reproductive stage. Plant Signal Behav 14(4):e1581557

Morran S, Eini O, Pyvovarenko T, Parent B, Singh R, Ismagul A, Eliby S, Shirley N, Langridge P, Lopato S (2011) Improvement of stress tolerance of wheat and barley by modulation of expression of DREB/CBF factors. Plant Biotechnol J 9(2):230–249

Nakamura J, Yuasa T, Huong TT, Harano K, Tanaka S, Iwata T, Phan T, Iwaya M (2011) Rice homologs of inducer of CBF expression (OsICE) are involved in cold acclimation. Plant Biotechnol 28(3):303–309

Novillo F, Alonso JM, Ecker JR, Salinas J (2004) CBF2/DREB1C is a negative regulator of CBF1/ DREB1B and CBF3/DREB1A expression and plays a central role in stress tolerance in *Arabidopsis*. Proc Natl Acad Sci 101(11):3985–3990

Oh SJ, Kwon CW, Choi DW, Song SI, Kim JK (2007) Expression of barley HvCBF4 enhances tolerance to abiotic stress in transgenic rice. Plant Biotechnol J 5(5):646–656

Pan YH, Wu WP, Xiong XD (2020) Circular RNAs: promising biomarkers for age-related diseases. Aging Dis 11(6):1585

Pandey R, MuÈller A, Napoli CA, Selinger DA, Pikaard CS, Richards EJ, Bender J, Mount DW, Jorgensen RA (2002) Analysis of histone acetyltransferase and histone deacetylase families of *Arabidopsis thaliana* suggests functional diversification of chromatin modification among multicellular eukaryotes. Nucleic Acids Res 30(23):5036–5055

Park J, Lim CJ, Shen M, Park HJ, Cha JY, Iniesto E, Rubio V, Mengiste T, Zhu JK, Bressan RA, Lee SY (2018) Epigenetic switch from repressive to permissive chromatin in response to cold stress. Proc Natl Acad Sci 115(23):E5400–E5409

Qian C, Li L, Guo H, Zhu G, Yang N, Tan X, Zhao H (2023) Genome-wide analysis of DREB family genes and characterization of cold stress responses in the woody plant *Prunus nana*. Gene 14(4):811

Ramakrishnan M, Papolu PK, Satish L, Vinod KK, Wei Q, Sharma A, Emamverdian A, Zou LH, Zhou M (2022) Redox status of the plant cell determines epigenetic modifications under abiotic stress conditions and during developmental processes. J Adv Res 42:99–116

Ramamoorthy R, Jiang SY, Kumar N, Venkatesh PN, Ramachandran S (2008) A comprehensive transcriptional profiling of the WRKY gene family in rice under various abiotic and phytohormone treatments. Plant Cell Physiol 49(6):865–879

Roy D, Paul A, Roy A, Ghosh R, Ganguly P, Chaudhuri S (2014) Differential acetylation of histone H3 at the regulatory region of OsDREB1b promoter facilitates chromatin remodelling and transcription activation during cold stress. PLoS One 9(6):e100343

Roychoudhury A, Paul S, Basu S (2013) Cross-talk between abscisic acid-dependent and abscisic acid-independent pathways during abiotic stress. Plant Cell Rep 32:985–1006

Salzman J, Gawad C, Wang PL, Lacayo N, Brown PO (2012) Circular RNAs are the predominant transcript isoform from hundreds of human genes in diverse cell types. PLoS One 7(2):e30733

Sanghera GS, Wani SH, Hussain W, Singh NB (2011) Engineering cold stress tolerance in crop plants. Curr Genomics 12(1):30

Sangwan V, Foulds I, Singh J, Dhindsa RJ (2001) Cold activation of *Brassica napus* BN115 promoter is mediated by structural changes in membranes and cytoskeleton, and requires Ca^{2+} influx. Plant J 27:1–12

Santer L, Bär C, Thum T (2019) Circular RNAs: a novel class of functional RNA molecules with a therapeutic perspective. Mol Ther 27(8):1350–1363

Sharma M, Kumar P, Verma V, Sharma R, Bhargava B, Irfan M (2022) Understanding plant stress memory response for abiotic stress resilience: molecular insights and prospects. Plant Physiol Biochem 179:10–24

Shen Q, Zhang P, Ho TH (1996) Modular nature of abscisic acid (ABA) response complexes: composite promoter units that are necessary and sufficient for ABA induction of gene expression in barley. Plant Cell 8(7):1107–1119

Shi Y, Yang S (2014) ABA regulation of the cold stress response in plants. In: Abscisic acid: metabolism, transport and signaling, pp 337–363

Su CF, Wang YC, Hsieh TH, Lu CA, Tseng TH, Yu SM (2010) A novel MYBS3-dependent pathway confers cold tolerance in rice. Plant Physiol 153(1):145–158

Su LT, Li JW, Liu DQ, Zhai Y, Zhang HJ, Li XW, Zhang QL, Wang Y, Wang QY (2014) A novel MYB transcription factor, GmMYBJ1, from soybean confers drought and cold tolerance in *Arabidopsis thaliana*. Gene 538(1):46–55

Tuteja N, Sopory SK (2008) Chemical signaling under abiotic stress environment in plants. Plant Signal Behav 3(8):525–536

Vannini C, Campa M, Iriti M, Genga A, Faoro F, Carravieri S, Rotino GL, Rossoni M, Spinardi A, Bracale M (2007) Evaluation of transgenic tomato plants ectopically expressing the rice *Osmyb4* gene. Plant Sci 173(2):231–239

Verma N, Giri SK, Singh G, Gill R, Kumar A (2022) Epigenetic regulation of heat and cold stress responses in crop plants. Plant Gene 29:100351

Wang F, Hou X, Tang J, Wang Z, Wang S, Jiang F, Li Y (2012) A novel cold-inducible gene from Pak-choi (*Brassica campestris* ssp. chinensis), BcWRKY46, enhances the cold, salt and dehydration stress tolerance in transgenic tobacco. Mol Biol Rep 39:4553–4564

Wang ST, Sun XL, Hoshino Y, Yu Y, Jia B, Sun ZW, Sun MZ, Duan XB, Zhu YM (2014) MicroRNA319 positively regulates cold tolerance by targeting OsPCF6 and OsTCP21 in rice (Oryza sativa L.). PLoS One 9(3):e91357

Wang P, Dai L, Ai J, Wang Y, Ren F (2019) Identification and functional prediction of cold-related long non-coding RNA (lncRNA) in grapevine. Sci Rep 9(1):6638

Wang MQ, Huang QX, Lin P, Zeng QH, Li Y, Liu QL, Zhang L, Pan YZ, Jiang BB, Zhang F (2020a) The overexpression of a transcription factor gene *VbWRKY32* enhances the cold tolerance in *Verbena bonariensis*. Front Plant Sci 10:1746

Wang X, Chang X, Jing Y, Zhao J, Fang Q, Sun M, Zhang Y, Li W, Li Y (2020b) Identification and functional prediction of soybean CircRNAs involved in low-temperature responses. J Plant Physiol 250:153188

Wang H, Lu S, Guan X, Jiang Y, Wang B, Hua J, Zou B (2022) Dehydration-responsive element binding protein 1C, 1E, and 1G promote stress tolerance to Chilling, Heat, Drought, and Salt in Rice. Front Plant Sci 13:851731

Wang L, Chen H, Chen G, Luo G, Shen X, Ouyang B, Bie Z (2023) Transcription factor SlWRKY50 enhances cold tolerance in tomato by activating the jasmonic acid signaling. Plant Physiol:kiad578

Waseem M, Yang X, Aslam MM, Li M, Zhu L, Chen S, Li Y, Liu P (2022) Genome-wide identification of long non-coding RNAs in two contrasting rapeseed (*Brassica napus* L.) genotypes subjected to cold stress. Environ Exp Bot 201:104969

Wei X, Liu S, Sun C, Xie G, Wang L (2021) Convergence and divergence: signal perception and transduction mechanisms of cold stress in *Arabidopsis* and Rice. Plants 10(9):1864

Welling A, Palva ET (2008) Involvement of CBF transcription factors in winter hardiness in birch. Plant Physiol 147(3):1199–1211

Wisniewski M, Willick IR, Duman JG, Livingston D, Newton SS (2020) Plant antifreeze proteins. In: Antifreeze proteins volume 1: environment, systematics and evolution, pp 189–226

Wu KL, Guo ZJ, Wang HH, Li J (2005) The WRKY family of transcription factors in rice and *Arabidopsis* and their origins. DNA Res 12(1):9–26

Wu R, Wang Y, Wu T, Xu X, Han Z (2017) MdMYB4, an R2R3-type MYB transcription factor, plays a crucial role in cold and salt stress in apple calli. J Am Soc Hortic Sci 142(3):209–216

Xie Z, Zhang ZL, Zou X, Huang J, Ruas P, Thompson D, Shen QJ (2005) Annotations and functional analyses of the rice *WRKY* gene superfamily reveal positive and negative regulators of abscisic acid signalling in aleurone cells. Plant Physiol 137(1):176–189

Xie JH, Wei TANG, Lu GD, Hong YH, Zhong ZH, Wang ZH, Zheng HK (2023) Histone H3K27me3 methylation regulates the expression of secreted proteins distributed at fast-evolving regions through transcriptional repression of transposable elements. J Integr Agric 22(10):3059–3068

Xing C, Liu Y, Zhao L, Zhang S, Huang X (2019) A novel MYB transcription factor regulates ascorbic acid synthesis and affects cold tolerance. Plant Cell Environ 42(3):832–845

Xu Y, Hu W, Song S, Ye X, Ding Z, Liu J, Wang Z, Li J, Hou X, Xu B, Jin Z (2023) *MaDREB1F* confers cold and drought stress resistance through common regulation of hormone synthesis and protectant metabolite contents in banana. Horticult Res 10(2):uhac275

Xue-Xuan X, Hong-Bo S, Yuan-Yuan M, Gang X, Jun-Na S, Dong-Gang G, Cheng-Jiang R (2010) Biotechnological implications from abscisic acid (ABA) roles in cold stress and leaf senescence as an important signal for improving plant sustainable survival under abiotic-stressed conditions. Crit Rev Biotechnol 30(3):222–230

Yamaguchi-Shinozaki K, Shinozaki K (2006) Transcriptional regulatory networks in cellular responses and tolerance to dehydration and cold stresses. Annu Rev Plant Biol 57:781–803

Yang W, Liu XD, Chi XJ, Wu CA, Li YZ, Song LL, Liu XM, Wang YF, Wang FW, Zhang C, Liu Y (2011) Dwarf apple MbDREB1 enhances plant tolerance to low temperature, drought, and salt stress via both ABA-dependent and ABA-independent pathways. Planta 233:219–229

Yang X, Liu Y, Zhang H, Wang J, Zinta G, Xie S, Zhu W, Nie WF (2020) Genome-wide identification of circular RNAs in response to low-temperature stress in tomato leaves. Front Genet 11:591806

Yang X, Luo Y, Bai H, Li X, Tang S, Liao X, Zhang L, Liu Q (2022) DgMYB2 improves cold resistance in chrysanthemum by directly targeting DgGPX1. Horticult Res 9:uhab028

Yang S, Zhou J, Li Y, Wu J, Ma C, Chen Y, Sun X, Wu L, Liang X, Fu Q, Xu Z (2023) AP2/EREBP pathway plays an important role in Chaling wild rice tolerance to cold stress. Int J Mol Sci 24(19):14441

Yao C, Li X, Li Y, Yang G, Liu W, Shao B, Zhong J, Huang P, Han D (2022) Overexpression of a *Malus baccata* MYB transcription factor gene *MbMYB4* increases cold and drought tolerance in *Arabidopsis thaliana*. Int J Mol Sci 23(3):1794

Yokotani N, Sato Y, Tanabe S, Chujo T, Shimizu T, Okada K, Yamane H, Shimono M, Sugano S, Takatsuji H, Kaku H (2013) WRKY76 is a rice transcriptional repressor playing opposite roles in blast disease resistance and cold stress tolerance. J Exp Bot 64(16):5085–5097

Yuan P, Yang T, Poovaiah BW (2018) Calcium signaling-mediated plant response to cold stress. Int J Mol Sci 19(12):3896

Zeng Z, Zhang W, Marand AP, Zhu B, Buell CR, Jiang J (2019) Cold stress induces enhanced chromatin accessibility and bivalent histone modifications H3K4me3 and H3K27me3 of active genes in potato. Genome Biol 20(1):1–17

Zhang F, Huang L, Wang W, Zhao X, Zhu L, Fu B, Li Z (2012) Genome-wide gene expression profiling of introgressed indica rice alleles associated with seedling cold tolerance improvement in a *japonica* rice background. BMC Genomics 13(1):1–15

Zhang ZW, Feng LY, Cheng J, Tang H, Xu F, Zhu F, Zhao ZY, Yuan M, Chen YE, Wang JH, Yuan S (2013) The roles of two transcription factors, ABI4 and CBFA, in ABA and plastid signalling and stress responses. Plant Mol Biol 83:445–458

Zhang Y, Yu H, Yang X, Li Q, Ling J, Wang H, Gu X, Huang S, Jiang W (2016) CsWRKY46, a WRKY transcription factor from cucumber, confers cold resistance in transgenic-plant by regulating a set of cold-stress responsive genes in an ABA-dependent manner. Plant Physiol Biochem 108:478–487

Zhang Q, Liang Z, Cui X, Ji C, Li Y, Zhang P, Liu J, Riaz A, Yao P, Liu M, Wang Y (2018) N6-methyladenine DNA methylation in *Japonica* and *Indica* rice genomes and its association with gene expression, plant development, and stress responses. Mol Plant 11(12):1492–1508

Zhang P, Li S, Chen M (2020) Characterization and function of circular RNAs in plants. Front Mol Biosci 7:91

Zhao H, Bughrara SS (2008) Isolation and characterization of cold-regulated transcriptional activator *LpCBF3* gene from perennial ryegrass (*Lolium perenne* L.). Mol Gen Genomics 279:585–594

Zhao C, Wang P, Si T, Hsu CC, Wang L, Zayed O, Yu Z, Zhu Y, Dong J, Tao WA, Zhu JK (2017a) MAP kinase cascades regulate the cold response by modulating ICE1 protein stability. Dev Cell 43(5):618–629

Zhao W, Cheng Y, Zhang C, You Q, Shen X, Guo W, Jiao Y (2017b) Genome-wide identification and characterization of circular RNAs by high throughput sequencing in soybean. Sci Rep 7(1): 5636

Zhou QY, Tian AG, Zou HF, Xie ZM, Lei G, Huang J, Wang CM, Wang HW, Zhang JS, Chen SY (2008) Soybean WRKY-type transcription factor genes, *GmWRKY13*, *GmWRKY21*, and *GmWRKY54*, confer differential tolerance to abiotic stresses in transgenic *Arabidopsis* plants. Plant Biotechnol J 6(5):486–503

Zilberman D, Henikoff S (2007) Genome-wide analysis of DNA methylation patterns. pp 3959–3965

Zou C, Jiang W, Yu D (2010) Male gametophyte-specific WRKY34 transcription factor mediates cold sensitivity of mature pollen in *Arabidopsis*. J Exp Bot 61(14):3901–3914

Zuo J, Wang Q, Zhu B, Luo Y, Gao L (2016) Deciphering the roles of circRNAs on chilling injury in tomato. Biochem Biophys Res Commun 479(2):132–138

Chapter 10
Applications of Microbial Biotechnology for Improving Toxic Heavy Metal Stress Tolerance and Yield in Crops

Susana Aylin Castillo-Cortés, Mía López-Portillo Ontiveros, María Fernanda Fonseca-Fonseca, María Fernanda Padilla-Nuño, Karina Franco-Gutiérrez, Andrea Del Rivero-Acuña, Marián Fenton-Aguilar, and Jorge Donato García-García

Abstract Essential heavy metals are vital for crop development, but high concentrations and prolonged exposure can be detrimental. Non-essential heavy metals, even at trace levels, can act as potent toxins for plant cells by mimicking essential metal ions and disrupting cellular functions. This similarity between essential and non-essential heavy metals complicates efforts to mitigate their toxic effects and to prevent the bio-magnification of harmful metals like cadmium, lead, chromium, nickel, and arsenic. Heavy metal toxicity not only affects plant yields, but also poses significant risks to human health through the consumption of contaminated crops. Here we analyze the role of microorganisms in ameliorating the accumulation of toxic heavy metals in crop fruits, either by sequestering metals in root cells or by immobilizing them in the soil. This summary highlights the similarities between heavy metal transporters and phytochelatin synthesis in plants and microorganisms, advocating for the use of microbial genes to engineer plants and enhance food security. Additionally, we explore the trends and perspectives in the field of plant growth-promoting bacteria (PGPB), which aid the adsorption of toxic metals by plants. Microbial biotechnology holds immense potential for tailoring metal transport processes, storing metals in vacuoles, inactivating metals, and limiting the availability of undesirable metals to plants. However, many opportunities remain unexplored in engineering plants with microbial elements. This emerging research area warrants further investigation, along with efforts to communicate the benefits of engineered organisms, addressing concerns and demonstrating controlled management and application of these technologies to ensure their safe and effective use.

S. A. Castillo-Cortés · M. L.-P. Ontiveros · M. F. Fonseca-Fonseca · M. F. Padilla-Nuño · K. Franco-Gutiérrez · A. Del Rivero-Acuña · M. Fenton-Aguilar · J. D. García-García (✉)
Tecnologico de Monterrey, Escuela de Ingenieria y Ciencias, Zapopan, Jalisco, Mexico
e-mail: jdonatogg@tec.mx

S. K. Panda, M. A. Khan (eds.), *Plant Functional Genomics for Abiotic Stress Resilience*, Plant in Challenging Environments 7,
https://doi.org/10.1007/978-3-032-01704-8_10

Keywords Heavy metal pollution · Bioaccumulation of metals · Protein engineering · Microbial communities · Microbial platforms

10.1 Introduction

Heavy metal pollution in lakes, rivers, streams, and freshwater wells, as well soil contamination, continues to pose a threat to agricultural activities and human health, primarily because metals are non-biodegradable. Additionally, the increase in industrial activity, including mining, oil production, the use of fertilizers and pesticides, as well as natural geological processes and human activities, contributes to this problem (Qin et al. 2021). In environmental science and research, “heavy metal” typically refers to metals and metalloids that possess toxic properties and a tendency to accumulate in ecosystems, posing risks to both human and environmental health. The term also encompasses metallic elements with high atomic weights and densities, usually exceeding 5 g/cm^3. Given their impact on health, various scientific disciplines are studying the behavior, fate, and translocation of heavy metals within ecosystems to develop strategies for mitigating or reversing their effects (Ali and Khan 2018). These contaminants may disrupt critical physiological processes in plants and alter microbial communities, thereby impacting food security. This threat necessitates the search for viable solutions to enhance crop yield, prevent fruit contamination, and avoid illness in humans who consume these fruits (Feng et al. 2023). Essential heavy metals play key roles in all cells throughout nature. In plant biological processes, low concentrations of heavy metals such as boron (B), copper (Cu), iron (Fe), manganese (Mn), molybdenum (Mo), nickel (Ni), and zinc (Zn) are necessary for plant health in specific concentrations for processes encompassing photosynthesis, the electron transport chain, and the synthesis of several cofactors. However, elevated levels of micronutrient heavy metals can have detrimental effects on organisms. Conversely, the presence of non-micronutrient heavy metals such as arsenic (As), cadmium (Cd), chromium (Cr), mercury (Hg), thallium (Tl), titanium (Ti), or vanadium (V) at trace concentrations can disrupt cell metabolism, inhibit growth, compromise quality, induce cellular damage, and even lead to plant mortality (Noor et al. 2022). While different heavy metals exhibit varying effects, overall, they impede the normal function of structural proteins, enzymes, metabolic pathways, signaling networks, and organelles (e.g., mitochondria and chloroplasts) within the plant. This leads to functional alterations, oxidative stress, and other adverse effects through the impairment of key components in biological molecules (Feng et al. 2023; Noor et al. 2022).

In humans, it is well-known that the intake of these toxic metals is associated with disruptions in metabolic activities because they can replace essential heavy metals, cause gastrointestinal issues, lead to renal failure, induce osteoporosis, compromise the immune system, and trigger cancer, depending on the duration and doses of exposure. Concerningly, one way to ingest these metals is through contaminated crops. In Nigeria, heavy metals such as Cu, Fe, Zn, As, Cd, Hg, Cr, silver (Ag), lead (Pb), and platinum (Pt) are prevalent in the environment and food crops, primarily due to anthropogenic activities like the use of fertilizers, pesticides, and industrial emissions (Onakpa et al. 2018). These metals are persistent environmental contaminants that bioaccumulate in the food chain, posing severe health risks (Onakpa et al. 2018). Heavy metals in rice and vegetables from Bangladesh pose significant health risks. Proshad et al. (2019) found high levels of Cr, Ni, Cu, As, Cd, and Pb in okra, chili, bitter gourd, papaya, brinjal, bean, bottle gourd, rice, cucumber, sponge gourd, and Indian spinach, indicating carcinogenic and non-carcinogenic risks for consumers. Six years earlier, another study in Bangladesh reported high As levels in drinking water and vegetables, among other non-essential heavy metals, highlighting the necessity of managing this environmental problem to avoid health risks in this country (Rahman et al. 2013). In Ponce Enriquez, Ecuador, the impact of gold mining activity on the metal content in celery, chives, corn, herbs, lettuce, turnips, green beans, cassava, and carrots was analyzed. The findings revealed that the contents of As, Cr, Cu, and Ni in soils exceeded the Ecuadorian quality guidelines for agricultural soils. Additionally, the trace elements concentration in local crops was higher than the maximum permissible levels set by the Food and Agriculture Organization of the United Nations (FAO). The study concluded that inhabitants of the area are at risk of developing carcinogenic and non-carcinogenic effects due to long-term consumption of food with high trace element concentrations (Romero-Crespo et al. 2023).

Regarding heavy metal content in rice, Ali et al. (2020) examined the metal contamination of paddy soil-rice systems in major rice-growing countries (i.e. China, Bangladesh, India, Malaysia, Japan, USA, Korea, Pakistan, and Thailand). The study revealed that the presence of several trace elements, including As, Se, Cd, Pb, Zn, Mn, Ni, and Cr, are linked to both natural and anthropogenic sources, such as industrial activities, agricultural practices, and environmental pollution. Industries contributing to this contamination include mining, smelting, and manufacturing, which release these metals into the environment. Additionally, the use of contaminated water for irrigation and the application of fertilizers and pesticides containing these metals further exacerbate the problem. The conclusions of the review emphasized the severe human health risks posed by the consumption of rice contaminated with these toxic metals. The paper called for rapid action to address this issue, highlighting the need for effective management and remediation measures (Ali et al. 2020). Meanwhile, Baghaie and Aghili (2019) investigated the contamination of Pb and Cd in soil, wheat, and barley in samples collected from 60 agricultural fields around the Shahin mine, Shazand County, Central Iran. The findings revealed that the mean concentrations of Pb (234 mg kg^{-1}) and Cd (1.9 mg kg^{-1}) in soil samples were significantly higher than the background values. In wheat grains, the mean concentrations of Pb and Cd were 0.44 and 0.23 μg g^{-1}, respectively, which were 6.5 and 5.8 times higher for Pb and Cd than those found in barley grains. The study assessed the total non-cancer hazard quotient (THQ) and excess lifetime cancer risks of Pb and Cd for different population groups through various exposure pathways, including soil ingestion, dermal contact, and inhalation of soil particulates. The THQ values for Pb and Cd were greater than one for all groups, indicating potential health risks (Baghaie and Aghili 2019).

Overall, these studies underscore the urgent need for monitoring and managing heavy metal contamination in food crops to protect public health, particularly in regions with high levels of industrial and agricultural pollution; therefore, the authors above (Baghaie and Aghili 2019; Ali et al. 2020) also call for future research to better understand and mitigate the contamination of rice in the global food supply. In this context, microbial biotechnology offers different solutions to overcome the challenges of heavy metal pollution. Microbial biotechnology encompasses the application of microorganisms, including bacteria, yeast, and fungi, to tackle a diverse array of challenges across industries, healthcare, agriculture, and environmental management. In the realm of environmental pollution, microbial biotechnology can directly contribute to ecosystem restoration through mechanisms such as adsorption and bioaccumulation. Additionally, it can indirectly improve plant tolerance to polluted soils, thus facilitating the recovery of degraded environments. In agricultural contexts, microbial biotechnology offers significant benefits by optimizing soil fertility and enhancing crop yields through the exploitation of microorganisms' metabolic capabilities. These technological advancements hold the potential to transform food production by enhancing quality, nutrition, and preservation (Yu et al. 2019). Furthermore, the integration of microbial biotechnology with cutting-edge molecular tools and synthetic biology enables the creation of innovative platforms for the precise engineering of crops. This approach allows for the development of cultivars with improved yield potential and increased resilience to both biotic and abiotic stressors. Such precision-driven methodologies accelerate crop breeding while preserving beneficial traits that have evolved naturally (Roell and Zurbriggen 2020). It is noteworthy that both plants and microorganisms employ various resistance strategies to handle stress from excessive essential heavy metals or harmful toxic metals. This opens opportunities to use enzymes or transporters from microorganisms, which have advantageous kinetic characteristics, to help manage and sequester metals in plant roots and fruits. One of the primary goals of plant genetic engineering is to prevent metal accumulation in fruits by (i) reducing root-to-shoot translocation and (ii) compartmentalizing metals within root vacuoles (Fig. 10.1). Another critical strategy in plant science is to elucidate the dynamics of plant growth-promoting bacteria (PGPB) to enhance their ability to reduce the bioavailability of toxic heavy metals in soils, thereby limiting plant uptake (Fig. 10.1). Accordingly, this review focuses on the role of heavy metal transporters, the synthesis of phytochelatins as a pivotal step in the detoxification and sequestration of metals within root vacuoles, and recent advancements in PGPB research that can be leveraged to mitigate metal accumulation in crops.

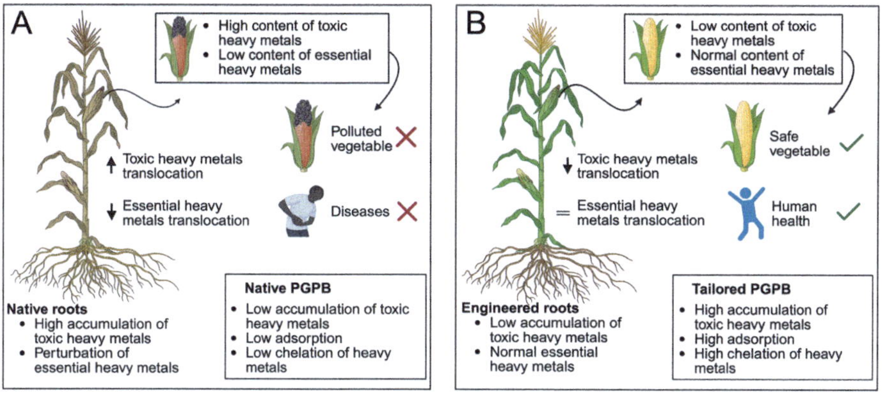

Fig. 10.1 Microbial biotechnology's role in safer food production through plasma membrane and vacuolar transport processes, metal inactivation via phytochelatins, and the reduction of metal availability in soils by PGPB. (**a**) Plant and PGPB interaction without engineering interventions, and (**b**) with engineering interventions to limit root-to-shoot translocation of toxic heavy metals while maintaining the homeostasis of essential metals. The goal of enhancing these mechanisms for toxic metal acquisition is to generate safer crops that can thrive in polluted soils, thereby reducing the risk of human diseases associated with heavy metal exposure

10.2 Contribution of Microbial Biotechnology in Reducing Heavy Metal Accumulation in Crop Fruits

The investigation into microbial mechanisms for mitigating heavy metal stress has become a promising area of research. This is because microorganisms have shown a remarkable ability to reduce heavy metal toxicity through various biochemical and physiological processes (García-García et al. 2016). Most of these mechanisms give microorganisms an advantage in resisting and accumulating heavy metals, which can be beneficial for the remediation of soils and water bodies. Therefore, this review focuses on the biochemical mechanisms used by microorganisms and how microbial technology can help address heavy metal stress in crops and plants, specifically the processes of transport through cellular and subcellular compartment membranes and bioaccumulation. In this specific section, we leave out the mechanism of biosorption because the focus is on bioaccumulation mechanisms of microorganisms that can be used in storage processes into the vacuole of roots of crops. However, we will revisit the biosorption mechanism on PGPB due to the impact that the soil bacterial community has on the availability of heavy metals for plant absorption.

In this section, we will analyze recent findings on transporters from plants and microorganisms that can be utilized in plant engineering, as well as the biosynthesis of glutathione (GSH) and phytochelatins (PCs). We will then review reports on engineering these processes in plants using components from microorganisms. This is significant because enhancing these pathways could help reduce the accumulation of harmful heavy metals in crop fruits while maintaining essential heavy metal levels (Fig. 10.1).

10.2.1 Family of Transporters for Bioaccumulation of Toxic Heavy Metals in Plants and Microorganisms

Plants have developed various mechanisms to cope with heavy metal stress, safeguarding against metal toxicity and ensuring survival in harsh conditions. Strategic interventions, such as soil amendments and agronomic practices, aim to sequester heavy metals within the soil matrix or reducing their bioavailability to crops. However, these intrinsic defense methods exhibit limited efficiency, especially with increasing pollutant levels in soil, water, and air. Therefore, enhancing resistance to heavy metals in plants is crucial.

In addition, plants that form symbiotic relationships with various communities of microorganisms can develop resistance to heavy metals. This resistance is achieved through the advantages gained from the symbiotic population, which we will discuss in detail later. It is important to note that free-living microorganisms play a crucial role in developing resistance capabilities through mutations and gene transfer mechanisms at a significantly faster rate than plants (Manríquez et al. 2021; Zilber-Rosenberg and Tosenberg 2008). This rapid adaptability is influenced by factors such as soil and water conditions, as well as abiotic and biotic stressors. Therefore, understanding the metabolic adaptability of microorganisms and enhancing it through genetic and protein engineering can provide alternative solutions for crops to improve their resistance against heavy metals and prevent the accumulation of toxic metals in fruits. Here, we will detail some key heavy metal transports and bioaccumulation mechanisms in plants and microorganisms, which could be targets for future engineering proposals.

Heavy metal ATPase (HMA) proteins are crucial for Cd transport and physiological ion traffic in plants. HMAs, or P_{1B}-ATPases, have 6–8 transmembrane domains (TDMs), a nucleotide-binding domain (NBD), a phosphorylation domain, and at least one metal-binding domain for interacting with toxic heavy metals like Cd, Ag, and Pb, as well as essential metals like Zn, Cu and Co (Chen et al. 2023; Takahashi et al. 2012; Zhang et al. 2021). The HMAs from *Arabidopsis thaliana* are well-studied, with HMA1 to HMA4 involved in Zn, Co, Cd, and Pb transport, and HMA5 to HMA8 linked to Cu and Ag transport. Notably, HMA2 and HMA4 are vital for Cd translocation from roots to shoots, as knockout plants for these proteins interrupt this process (Wong and Cobbett 2009). HMA3 manages Cd, Zn, and other divalent ion transport into the vacuole (Pita-Barbosa and Ricachenevsky 2022). In *Oryza sativa*, multiple HMA proteins have been identified. OsHMA1 to OsHMA3 translocate Cd, Co, Pb, and Zn, while OsHMA4 to OsHMA9 are associated with Cu and Ag. Similar to AtHMA3, OsHMA3 specializes in sequestering Cd in root vacuoles (Takahashi et al. 2012). Although fully characterized in *A. thaliana* and *O. sativa*, recent genomic projects have expanded HMA understanding in crops like maize, barley, wheat, peanut, and flax (Batool et al. 2023; Khan et al. 2020; Li et al. 2024; Liao et al. 2023; Zhang et al. 2021). In microorganisms, HMA studies are limited. In *Chlamydomonas reinhardtii* and *Saccharomyces cerevisiae*, P-type ATPases called CTRs are responsible for copper uptake without ATP consumption.

CTR1 and CTR2 show higher copper uptake contributions than an HMA1 isoform in *C. reinhardtii* (Page et al. 2009). However, their role in toxic metal accumulation is less studied. In *Dunaliella tertiolecta*, transcriptomics identified HMA1 to HMA4, but they remained uncharacterized (Popova et al. 2018). *S. cerevisiae*'s P_{1B}-type ATPase CAD2 and PCA1 mutants correlate with Cd resistance, reducing Cd concentration in yeast (Adle et al. 2007; Shiraishi et al. 2000). Other vacuolar transporters, such as MTP, VIT, and ZIP, regulate essential and toxic metal homeostasis in plants and microorganisms, primarily for Fe, Zn, and Cd (Chakravorty et al. 2023; Chen et al. 2023; Tripathi and Poluri 2021). The NRAMP family, with 10–12 TMDs, is considered low-affinity metal transporters for Mn and is involved in Fe, Mn, Al, and Cd homeostasis in crops and microorganisms (Pottier et al. 2015; Kanwal et al. 2024). Despite the complexities of microbial isoforms, NRAMP proteins in plants like TjNRAMP4 from *Thlaspi japonicum* exhibit specialized transport capacities for metals like Ni (Mizuno et al. 2005). In *C. reinhardtii*, there are identified isoforms NRAMP1 to NRAMP4, where NRAMP1 and NRAMP2 have similarities with NRAMP from bacteria, and NRAMP4 have high similarities with AtNRAMP3 and AtNRAMP4 (Blaby-Haas and Merchant 2012). Interestingly, a transcriptomic study of *C. reinhardtii* showed that none of the four NRAMP members were overexpressed under cadmium stress, suggesting a low contribution to managing this toxicity (Zhang et al. 2023). In contrast, Smf1p (an NRAMP isoform) showed a high contribution to Cd accumulation in *S. cerevisiae* (Liu et al. 1997). Therefore, many differences in kinetic properties govern microbial isoforms.

The ABC transporters are classified into eight subfamilies in plants: A, B, C, D, E, F, G, and I (Naaz et al. 2023). Typically, ABC transporters have four structures conserved: two cytoplasmic nucleotide-binding domains (NBD) and two highly hydrophobic transmembrane domains (TMD) with highly conserved motifs. The energy required to move substrates from one cellular compartment to another comes from an ATP molecule that binds in one of the NBD (Brunetti et al. 2015; Naaz et al. 2023). Particularly, C-type ABC transporters have shown the capacity to sequester heavy metals in complex with metabolites into the vacuole. In *Arabidopsis*, of the 15 proteins that constitute ABCC family, AtABCC1, AtABCC2, AtABCC3, AtABCC4, and AtABCC7 have been characterized for transporting cadmium or arsenic complexed with phytocheatins (PCs, polymers of glutathione units) (Brunetti et al. 2015; Park et al. 2012; Song et al. 2010). Importantly, these ABC transporters play a crucial role in compartmentalizing metals into vacuoles, similar to HMA and NRAMP transporters. However, the difference is that ABC transporters translocate Cd in the form of complexes with PCs, which is an important mechanism for accumulating toxic metals in shoots and roots (Brunetti et al. 2015; Zhang et al. 2018). Kinetic characterization of the vacuolar transport of Cd-PC_2 (PC_2 = (γGlu-Cys)$_2$-Gly) and Cd-PC_3 (PC_3 = (γGlu-Cys)$_3$-Gly) in tonoplasts of *Avena sativa* showed a $V_{max} = 0.7$–1 nmol × min/mg protein, in an ATP-dependent process (Salt and Rauser 1995). The engineering of ABC transporters to develop crops free of toxic heavy metals is promising through the promotion of the metal accumulation in vacuoles of roots (Naaz et al. 2023; Zhang et al. 2018). In fungi, ABC transporters are classified into the same family groups as in plants. However, B-type ABC transportes include Atm1p from *S. cerevisiae* which is associated with the export of iron-sulfur clusters from the mitochondrial matrix, and Hmt1 (Heavy metal tolerance 1) from *Schizosaccharomyces pombe*, which is localized in the vacuole to translocate Cd-PCs complexes and contribute to heavy metal resistant mechanisms (Kovalchuk and Driessen 2010). Additionally, *S. cerevisiae*, *S. pombe*, *Candida albicans*, *Kluyveromyces lactis*, *Yarrowia lipolytica*, and *Puccinia graminis* have the C-type ABC transporter called Ycf1 (yeast cadmium factor protein), which is linked to the vacuolar transport of heavy metals complexed with glutathione (Kovalchuk and Driessen 2010). Cd-PCs complexes have been described in *S. pombe*, *C. reinhardtii* and the alga-like *Euglena gracilis* (Mendoza-Cózatl et al. 2006; Murasugi et al. 1981; Shengxi et al. 2001). However, the molecular and kinetic characterization of the ABC transporters in vacuoles remains unknown. Interestingly, the storage of Cd-PC complexes has been described in the chloroplast of *E. gracilis*, which is suggested to be associated with ABC transporters (Khatiwada et al. 2020a; Mendoza-Cózatl et al. 2006). The heavy metal resistance and accumulation capabilities of this alga-like organism are promising for use as a model of bioaccumulation due to its ability to accumulate Zn, Pb, nickel (Ni), and chromium (Cr) (García-García et al. 2018; García-García et al. 2009; Hernández-Garnica et al. 2021; Sánchez-Thomas et al. 2006). Recent and future genomic and transcriptomic projects will provide information about the molecular components of metal resistant mechanisms, including ABC transporters, facilitating their use in engineering applications (Geimer et al. 2009; Khatiwada et al. 2020b; Krüger et al. 2019).

10.2.2 Phytochelatins as a Key Metabolite for Bioaccumulation of Toxic Heavy Metals

Heavy metals are stored in intracellular compartments to regulate their homeostasis and prevent toxicity in all eukaryotic cells. The plant vacuole, an intracellular membrane-bound organelle, plays a multifaceted role as a reservoir for various physiological ions, metabolites, and toxic compounds (Naaz et al. 2023; Zhang et al. 2018). Within the vacuole, diverse families of transporters orchestrate the uptake and sequestration of heavy metals, thereby enhancing the plant's resilience against metal-induced stress. This mechanism also governs metal homeostasis in microorganisms (Chakravorty et al. 2023; Chen et al. 2023; Khatiwada et al. 2020a; Tripathi and Poluri 2021; Zhang et al. 2023). In some algae, such as *E. gracilis*, the chloroplast also serves as a storage compartment (Mendoza-Cózatl et al. 2006). Interestingly, the storage mechanisms might be complemented by chelation mechanism to inactivate the reactivity of toxic heavy metals. Phytochelatins (PCs) represent one of the main cellular detoxification mechanisms via the sequestration of metals. PCs possess the capability to form metal-PC complexes, facilitating their transport via ABC-type transporters across cellular membranes toward the vacuole or chloroplasts in plants. PCs, which are polymers of glutathione (GSH), have also been identified in bacteria, fungi, algae, protists, plants, and some worms (Mendoza-Cózatl et al. 2006; Moreno-Sánchez et al. 2017; Vatamaniuk et al. 2004). Unlike metallothioneins, PCs are not encoded in the genome and have the general formula of (γ-Glu-Cys)n-Gly, where "n" commonly ranges from two to six (Moreno-Sánchez et al. 2017). The biosynthesis of PCs depends on the GSH biosynthesis. GSH, an antioxidant metabolite composed of the three amino acids γ-Glu-Cys-Gly, plays an important role against heavy metals, as it is highly abundant (~10 mM) and has the capacity to form metal-glutathione complexes (Pearson and Cowan 2021). The GSH molecule is oxidized to GSSG when reactive oxygen species interact with its thiol group (-SH) in the Cys residue. However, the enzyme GSH reductase (GR) regenerates GSH by consuming NADPH (Asthir et al. 2020). The first reaction of GSH biosynthesis depends on the pool of cysteine produced from the sulfur-assimilation pathway and the pool of glutamate, as well as the activity of the enzyme γ-glutamylcysteine synthetase (γ-ECS). Then, the intermediary γ-glutamylcysteine and glycine are taken as substrates by glutathione synthetase (GS) to yield GSH. Both reactions consume ATP and occur in all cell lineages, from prokaryotes to eukaryotes (Asthir et al. 2020; Pearson and Cowan 2021). GSH has distinctive thiol and carboxyl functional groups, enabling it to form complexes with heavy metals. This increases the resistance of plants and microorganisms, as the metal-GSH complexes are transported across vacuolar membrane using ABC-type transporters for storage (Pearson and Cowan 2021; Fan et al. 2023; García-García et al. 2016).

Although PC biosynthesis has been observed upon exposure to various toxic metals, this mechanism appears to have evolved to specialize in Cd stress, the reason for which remains unknown (García-García et al. 2016; García-García et al. 2020; Vatamaniuk et al. 2004). In 2005, Vivares et al. presented the crystal structure for a PC synthase (PCS) of the cyanobacterium *Nostoc* sp. (NsPCS), which contains only one catalytic domain of the eukaryotic synthase. The significance of this crystal structure lies in the fact that no other PCS from plants or microorganisms has been crystallized, making this enzyme an engineering challenge to improve its catalytic performance and metal-affinity preference through rational design. The NsPCS crystal showed that the enzyme is a dimer belonging to the papain superfamily of cysteine proteases, with a conserved catalytic site that binds a glutamyl cysteine acyl-enzyme intermediate derived from GSH (the enzyme's first substrate). However, the second site that binds the metallic substrate Metal-GS_2 (bis-glutathionate of metal) remains completely unknown. For Cd and Zn exposure, the second substrates are Cd-GS_2 and Zn-GS_2. It is worth mentioning that the structures of other metals should be different, which could determine the difference in affinity for the second site of PCSs (García-García et al. 2016; García-García et al. 2020).

PCs play a crucial role in heavy metal resistance in maize, wheat, rice, and avena, among other crops, by chelating and sequestering toxic metal ions, thereby reducing their harmful effects. In maize, the upregulation of the phytochelatin synthase gene *ZmPCS1* under Cd stress enhances Cd tolerance. Overexpression of *ZmPCS1* in *Arabidopsis* and transient expression in tobacco leaves increase Cd tolerance by enhancing GSH and PC contents, which in turn reduce reactive oxygen species

(ROS) accumulation (Jin et al. 2022). In wheat, the *Rht-B1c* dwarf mutant exhibits higher Cd tolerance compared to the wild type due to a metabolic shift from proline and polyamine synthesis to phytochelatin synthesis, leading to higher accumulation of PCs, particularly PC_4, which detoxifies Cd via vacuolar compartmentalization (Szalai et al. 2020). Analysis of two rice mutants defective in *OsABCC1* and *OsPCS1*, known as *has1* and *has2* mutants, respectively, showed that activity of OsPCS1 and the vacuolar transporter OsABCC1 are essential for reducing As levels in grains (Hayashi et al. 2017). The study suggested that OsPCS1, which functions similarly in the presence of As and Cd, works with the OsABCC1 transporter to sequester As-PC complexes into vacuoles of nodes in wild-type rice plants. As accumulation in grains was approximately nine times lower in the nodes of *has1* and *has2* mutants than in wild-type plants, while As accumulation was approximately five times higher in the grains of *has1* and *has2* mutants compared to wild-type plants. Interestingly, complementation of *has2* with OsPCS1 decreased the As content in grains, achieving metal levels similar to those in wild-type plants. (Hayashi et al. 2017). In avena (black oat), the high Cd-accumulating cultivar "New oat" shows strong Cd tolerance primarily through ascorbate peroxidase and superoxide dismutase activities than PCs, compared to the Cd-sensitive cultivar "Soil saver". Although Cd exposure increases PC levels in both 'New oat' and "Soil saver", the induction of PCs is not superior in 'New oat,' indicating that antioxidative activities play more significant roles in Cd hyperaccumulator oats (Uraguchi et al. 2009). Overall, while PCs are important for heavy metal detoxification in these crops, their relative contribution varies, with other mechanisms such as antioxidative enzyme activities also playing crucial roles.

10.2.3 ABC Transporters and Phytochelatin Metabolism Using Microbial Proteins to Favor Bioaccumulation of Metals in Plant Roots

Heavy metal contamination in agricultural soils poses a significant threat to food security, as it leads to the accumulation of toxic metals in crops, which can enter the food chain and endanger human health. To address this issue, genetic engineering offers a promising solution to develop crops with reduced heavy metal uptake and enhanced tolerance. Targeting plasmatic membrane and vacuolar transporters of heavy metals and metal complexes is crucial for developing plants with a high capacity for metal accumulation in roots, thereby preventing metal translocation from root to shoot. Additionally, enhancing the capacity to produce PCs could be beneficial, as PCs are highly efficient metal chelators and substrates for vacuolar transporters. This approach can be extended to various crop plants, potentially reducing heavy metal content in food crops and ensuring safer food production. Enzymes from microorganisms could be used as genetic components to achieve these goals. Here, we summarize representative experimental trials demonstrating the potential of these genetic engineering strategies.

The paper by Lee et al. (2003) investigated the functional expression of the *ZntA* gene—a Zn(II)-translocating P-type ATPase from *E. coli*—in *A. thaliana* and its consequent impact on Cd accumulation. Initially, the *ZntA* conferred enhanced resistance to Cd and Pb in the YCF1-deficient mutant of *S. cerevisiae* strain DTY167, as evidenced by the growth of transformed yeast cells in the presence of 2 mM Pb or 50 μM Cd. Subsequently, *ZntA* was introduced into Arabidopsis to assess its efficacy in conferring tolerance to and reducing the accumulation of Cd and Pb. When cultivated in media containing 700 μM Pb^{2+} or 70 μM Cd^{2+}, the transgenic lines displayed superior growth metrics, including broader and greener leaves, higher chlorophyll content, and increased fresh weights compared to wild-type plants. Quantitative analysis revealed that the Pb content in the shoots of transgenic plants was approximately 45% of that in wild-type plants, while Cd content ranged from 46% to 67% of the control level. Intriguingly, the root content of both metals remained comparable to that in control plants. Moreover, experiments utilizing $^{109}Cd^{2+}$ medium demonstrated a more rapid extrusion of Cd in ZntA-transgenic protoplasts compared to wild-type protoplasts, underscoring the enhanced Cd efflux capacity conferred by *ZntA* expression (Lee et al. 2003). These findings illustrate that the expression of the *ZntA* gene in Arabidopsis not only mitigates Cd accumulation in shoots but also enhances Cd resistance, highlighting the promising potential of employing bacterial transporters to engineer plants with reduced heavy metal uptake. Nevertheless, further research is imperative to evaluate a broader array of transporter families from diverse microorganisms, including bacteria, microalgae, protists, and yeast, and to extend these experimental approaches to agriculturally significant crops such as maize, rice, wheat, and barley.

PCS engineering is a promising field of research with the potential to enhance our understanding of the enzyme's mechanism of action and improve its activity. One primary motivation for PCS engineering is to elucidate why cadmium, as Cd-GS_2, is the optimal substrate, while other toxic metals are not (García-García et al. 2016). This increased promiscuity could expand PCS's ability to chelate a broader range of heavy metals, thereby promoting the formation of metal-PC complexes with Ni, Cr, Pb, and others to inactivate these toxic metals and store them in the vacuoles of crop roots. As mentioned previously, there is still no complete crystal structure of PCS from bacteria, yeast, algae, plants, or animals. Early studies in the 2000s aimed to understand PCS's enzymatic properties using truncated enzymes generated by site-directed mutagenesis. In 2004, Vatamaniuk et al. confirmed that AtPCS1 carries out dipeptidyltransferase activity and undergoes γ-Glu-Cys acylation at site one during catalysis, associated with Cys56, although the metal-binding site remained unidentified. Later, Cahoon et al. (2015) employed directed evolution to enhance AtPCS performance by screening Cd tolerant mutants through a random mutagenesis approach over a specific open reading frame. Despite these efforts, the mutants did not surpass the wild-type enzyme's characteristics, underscoring the challenge in evolving AtPCS1 successfully. More recently, Li et al. (2020) isolated and characterized the first *PCS* gene from the liverwort *Marchantia polymorpha* (*MpPCS*). Site-directed mutagenesis revealed twin-cysteine motifs in the C-terminus that inhibit enzyme activation by heavy metals, suggesting a regulatory feedback mechanism. Regarding PCS from microorganisms, Garcia-Garcia et al. (2020) conducted site-directed mutagenesis on *Euglena gracilis* PCS (EgPCS), identifying Cys421 and Cys422 as crucial for catalysis. Despite the metal-glutathione substrate-binding site remaining elusive, this study showed that EgPCS is more active in the presence of Zn than Cd, suggesting its potential for engineering in crops. These findings highlight the intricate nature of PCS and the ongoing efforts to fully understand and enhance its capabilities through engineering.

Nonetheless, despite PCS from microorganisms showing different kinetic properties compared to PCS from plants (e.g., most plant PCS used Cd-GS_2 as the best substrate), there is an opportunity in engineering plants with PCS from microorganisms. Researchers have assessed the impact of PCS from plant to engineer other plants. For instance, *A. thaliana* was transformed to express the PCS from ramie (*Boehmeria nivea*) to test Cd tolerance and Cd translocation (Zhu et al. 2021). Overexpression of *BnPCS1* enhanced root growth and fresh weight, while reducing oxidative damage markers like H_2O_2 and malondialdehyde (MDA). Regarding Cd accumulation, most Cd was accumulated in roots, but no differences were reported linked to BnPCS1; however, Cd content increased in shoots of transformed plants. This behavior correlates with the eight-fold higher *BnPCS* expression in shoots than in roots. Similar results were observed in a previous study focused on *A. thaliana* plants overexpressing PCS1 from garlic (*Allium sativum*); these plants showed significantly higher tolerance to cadmium (Cd) and accumulated more Cd compared to wild-type plants. Interestingly, a dual-gene transgenic line expressing *AsPCS* and *GSH1* (which codes for the bottleneck enzyme γ-ECS from the yeast *S. cerevisiae*) accumulated twice the amount of Cd as the single-gene transgenic lines. This increased accumulation was observed in both the roots and shoots of the transformed plants, indicating that the simultaneous overexpression of AsPCS1 and ScGSH1 enhances the plant's ability to uptake and sequester Cd throughout its tissues. The elevated total phytochelatin (PC) production in these transgenic lines is likely responsible for the increased Cd accumulation and tolerance (García-García et al. 2016). These findings exemplify the potential of PCS for enhancing Cd tolerance and phytoremediation. However, the challenge remains to favor cadmium accumulation in roots while preventing its translocation to shoots. The solution might lie in developing expression systems that enable tissue-specific PCS interventions or employing PCS from microorganisms as genetic components in crop engineering. In any scenario, complementing PCS with other enzymes involved in GSH biosynthesis, such as γ-ECS, as well as integrating ABC transporters or other transporters from different families, is essential. These components would facilitate the storage of toxic heavy metal complexes in the vacuoles of roots, thereby optimizing the phytoremediation process and enhancing plant resilience to heavy metal stress (Fig. 10.1).

Finally, it is worth mentioning that PCS has been used for genetic intervention of PGPB. The study by Yong et al. (2014) investigated the effects of exposing wheat seeds to cadmium in the presence of the engineered strain *Pseudomonas putida* KT2440-SpPCS, capable of expressing the PCS from *S. pombe* (ScPCS). This study is relevant for crops because *P. putida* KT2440 is a PGPB. The engineered *P. putida*

KT2440-SpPCS demonstrated enhanced resistance to Cd, Ag, and Hg and showed a five-fold increase in Cd accumulation capacity compared to the control cells when grown in the presence $CdCl_2$. Furthermore, wheat seeds inoculated with *P. putida* KT2440-SpPCS exhibited significantly enhanced germination rates and growth when exposed to Cd. This suggests that the presence of the modified *P. putida*-SpPCS, as an engineered PGPB strain, can mitigate the toxic effects of cadmium on wheat seed germination and growth (Yong et al. 2014). In the next section, the role of PGPB, their mechanisms, and future perspectives are discussed.

10.3 Contribution of PGPB to Avoid the Accumulation of Toxic Heavy Metals in Fruits of Crops

In addition to advanced plant engineering techniques, harnessing the inherent capabilities of soil microorganisms holds significant potential for enhancing crop yield. A specific subset of free-living soil organisms, known as plant growth-promoting bacteria (PGPB), colonizes the root rhizosphere and improves crop productivity by helping plants cope with biotic and abiotic stress (Cao et al. 2023; Ramakrishna et al. 2019). PGPB can be found on the rhizoplane (attached to the root surface), in the rhizosphere (the area surrounding the roots), and within the plant interior (Glick 2020). These microorganisms can improve crop yield in heavy metal-contaminated soils and potentially decrease the accumulation of these pollutants in the edible parts of crops by both mitigating the uptake of metals and increasing crop biomass (Bai et al. 2024). PGPB contribute to heavy metal tolerance through various mechanisms. They can interact with metals in the soil either directly (by contact with the cell surface) or indirectly (by secreting certain compounds), reducing the bioavailability of toxic metal ions in the soil and preventing their uptake (Braud et al. 2006). Additionally, PGPB can promote plant growth by solubilizing phosphates, improving nitrogen fixation, and producing phytohormones and enzymes that stimulate growth and antioxidative mechanisms (Gupta et al. 2020). Here, we will summarize recent advances on these mechanisms.

10.3.1 Biosorption and Bioaccumulation of Heavy Metals Mediated by PGPB

Biosorption is the process by which biological materials adsorb substances from a solution through physicochemical pathways, such as electrostatic forces and ion/proton displacement. This process relies on ionic interactions between the extracellular surfaces of dead biomass, living cells, and metal ions. Consequently, contaminants bind to the cellular surface rather than being oxidized through aerobic or anaerobic metabolism. The functional groups in bacterial cell walls, including carboxyl, phosphonate, amine, hydroxyl, and sulfhydryl groups, play a key role in this process, as well as exopolysaccharides (EPS) which can entrap cationic heavy metals (Lal et al. 2018; Narayanan and Ma 2023). Therefore, the success of biosorption depends on the diversity of cell wall structures. For example, zinc-tolerant bacteria isolated from contaminated soils in India, some of which belong to the *Serratia sp.*, demonstrated high biosorption of zinc (Table 10.1). This could be attributed to biosorption through the secretion of exopolysaccharides, for which they tested positive, along with other plant growth-promoting properties. Inoculation of maize seedlings with these zinc-tolerant bacterial (ZTB) strains under zinc stress resulted in increased shoot length, root length, and total chlorophyll content compared to the uninoculated control (Kour et al. 2019). In the case of *Pseudomonas taiwanensis* WRS8, this strain showed a reduction of Cd uptake in wheat by increasing root surface Cd adsorption, decreasing Cd content in roots and above-ground tissues, and downregulating Cd transport-related genes *LCT1* (*Low-affinity Cation Transporter*) and HMA2 by 46% and 30%, respectively, compared to the controls (Table 10.1). This strain effectively mitigates Cd contamination in wheat grown in Cd-polluted environments (Cheng et al. 2021)

Bioaccumulation refers to the process by which organisms, such as plants, absorb and accumulate contaminants, including metals, from their environment into their tissues over time. This process is energy-dependent, requiring metabolic energy for the uptake and internal transport of metals within the organism. In contrast, biosorption is a passive process that does not require metabolic energy. Thus, the key differences between bioaccumulation and biosorption lie in the energy requirements and the mechanisms involved (Alves et al. 2022). Bioaccumulation requires energy-driven respiration within the cell's metabolic cycle and occurs in two steps: first, heavy metals are transported inside the cells, where they can be recognized as a substrate by membrane transporters and biomolecule ligands as a storage. During intracellular sequestration, metal ions are taken up by various compounds (e.g., peptides and organic acids) within the cell cytoplasm (Pham et al. 2022). This mechanism is also considered a plant growth-promoting process because it reduces the availability of toxic metals in soils. For example, Mitra et al. (2018) isolated a cadmium-resistant *Enterobacter* sp. strain (S2) from a contaminated rice rhizosphere. The strain showed high cadmium removal efficiency (>95%) through intracellular accumulation and exhibited plant growth-promoting traits such as 1-aminocyclopropane-1-carboxylate (ACC) deaminase, IAA (indole-3-acetic acid) production, phosphate solubilization, and nitrogen fixation. Inoculation with S2 enhanced rice seedling growth under cadmium stress, alleviated oxidative stress, reduced stress ethylene, and decreased cadmium accumulation in seedlings. This strain shows promising for bioremediation and promoting plant growth in contaminated soils (Table 10.1).

Table 10.1 Case studies on PGPB-mediated amelioration of heavy-metal (HM) stress in edible crops

Crop	HM	Bacteria	Identified mechanisms	Mode of application	Effect	References
Zea mays (Maize)	Zn	*Serratia sp.*	Zn biosorption and nutrient solubilization, including zinc Identification of Zinc-Tolerant Bacteria (ZTB)	ZTB can solubilize up to 0.25% insoluble ZnO in the medium; at higher concentrations (2% ZnO in MSM), 12 isolates were effective. The highest zinc biosorption (< 90%) was recorded with the bacterial strain Z-15	The ZTB strains exhibited multiple plant growth-promoting activities, including the production of indole-3-acetic acid, gibberellic acid, ammonia, and in some cases, ACC deaminase, siderophores, and hydrogen cyanide. Additionally, many of these strains were capable of solubilizing phosphorus, potash, and silica, and producing phytase, further supporting plant growth	Kour et al. (2019)
	Pb	Rhizobacterial isolate AN8 originated from tomato crops	ACC deaminase and nitrogen-fixing activity	Pot experiment in soil with different Pb concentrations. Maize seeds were dipped in broth with each isolate for inoculation	The tested rhizobacterial isolates generally improved maize growth in Pb-contaminated soil. The AN8 isolate was the most efficient, leading to the highest increases in shoot and root length, seedling fresh and dry weights,	Hassan et al. (2014)

(continued)

Table 10.1 (continued)

Crop	HM	Bacteria	Identified mechanisms	Mode of application	Effect	References
					and the most favorable chemical and enzymatic growth parameters, while causing the lowest Pb uptake in the shoot and root, even at the highest Pb level (400 Pb mg kg^{-1} soil)	
Triticum aestivum (Wheat)	Cd	*Pseudomonas taiwanensis* WRS8 isolated from wheat crops	Cd adsorption and accumulation, IAA production, and ACC deaminase activity	Hydroponic experiment on media with Cd The plants were inoculated by immersing their roots in a bacterial suspension	The strain reduced Cd concentration in the media. Plants grown hydroponically in the presence of Cd had decreased Cd content in roots and shoots, and the Cd bioconcentration and translocation factors were reduced The root surface-adsorbed Cd of the wheat plants was increased	Cheng et al. (2021)
	Cd	Rhizobacterial isolated from sugarcane rhizosphere	IAA production, ACC deaminase, and nitrogen fixation activity	Pot experiment in soil with different Cd concentrations (0–200 mg kg^{-1}). Wheat seedlings were inoculated in broth with each isolate	An increase in root and shoot lengths, fresh and dry weights of wheat seedlings under Cd contamination was observed along with increased chlorophyll	Hassan et al. (2016)

					a, b and carotenoid contents, as well as elevated levels of plant protein, proline, and antioxidant and enzyme production. A significant decrease in Cd uptake in roots and shoots was also noted	
Oryza sativa (Rice)	Cd	*Enterobacter sp.*	Cd accumulation, IAA production, ACC deaminase, and phosphate solubilization	Bacterial strains (S1, S2, S3, and S5) isolated from a metal-contaminated rice rhizosphere were evaluated for their plant growth-promoting traits in rice under Cd stress	The S2 strain demonstrated high cadmium removal efficiency (>95%) from the growth medium, primarily through intracellular cadmium accumulation. Additionally, the S2 strain produced ACC deaminase (236 ng α-ketobutyrate/mg protein/h), IAA (726 μg/ml), solubilized phosphate (73 ppm), and fixed nitrogen (4.4 μg of nitrogen fixed/h/mg protein), all of which are beneficial for plant growth. Inoculation of rice seedlings with the S2 strain significantly enhanced various	Mitra et al. (2018)

(continued)

Table 10.1 (continued)

Crop	HM	Bacteria	Identified mechanisms	Mode of application	Effect	References
					morphological and biochemical characteristics of seedling growth under cadmium stress compared to uninoculated seedlings	
	Cd and Zn	*Bacillus* sp. ZC3-2-1	IAA, EPS, siderophore production, phosphate solubilization, and ACC deaminase activity	Pot experiment in soil with Cd, Zn, co-contamination Rice seedlings were soaked in bacterial culture medium with the inoculum	ZC3-2-1 significantly increased the rice biomass while also increasing the phytoextraction of Cd^{2+} and Zn^{2+}; suggesting that rice could be useful both for phytoremediation and as food, as the concentration of the heavy metals per biomass unit of rice did not change significantly ZC3-2-1 also decreased the bioavailability of Cd and Zn in the soil	Liu et al. (2022)
Glycine max (Soybean)	As	*Bradyrhizobium japonicum* E109 and *Azospirillum brasilense* Az39 co-inoculation	IAA production, metal chelation, and biofilm formation	Pot experiments in the presence of As stress (AsV 25 µM and AsIII 25 µM). Soybean seeds were	The co-inoculation led to improved plant growth in AsV treated plants, increased nodule number and	Armendariz et al. (2019)

				inoculated with the bacteria both separately and together	nitrogen content, and reduced As translocation to aerial parts of the plants	
Vigna mungo (Black gram)	Cd	*Pseudomonas aeruginosa* isolated from agricultural fields of different plant species in India	Siderophore production, ACC deaminase activity, phosphate solubilization, and auxin synthesis	Pot experiment. Microbial seed coating of black gram	Increased root and shoot lengths and weight were observed, along with decreased Cd accumulation in soil amended with up to 200 µM $CdCl_2$. Lower levels of Cd were detected in the leaves, indicating effective protection of the plants from growth inhibition due to Cd toxicity	Ganesan (2008)
Solanum lycopersicum (Tomato)	Cr	*Klebsiella* sp. CPSB4 and *Enterobacter* sp. CPSB49, isolated from polluted soils	Siderophore and IAA production, phosphate solubilization, and enhancement of antioxidative enzymes	Pot experiment under different levels of Cr contamination (0–200 mg kg^{-1}). Tomato seeds were dipped in a bacterial solution for inoculation	The strains promoted plant growth and biomass, with CPSB49 notably enhancing root and shoot length, fresh and dry weight, and nutrient uptake. Both strains increased levels of enzymes that combat oxidative stress. Inoculation with CPSB4 resulted in higher chromium accumulation in roots but limited translocation to shoots	Gupta et al. (2019, 2020)

10.3.2 Excretion of Molecules from PGPB for Immobilization of Heavy Metals in Soils

PGPB can secrete natural organic chelating agents, such as biosurfactants, organic acids (OAs) and siderophores, which decrease the bioavailability and toxicity of heavy metals. Biosurfactants are amphiphilic molecules that preferentially bind to toxic heavy metals, sometimes showing a stronger affinity for these metals compared to essential metals in the soil (Ahemad 2015). Organic acids react with metal cations and aid in the release of inorganic phosphate, while siderophores help regulate the absorption and transport of heavy metals by enhancing the availability of essential metals such as iron (Panchal et al. 2021). In this context, the biosynthesis of phosphates has been identified as a key resistance mechanism under heavy metal stress in *Enterobacter* sp., *Bacillus* sp. ZC3-2-1, *Pseudomonas aeruginosa*, and *Klebsiella* sp. This mechanism is also linked to the enhancement of heavy metal tolerance in crops such as rice, black gram, and tomato, where the aforementioned bacteria are part of the PGPB (Table 10.1).

The synthesis of phytohormones can help plants adapt to metal toxicity; in particular, IAA is known to modulate the plant auxin pool, mitigating the effects of stress, and is commonly tested for in potential PGPBs. For instance, the rhizobacterial isolate SAN1 (Strain 1 exhibiting ACC-deaminase activity and Nitrogen fixing ability), from the sugarcane rhizosphere, which also has an increased glutathione S- transferase, peroxidase and catalase activities, produced more IAA (14 mg L^{-1}) than *Azotobacter* (7.6 mg L^{-1}) in the presence of L-tryptophan. This metabolic capability was correlated with substantial improvements in wheat growth and physiological parameters under Cd stress (Hassan et al. 2016). Plants inoculated with SAN1 showed significant increases in shoot and root lengths compared to control plants (13- and 11-fold increases), as well as enhanced of chlorophyll a/b, carotenoids, protein, proline, and antioxidant enzyme content under cadmium stress (50–200 mg Kg^{-1}). Interestingly, Hassan et al. (2016) found that SAN1 significantly reduced Cd uptake in wheat seedlings' shoots and roots, with values of 0.47 and 0.38 mg Cd Kg^{-1}, respectively, in plants inoculated with SAN1, compared to 4.5 and 2.8 mg Cd Kg^{-1} in roots and shoots of control plants. Overall, this study, along with others analyzing the role of IAA production in *Pseudomonas taiwanensis*, *Enterobacter* sp., *Bacillus* sp. ZC3-2-1, *Bradyrhizobium japonicum* E109, *Azospirillum brasilense* Az39, *Klebsiella* sp. CPSB4, and *Enterobacter* sp. CPSB49, demonstrates that PGPR effectively promotes plant growth and reduces Cd toxicity in crops. (Table 10.1).

ACC deaminase is another common mechanism employed by PGPB due to its ability to hydrolyze ACC, the immediate precursor of ethylene, which is often produced in excess under metal stress, leading to several negative effects (Etesami and Maheshwari 2018). Rhizobacteria such as *Enterobacter* sp., *Bacillus* sp. ZC3-2-1, *P. auruginosa*, *Kleibsella* sp, and *Enterobacter* sp. are among the PGPB capable of using this mechanism to mitigate the toxicity of As, Cd, Pb, Zn, and Cr in crops like maize, wheat, rice, soybean, black gram, and tomato (Table 10.1). Additionally,

PGPB can upregulate the expression of plant genes related to stress response and activate antioxidant defense systems, including enzymes like peroxidase (POD) and superoxide dismutase (SOD). This activation is often linked to signal transduction mediated by the induction of phytohormones or reactive species (Bai et al. 2024).

10.3.3 Future Perspectives in PGPB

While the efficacy of PGPB in alleviating heavy metal stress in crops can be substantial, outcomes are highly variable and depend on multiple factors, including crop type, soil conditions, and the native rhizosphere microbial community. Additionally, results can vary across experimental setups, such as hydroponic, pot, or field trials. For example, PGPB might exhibit different levels of effectiveness in controlled laboratory conditions compared to field environments due to variations in soil composition and microbial interactions. Understanding these variables is crucial for optimizing PGPB applications to achieve consistent benefits across diverse agricultural settings. Advances in biotechnology and 'omics' technologies can enhance our understanding of PGPB mechanisms and interactions. Techniques such as meta-transcriptomics can identify which microbial groups are functionally active and how they contribute to plant health under heavy metal stress (Ramakrishna et al. 2019). Furthermore, proteomics and metabolomics can identify specific bacterial strains and metabolic pathways that enhance plant resilience and soil health. These insights will enable the development of targeted microbial treatments and more precise rhizosphere engineering strategies. As our ability to analyze and interpret complex microbial interactions improves, more effective and tailored applications of PGPB can be expected, leading to better management of soil pollution and improved crop productivity.

In this regard, genetic modifications of bacteria have shown significant promise in enhancing phytoremediation and plant growth. Wu et al. (2006) explore the use of engineered plant-microbe symbiosis for the rhizoremediation of Cd. The study focuses on the bacterium *P. putida* 06909, which was genetically modified to express a metal-binding peptide with a high affinity for cadmium, known as EC20. Sunflower seedlings were inoculated with both control and EC20-engineered *P. putida* 06909. After a colonization period, plants were exposed to 80 μM $CdCl_2$ for three days. Results showed that Cd accumulation in the roots was significantly enhanced by inoculation with the engineered *P. putida* 06909 expressing the EC20 peptide. Specifically, sunflower plants inoculated with the engineered strain bound 1.6 times more Cd than plants inoculated with control *P. putida*. This increased cadmium accumulation in the roots was due to the enhanced cadmium-binding capacity of the EC20-expressing bacteria (Wu et al. 2006). Although there are few studies about the use of engineered bacteria for improving heavy metal resistance in crops, engineered bacteria developed for other porpoises can potentially be applied in this. For instance, *Pseudomonas protegens* Pf-5 was engineered to enhance its antifungal activity against the *A. thaliana* pathogen *Rhizoctonia solani* (Jing et al.

2020). The *retS* gene was deleted in *P. protegens* to increase the production of antifungal secondary metabolites, particularly 2,4-diacetylphloroglucinol (2,4-DAPG), which showed strong inhibition of *R. solani*. Additionally, the strategy included introducing the entire *nif* island from *Pseudomonas stutzeri* DSM4166 into the retS mutant strain to increase nitrogen fixation capabilities. The engineered *P. protegens*, named *retS* mutant-*nif*, displayed significant benefits for *A. thaliana*. Under nitrogen-limited conditions, plants inoculated with the recombinant retS mutant-*nif* strain had increased rosette diameter and enhanced seedling fresh weights under pathogen and nitrogen-limited conditions. This indicates that the *nif* genes in the 2,4-DAPG producing *P. protegens* strain promoted plant growth and health in challenging environments (Jing et al. 2020). Although heavy metal resistance was not assessed in this study, it is expected that bacteria with a high nitrogenase activity can tolerate heavy metals. For instance, rhizobacteria exposed to Pb have shown increased nitrogenase activity (Hassan et al. 2014). Additionally, the 2,4-DAPG overproducer strain *Pseudomonas fluorescens* 5.014 has demonstrated high glutathione content and Cd resistance (Hultberg 1998). Another example is the engineering of IAA biosynthesis from glucose using an *E. coli* strain harboring feedback inhibition-resistant genes $aroG^{fbr}$ (3-deoxy-7-phosphoheptulonate synthase), $trpE^{fbr}$ (anthranilate synthase), and $serA^{fbr}$ (phosphoglycerate dehydrogenase). This synthetic pathway was improved by coexpressing the *aro8* (aminotransferase) and *kdc* (decarboxylase) genes from *S. cerevisiae* and the *aldH* (aldehyde dehydrogenase) gene from *E. coli* (Guo et al. 2019). This genetic intervention facilitated the conversion of glucose to L-tryptophan and a downstream pathway converting L-tryptophan to IAA, achieving a yield of 744 mg IAA/L. Meanwhile, engineered *Bacillus amyloliquefaciens* SQR9 and *Bacillus subtilis* 168 (two PGPB strains) coexpressing *patB* (aminotransferase), *yclC* (decarboxylase), and *dhaS* (indole-3-acetaldehyde dehydrogenase) produced 39 and 22 mg/L of IAA, respectively (Guo et al. 2019). These studies suggest that further engineering of PGPB might enhance their IAA overproduction, potentially improving heavy metal resistance with benefits for crops such as wheat, rice, soybean, and tomato. (Table 10.1).

Therefore, these engineered bacteria also can be part of the strategy to tailor PGPB for alleviating heavy metal stress in crops. However, currently, PGPB consortia have been developed using bacteria with natural capabilities for any of the heavy metal-resistant mechanisms discussed previously (Flores-Duarte et al. 2023). The deployment of genetically modified organisms (GMOs) in this context presents several challenges. Technical issues such as metabolic burden can impact the efficiency and viability of engineered bacteria. Moreover, ethical concerns regarding the environmental release and utilization of GMOs need to be addressed (Somayaji et al. 2022). These challenges necessitate a balanced approach, incorporating responsible research and education to mitigate potential risks while allowing the advancement of synthetic biology solutions, as metabolic engineering holds great promise for addressing heavy metal contamination in crops.

10.4 Conclusions

In conclusion, ABC transporters from microorganisms emerge as promising candidates for enhancing the storage of toxic heavy metals in plant vacuoles. By harnessing these microbial transporters, plants could significantly improve their ability to sequester harmful metals within vacuoles, thus reducing their translocation to edible parts and mitigating toxic effects. Additionally, phytochelatin synthase (PCS) from microorganisms shows substantial promise in promoting metal accumulation in plant roots. However, optimizing this potential requires the development of evolved PCS versions that perform better with a broader range of metals beyond cadmium. Such targeted evolution could empower plants to manage and tolerate various toxic metals more effectively, contributing to safer and more sustainable agricultural practices.

Furthermore, plant growth-promoting bacteria (PGPB) play a crucial role in minimizing the availability of toxic metals for plant uptake. Engineering bacterial resistance mechanisms and tailoring rhizobacterial communities are pivotal microbial biotechnological strategies to overcome food security challenges in polluted soils. By optimizing microbial interactions and enhancing metal-immobilization capabilities, we can mitigate heavy metal contamination's adverse effects on crops. This approach not only improves plant health and yields but also ensures safer food production in contaminated environments. Continued research and development in these areas have the potential to revolutionize plant engineering for improved heavy metal tolerance and accumulation, ultimately safeguarding food security and human health.

Acknowledgements The present work was supported by Secretaría de Innovación, Ciencia y Tecnología de Jalisco, Consejo Estatal de Ciencia y Tecnología de Jalisco (COECYTJAL), and Tecnológico de Monterrey (Project: FODECIJAL No. 10307-2022).

References

Adle DJ, Sinani D, Kim H, Lee J (2007) A cadmium-transporting P1B-type ATPase in yeast Saccharomyces cerevisiae. J Biol Chem 282(2):947–955. https://doi.org/10.1074/jbc.M609535200

Ahemad M (2015) Enhancing phytoremediation of chromium-stressed soils through plant-growth-promoting bacteria. J Genet Eng Biotechnol 13(1):51–58. https://doi.org/10.1016/j.jgeb.2015.02.001

Ali H, Khan E (2018) What are heavy metals? Long-standing controversy over the scientific use of the term "heavy metals"—proposal of a comprehensive definition. Toxicol Environ Chem 101(1):6–19. https://doi.org/10.1080/02772248.2017.1413652

Ali W, Mao K, Zhang H, Junaid M, Xu N, Rasool A, Feng X, Yang Z (2020) Comprehensive review of the basic chemical behaviours, sources, processes, and endpoints of trace element contamination in paddy soil-rice systems in rice-growing countries. J Hazard Mater 397:122720. https://doi.org/10.1016/j.jhazmat.2020.122720

Alves ARA, Yin Q, Oliveira RS, Silva EF, Novo LAB (2022) Plant growth-promoting bacteria in phytoremediation of metal-polluted soils: current knowledge and future directions. Sci Total Environ 838(Pt 4):156435. https://doi.org/10.1016/j.scitotenv.2022.156435

Armendariz AL, Talano MA, Olmos Nicotra MF, Escudero L, Breser ML, Porporatto C, Agostini E (2019) Impact of double inoculation with Bradyrhizobium japonicum E109 and Azospirillum brasilense Az39 on soybean plants grown under arsenic stress. Plant Physiol Biochem PPB 138: 26–35. https://doi.org/10.1016/j.plaphy.2019.02.018

Asthir B, Kaur G, Kaur B (2020) Convergence of pathways towards ascorbate–glutathione for stress mitigation. J Plant Biol 63:243–257. https://doi.org/10.1007/s12374-020-09253-7

Baghaie AH, Aghili F (2019) Health risk assessment of Pb and Cd in soil, wheat, and barley in Shazand County, central of Iran. J Environ Health Sci Eng 17(1):467–477. https://doi.org/10.1007/s40201-019-00365-y

Bai X, Bol R, Chen H, Cui Q, Qiu T, Zhao S, Fang L (2024) A meta-analysis on crop growth and heavy metals accumulation with PGPB inoculation in contaminated soils. J Hazard Mater 471: 134370. https://doi.org/10.1016/j.jhazmat.2024.134370

Batool TS, Aslam R, Gul A, Paracha RZ, Ilyas M, De Abreu K, Munir F, Amir R, Williams LE (2023) Genome-wide analysis of heavy metal ATPases (HMAs) in Poaceae species and their potential role against copper stress in Triticum aestivum. Sci Rep 13(1):7551. https://doi.org/10.1038/s41598-023-32023-7

Blaby-Haas CE, Merchant SS (2012) The ins and outs of algal metal transport. Biochim Biophys Acta 1823(9):1531–1552. https://doi.org/10.1016/j.bbamcr.2012.04.010

Braud A, Jézéquel K, Vieille E, Tritter A, Lebeau T (2006) Changes in extractability of CR and PB in a polycontaminated soil after bioaugmentation with microbial producers of biosurfactants, organic acids and siderophores. Water Air Soil Pollut Focus/Water Air Soil Pollut Focus 6(3–4): 261–279. https://doi.org/10.1007/s11267-005-9022-1

Brunetti P, Zanella L, De Paolis A, Di Litta D, Cecchetti V, Falasca G, Barbieri M, Altamura MM, Costantino P, Cardarelli M (2015) Cadmium-inducible expression of the ABC-type transporter AtABCC3 increases phytochelatin-mediated cadmium tolerance in Arabidopsis. J Exp Bot 66(13):3815–3829. https://doi.org/10.1093/jxb/erv185

Cahoon RE, Lutke WK, Cameron JC, Chen S, Lee SG, Rivard RS, Rea PA, Jez JM (2015) Adaptive engineering of phytochelatin-based heavy metal tolerance. J Biol Chem 290(28):17321–17330. https://doi.org/10.1074/jbc.M115.652123

Cao M, Narayanan M, Shi X, Chen X, Li Z, Ma Y (2023) Optimistic contributions of plant growth-promoting bacteria for sustainable agriculture and climate stress alleviation. Environ Res 217: 114924. https://doi.org/10.1016/j.envres.2022.114924

Chakravorty M, Nanda M, Bisht B, Sharma R, Kumar S, Mishra A, Vlaskin MS, Chauhan PK, Kumar V (2023) Heavy metal tolerance in microalgae: detoxification mechanisms and applications. Aquat Toxicol (Amsterdam, Netherlands) 260:106555. https://doi.org/10.1016/j.aquatox.2023.106555

Chen X, Zhao Y, Zhong Y, Chen J, Qi X (2023) Deciphering the functional roles of transporter proteins in subcellular metal transportation of plants. Planta 258(1):17. https://doi.org/10.1007/s00425-023-04170-8

Cheng C, Wang Q, Wang QX, He LY, Sheng XF (2021) Wheat-associated Pseudomonas taiwanensis WRS8 reduces cadmium uptake by increasing root surface cadmium adsorption and decreasing cadmium uptake and transport related gene expression in wheat. Environ Pollut (Barking, Essex: 1987) 268(Pt A):115850. https://doi.org/10.1016/j.envpol.2020.115850

Etesami H, Maheshwari DK (2018) Use of plant growth promoting rhizobacteria (PGPRs) with multiple plant growth promoting traits in stress agriculture: action mechanisms and future prospects. Ecotoxicol Environ Saf 156:225–246. https://doi.org/10.1016/j.ecoenv.2018.03.013

Fan P, Wu L, Wang Q, Wang Y, Luo H, Song J, Yang M, Yao H, Chen S (2023) Physiological and molecular mechanisms of medicinal plants in response to cadmium stress: current status and future perspective. J Hazard Mater 450:131008. https://doi.org/10.1016/j.jhazmat.2023.131008

Feng D, Wang R, Sun X, Liu L, Liu P, Tang J, Zhang C, Liu H (2023) Heavy metal stress in plants: ways to alleviate with exogenous substances. Sci Total Environ 897:165397. https://doi.org/10.1016/j.scitotenv.2023.165397

Flores-Duarte NJ, Pajuelo E, Mateos-Naranjo E, Navarro-Torre S, Rodríguez-Llorente ID, Redondo-Gómez S, Carrasco López JA (2023) A culturomics-based bacterial synthetic community for improving resilience towards arsenic and heavy metals in the nutraceutical plant *Mesembryanthemum crystallinum*. Int J Mol Sci 24(8):7003. https://doi.org/10.3390/ijms24087003

Ganesan V (2008) Rhizoremediation of cadmium soil using a cadmium-resistant plant growth-promoting rhizopseudomonad. Curr Microbiol 56(4):403–407. https://doi.org/10.1007/s00284-008-9099-7

García-García JD, Rodríguez-Zavala JS, Jasso-Chávez R, Mendoza-Cozatl D, Moreno-Sánchez R (2009) Chromium uptake, retention and reduction in photosynthetic Euglena gracilis. Arch Microbiol 191(5):431–440. https://doi.org/10.1007/s00203-009-0469-8

García-García JD, Sánchez-Thomas R, Moreno-Sánchez R (2016) Bio-recovery of non-essential heavy metals by intra- and extracellular mechanisms in free-living microorganisms. Biotechnol Adv 34(5):859–873. https://doi.org/10.1016/j.biotechadv.2016.05.003

García-García JD, Peña-Sanabria KA, Sánchez-Thomas R, Moreno-Sánchez R (2018) Nickel accumulation by the green algae-like *Euglena gracilis*. J Hazard Mater 343:10–18. https://doi.org/10.1016/j.jhazmat.2017.09.008

García-García J, Sánchez-Thomas R, Saavedra E, Fernández-Velasco D, Romero-Romero S, Casanova-Figueroa K, Mendoza-Cózatl D, Moreno-Sánchez R (2020) Mapping the metal-catalytic site of a zinc-activated phytochelatin synthase. Algal Res 47:101890. https://doi.org/10.1016/j.algal.2020.101890

Geimer S, Belicová A, Legen J, Slávikovâ S, Herrmann RG, Krajcovic J (2009) Transcriptome analysis of the Euglena gracilis plastid chromosome. Curr Genet 55(4):425–438. https://doi.org/10.1007/s00294-009-0256-8

Glick BR (2020) Introduction to plant growth-promoting bacteria. In: Springer eBooks, pp 1–37. https://doi.org/10.1007/978-3-030-44368-9_1

Guo D, Kong S, Chu X, Li X, Pan H (2019) De novo biosynthesis of indole-3-acetic acid in engineered *Escherichia coli*. J Agric Food Chem 67(29):8186–8190. https://doi.org/10.1021/acs.jafc.9b02048

Gupta P, Kumar V, Usmani Z, Rani R, Chandra A, Gupta VK (2019) A comparative evaluation towards the potential of Klebsiella sp. and Enterobacter sp. in plant growth promotion, oxidative stress tolerance and chromium uptake in Helianthus annuus (L.). J Hazard Mater 377:391–398. https://doi.org/10.1016/j.jhazmat.2019.05.054

Gupta P, Kumar V, Usmani Z, Rani R, Chandra A, Gupta VK (2020) Implications of plant growth promoting Klebsiella sp. CPSB4 and Enterobacter sp. CPSB49 in luxuriant growth of tomato plants under chromium stress. Chemosphere 240:124944. https://doi.org/10.1016/j.chemosphere.2019.124944

Hassan W, Bano R, Bashir F, David J (2014) Comparative effectiveness of ACC-deaminase and/or nitrogen-fixing rhizobacteria in promotion of maize (Zea mays L.) growth under lead pollution. Environ Sci Pollut Res Int 21(18):10983–10996. https://doi.org/10.1007/s11356-014-3083-5

Hassan W, Bashir S, Ali F, Ijaz M, Hussain M, David J (2016) Role of ACC-deaminase and/or nitrogen fixing rhizobacteria in growth promotion of wheat (Triticum aestivum L.) under cadmium pollution. Environ Earth Sci 75(3). https://doi.org/10.1007/s12665-015-4902-9

Hayashi S, Kuramata M, Abe T, Takagi H, Ozawa K, Ishikawa S (2017) Phytochelatin synthase OsPCS1 plays a crucial role in reducing arsenic levels in rice grains. Plant J 91(5):840–848. https://doi.org/10.1111/tpj.13612

Hernández-Garnica M, García-García JD, Moreno-Sánchez R, Sánchez-Thomas R (2021) Lead accumulation in photosynthetic Euglena gracilis depends on polyphosphates and calcium. Environ Pollut (Barking, Essex: 1987) 272:116007. https://doi.org/10.1016/j.envpol.2020.116007

Hultberg M (1998) Rhizobacterial glutathione levels as affected by starvation and cadmium exposure. Curr Microbiol 37(5):301–305. https://doi.org/10.1007/s002849900382

Jin D, Zhang Q, Liu Y, Liang M, Li A, Wu J (2022) Overexpression of the maize phytochelatin synthase gene (zmpcs1) enhances CD tolerance in plants. Acta Physiol Plant 44(11). https://doi.org/10.1007/s11738-022-03451-1

Jing X, Cui Q, Li X, Yin J, Ravichandran V, Pan D, Fu J, Tu Q, Wang H, Bian X, Zhang Y (2020) Engineering Pseudomonas protegens Pf-5 to improve its antifungal activity and nitrogen fixation. Microb Biotechnol 13(1):118–133. https://doi.org/10.1111/1751-7915.13335

Kanwal F, Riaz A, Ali S, Zhang G (2024) NRAMPs and manganese: magic keys to reduce cadmium toxicity and accumulation in plants. Sci Total Environ 921:171005. https://doi.org/10.1016/j.scitotenv.2024.171005

Khan N, You FM, Datla R, Ravichandran S, Jia B, Cloutier S (2020) Genome-wide identification of ATP binding cassette (ABC) transporter and heavy metal associated (HMA) gene families in flax *(Linum usitatissimum* L.). BMC Genomics 21(1):722. https://doi.org/10.1186/s12864-020-07121-9

Khatiwada B, Hasan MT, Sun A, Kamath KS, Mirzaei M, Sunna A, Nevalainen H (2020a) Probing the role of the chloroplasts in heavy metal tolerance and accumulation in *Euglena gracilis*. Microorganisms 8(1):115. https://doi.org/10.3390/microorganisms8010115

Khatiwada B, Hasan MT, Sun A, Kamath KS, Mirzaei M, Sunna A, Nevalainen H (2020b) Proteomic response of *Euglena gracilis* to heavy metal exposure—identification of key proteins involved in heavy metal tolerance and accumulation. Algal Res 45:2211–9264. https://doi.org/10.1016/j.algal.2019.101764

Kour R, Jain D, Bhojiya AA, Sukhwal A, Sanadhya S, Saheewala H, Jat G, Singh A, Mohanty SR (2019) Zinc biosorption, biochemical and molecular characterization of plant growth-promoting zinc-tolerant bacteria. 3 Biotech 9(11):421. https://doi.org/10.1007/s13205-019-1959-2

Kovalchuk A, Driessen AJ (2010) Phylogenetic analysis of fungal ABC transporters. BMC Genomics 11:177. https://doi.org/10.1186/1471-2164-11-177

Krüger J, Richter P, Stoltze J, Strauch SM, Krüger M, Daiker V, Prasad B, Sonnewald S, Reid S, Lebert M (2019) Changes of gene expression in Euglena gracilis obtained during the 29th DLR parabolic flight campaign. Sci Rep 9(1):14260. https://doi.org/10.1038/s41598-019-50611-4

Lal S, Ratna S, Said OB, Kumar R (2018) Biosurfactant and exopolysaccharide-assisted rhizobacterial technique for the remediation of heavy metal contaminated soil: an advancement in metal phytoremediation technology. Environ Technol Innov 10:243–263. https://doi.org/10.1016/j.eti.2018.02.011

Lee J, Bae H, Jeong J, Lee JY, Yang YY, Hwang I, Martinoia E, Lee Y (2003) Functional expression of a bacterial heavy metal transporter in Arabidopsis enhances resistance to and decreases uptake of heavy metals. Plant physiology 133(2):589–596. https://doi.org/10.1104/pp.103.021972

Li M, Barbaro E, Bellini E, Saba A, Sanità di Toppi L, Varotto C (2020) Ancestral function of the phytochelatin synthase C-terminal domain in inhibition of heavy metal-mediated enzyme overactivation. J Exp Bot 71(20):6655–6669. https://doi.org/10.1093/jxb/eraa386

Li J, Zhang Z, Shi G (2024) Genome-wide identification and expression profiling of heavy metal ATPase (HMA) genes in peanut: potential roles in heavy metal transport. Int J Mol Sci 25(1):613. https://doi.org/10.3390/ijms25010613

Liao C, Li Y, Wu X, Wu W, Zhang Y, Zhan P, Meng X, Hu G, Yang S, Lin H (2023) *ZmHMA3*, a member of the heavy-metal-transporting ATPase family, regulates Cd and Zn tolerance in maize. Int J Mol Sci 24(17):13496. https://doi.org/10.3390/ijms241713496

Liu XF, Supek F, Nelson N, Culotta VC (1997) Negative control of heavy metal uptake by the Saccharomyces cerevisiae BSD2 gene. J Biol Chem 272(18):11763–11769. https://doi.org/10.1074/jbc.272.18.11763

Liu A, Wang W, Zheng X, Chen X, Fu W, Wang G, Ji J, Jin C, Guan C (2022) Improvement of the Cd and Zn phytoremediation efficiency of rice (Oryza sativa) through the inoculation of a metal-resistant PGPR strain. Chemosphere 302:134900. https://doi.org/10.1016/j.chemosphere.2022.134900

Manríquez B, Muller D, Prigent-Combaret C (2021) Experimental evolution in plant-microbe systems: a tool for deciphering the functioning and evolution of plant-associated microbial communities. Front Microbiol 12:619122. https://doi.org/10.3389/fmicb.2021.619122

Mendoza-Cózatl DG, Rodríguez-Zavala JS, Rodríguez-Enríquez S, Mendoza-Hernandez G, Briones-Gallardo R, Moreno-Sánchez R (2006) Phytochelatin-cadmium-sulfide high-molecular-mass complexes of Euglena gracilis. FEBS J 273(24):5703–5713. https://doi.org/10.1111/j.1742-4658.2006.05558.x

Mitra S, Pramanik K, Sarkar A, Ghosh PK, Soren T, Maiti TK (2018) Bioaccumulation of cadmium by Enterobacter sp. and enhancement of rice seedling growth under cadmium stress. Ecotoxicol Environ Saf 156:183–196. https://doi.org/10.1016/j.ecoenv.2018.03.001

Mizuno T, Usui K, Horie K, Nosaka S, Mizuno N, Obata H (2005) Cloning of three ZIP/Nramp transporter genes from a Ni hyperaccumulator plant *Thlaspi japonicum* and their Ni^{2+}-transport abilities. Plant Physiol Biochem PPB 43(8):793–801. https://doi.org/10.1016/j.plaphy.2005.07.006

Moreno-Sánchez R, Rodríguez-Enríquez S, Jasso-Chávez R, Saavedra E, García-García JD (2017) Biochemistry and physiology of heavy metal resistance and accumulation in euglena. Adv Exp Med Biol 979:91–121. https://doi.org/10.1007/978-3-319-54910-1_6

Murasugi A, Wada C, Hayashi Y (1981) Cadmium-binding peptide induced in fission yeast, *Schizosaccharomyces pombe*. J Biochem 90(5):1561–1564. https://doi.org/10.1093/oxfordjournals.jbchem.a133627

Naaz S, Ahmad N, Jameel MR, Al-Huqail AA, Khan F, Qureshi MI (2023) Impact of some toxic metals on important ABC transporters in soybean (*Glycine max* L.). ACS Omega 8(30): 27597–27611. https://doi.org/10.1021/acsomega.3c03325

Narayanan M, Ma Y (2023) Mitigation of heavy metal stress in the soil through optimized interaction between plants and microbes. J Environ Manag 345:118732. https://doi.org/10.1016/j.jenvman.2023.118732

Noor I, Sohail H, Sun J, Nawaz MA, Li G, Hasanuzzaman M, Liu J (2022) Heavy metal and metalloid toxicity in horticultural plants: tolerance mechanism and remediation strategies. Chemosphere 303(Pt 3):135196. https://doi.org/10.1016/j.chemosphere.2022.135196

Onakpa MM, Njan AA, Kalu OC (2018) A review of heavy metal contamination of food crops in Nigeria. Ann Glob Health 84(3):488–494. https://doi.org/10.29024/aogh.2314

Page MD, Kropat J, Hamel PP, Merchant SS (2009) Two Chlamydomonas CTR copper transporters with a novel cys-met motif are localized to the plasma membrane and function in copper assimilation. Plant Cell 21(3):928–943. https://doi.org/10.1105/tpc.108.064907

Panchal P, Miller AJ, Giri J (2021) Organic acids: versatile stress-response roles in plants. J Exp Bot 72(11):4038–4052. https://doi.org/10.1093/jxb/erab019

Park J, Song WY, Ko D, Eom Y, Hansen TH, Schiller M, Lee TG, Martinoia E, Lee Y (2012) The phytochelatin transporters AtABCC1 and AtABCC2 mediate tolerance to cadmium and mercury. Plant J 69(2):278–288. https://doi.org/10.1111/j.1365-313X.2011.04789.x

Pearson SA, Cowan JA (2021) Glutathione-coordinated metal complexes as substrates for cellular transporters. Metallomics 13(5):mfab015. https://doi.org/10.1093/mtomcs/mfab015

Pham VHT, Kim J, Chang S, Chung W (2022) Bacterial biosorbents, an efficient heavy metals green clean-up strategy: prospects, challenges, and opportunities. Microorganisms 10(3):610. https://doi.org/10.3390/microorganisms10030610

Pita-Barbosa A, Ricachenevsky FK (2022) Heavy metal ATPase 3 expression controls zinc homeostasis in natural accessions of *Arabidopsis thaliana*. Theor Exp Plant Physiol 34:509–515. https://doi.org/10.1007/s40626-022-00261-1

Popova LG, Belyaev DV, Shuvalov AV, Yurchenko AA, Matalin DA, Khramov DE, Orlova YV, Balnokin YV (2018) Mol Biol 52(4):601–615. https://doi.org/10.1134/S002689841804016X

Pottier M, Oomen R, Picco C, Giraudat J, Scholz-Starke J, Richaud P, Carpaneto A, Thomine S (2015) Identification of mutations allowing Natural Resistance Associated Macrophage Proteins

(NRAMP) to discriminate against cadmium. Plant J 83(4):625–637. https://doi.org/10.1111/tpj.12914
Proshad R, Kormoker T, Islam MS, Chandra K (2019) Potential health risk of heavy metals via consumption of rice and vegetables grown in the industrial areas of Bangladesh. Hum Ecol Risk Assess 26(4):921–943. https://doi.org/10.1080/10807039.2018.1546114
Qin G, Niu Z, Yu J, Li Z, Ma J, Xiang P (2021) Soil heavy metal pollution and food safety in China: effects, sources and removing technology. Chemosphere 267:129205. https://doi.org/10.1016/j.chemosphere.2020.129205
Rahman MM, Asaduzzaman M, Naidu R (2013) Consumption of arsenic and other elements from vegetables and drinking water from an arsenic-contaminated area of Bangladesh. J Hazard Mater 262:1056–1063. https://doi.org/10.1016/j.jhazmat.2012.06.045
Ramakrishna W, Yadav R, Li K (2019) Plant growth promoting bacteria in agriculture: two sides of a coin. Appl Soil Ecol 138:10–18. https://doi.org/10.1016/j.apsoil.2019.02.019
Roell MS, Zurbriggen MD (2020) The impact of synthetic biology for future agriculture and nutrition. Curr Opin Biotechnol 61:102–109. https://doi.org/10.1016/j.copbio.2019.10.004
Romero-Crespo P, Jiménez-Oyola S, Salgado-Almeida B, Zambrano-Anchundia J, Goyburo-Chávez C, González-Valoys A, Higueras P (2023) Trace elements in farmland soils and crops, and probabilistic health risk assessment in areas influenced by mining activity in Ecuador. Environ Geochem Health 45(7):4549–4563. https://doi.org/10.1007/s10653-023-01514-x
Salt DE, Rauser WE (1995) MgATP-dependent transport of phytochelatins across the tonoplast of oat roots. Plant Physiol 107(4):1293–1301. https://doi.org/10.1104/pp.107.4.1293
Sánchez-Thomas R, Moreno-Sánchez R, García-García JD (2006) Accumulation of zinc protects against cadmium stress in photosynthetic *Euglena gracilis*. Environ Exp Bot 131:19–31. https://doi.org/10.1016/j.envexpbot.2016.06.009
Shengxi H, Ken WKL, Madeline W (2001) Cadmium sequestration in *Chlamydomonas reinhardtii*. Plant Sci 161(5):987–996. https://doi.org/10.1016/S0168-9452(01)00501-5
Shiraishi E, Inouhe M, Joho M, Tohoyama H (2000) The cadmium-resistant gene, CAD2, which is a mutated putative copper-transporter gene (PCA1), controls the intracellular cadmium-level in the yeast S. cerevisiae. Curr Genet 37(2):79–86. https://doi.org/10.1007/s002940050013
Somayaji A, Sarkar S, Balasubramaniam S, Raval R (2022) Synthetic biology techniques to tackle heavy metal pollution and poisoning. Synthetic Syst Biotechnol 7(3):841–846. https://doi.org/10.1016/j.synbio.2022.04.007
Song WY, Park J, Mendoza-Cózatl DG, Suter-Grotemeyer M, Shim D, Hörtensteiner S, Geisler M, Weder B, Rea PA, Rentsch D, Schroeder JI, Lee Y, Martinoia E (2010) Arsenic tolerance in Arabidopsis is mediated by two ABCC-type phytochelatin transporters. Proc Natl Acad Sci USA 107(49):21187–21192. https://doi.org/10.1073/pnas.1013964107
Szalai G, Tajti J, Hamow KÁ, Ildikó D, Khalil R, Vanková R, Dobrev P, Misheva SP, Janda T, Pál M (2020) Molecular background of cadmium tolerance in Rht dwarf wheat mutant is related to a metabolic shift from proline and polyamine to phytochelatin synthesis. Environ Sci Pollut Res Int 27(19):23664–23676. https://doi.org/10.1007/s11356-020-08661-z
Takahashi R, Bashir K, Ishimaru Y, Nishizawa NK, Nakanishi H (2012) The role of heavy-metal ATPases, HMAs, in zinc and cadmium transport in rice. Plant Signal Behav 7(12):1605–1607. https://doi.org/10.4161/psb.22454
Tripathi S, Poluri KM (2021) Heavy metal detoxification mechanisms by microalgae: Insights from transcriptomics analysis. Environ Pollut (Barking, Essex: 1987) 285:117443. https://doi.org/10.1016/j.envpol.2021.117443
Uraguchi S, Kiyono M, Sakamoto T, Watanabe I, Kuno K (2009) Contributions of apoplasmic cadmium accumulation, antioxidative enzymes and induction of phytochelatins in cadmium tolerance of the cadmium-accumulating cultivar of black oat (Avena strigosa Schreb.). Planta 230(2):267–276. https://doi.org/10.1007/s00425-009-0939-x
Vatamaniuk OK, Mari S, Lang A, Chalasani S, Demkiv LO, Rea PA (2004) Phytochelatin synthase, a dipeptidyltransferase that undergoes multisite acylation with gamma-glutamylcysteine during catalysis: stoichiometric and site-directed mutagenic analysis of arabidopsis thaliana

PCS1-catalyzed phytochelatin synthesis. J Biol Chem 279(21):22449–22460. https://doi.org/10.1074/jbc.M313142200

Vivares D, Arnoux P, Pignol D (2005) A papain-like enzyme at work: native and acyl-enzyme intermediate structures in phytochelatin synthesis. Proc Natl Acad Sci USA 102(52): 18848–18853. https://doi.org/10.1073/pnas.0505833102

Wong CKE, Cobbett CS (2009) HMA P-type ATPases are the major mechanism for root-to-shoot Cd translocation in Arabidopsis thaliana. New Phytol 181(1):71–78. https://doi.org/10.1111/j.1469-8137.2008.02638.x

Wu CH, Wood TK, Mulchandani A, Chen W (2006) Engineering plant-microbe symbiosis for rhizoremediation of heavy metals. Appl Environ Microbiol 72(2):1129–1134. https://doi.org/10.1128/AEM.72.2.1129-1134.2006

Yong X, Chen Y, Liu W, Xu L, Zhou J, Wang S, Chen P, Ouyang P, Zheng T (2014) Enhanced cadmium resistance and accumulation in Pseudomonas putida KT2440 expressing the phytochelatin synthase gene of Schizosaccharomyces pombe. Lett Appl Microbiol 58(3): 255–261. https://doi.org/10.1111/lam.12185

Yu LP, Wu FQ, Chen GQ (2019) Next-generation industrial biotechnology-transforming the current industrial biotechnology into competitive processes. Biotechnol J 14(9):e1800437. https://doi.org/10.1002/biot.201800437

Zhang J, Martinoia E, Lee Y (2018) Vacuolar transporters for cadmium and arsenic in plants and their applications in phytoremediation and crop development. Plant Cell Physiol 59(7): 1317–1325. https://doi.org/10.1093/pcp/pcy006

Zhang C, Yang Q, Zhang X, Zhang X, Yu T, Wu Y, Fang Y, Xue D (2021) Genome-wide identification of the *HMA* gene family and expression analysis under Cd stress in barley. Plants (Basel, Switzerland) 10(9):1849. https://doi.org/10.3390/plants10091849

Zhang B, Tang Y, Yu F, Peng Z, Yao S, Deng X, Long H, Wang X, Huang K (2023) Translatomics and physiological analyses of the detoxification mechanism of green alga Chlamydomonas reinhardtii to cadmium toxicity. J Hazard Mater 448:130990. https://doi.org/10.1016/j.jhazmat.2023.130990

Zhu S, Shi W, Jie Y (2021) Overexpression of *BnPCS1*, a novel phytochelatin synthase gene from ramie (*Boehmeria nivea*), enhanced Cd tolerance, accumulation, and translocation in *Arabidopsis thaliana*. Front Plant Sci 12:639189. https://doi.org/10.3389/fpls.2021.639189

Zilber-Rosenberg I, Rosenberg E (2008) Role of microorganisms in the evolution of animals and plants: the hologenome theory of evolution. FEMS Microbiol Rev 32(5):723–735. https://doi.org/10.1111/j.1574-6976.2008.00123.x

GPSR Compliance

The European Union's (EU) General Product Safety Regulation (GPSR) is a set of rules that requires consumer products to be safe and our obligations to ensure this.

If you have any concerns about our products, you can contact us on ProductSafety@springernature.com

In case Publisher is established outside the EU, the EU authorized representative is:

Springer Nature Customer Service Center GmbH
Europaplatz 3
69115 Heidelberg, Germany

Batch number: 10755329

Printed by Printforce, the Netherlands